Biotechnology of fungi for improving plant growth

Biotechnology of fungi for improving plant growth

Symposium of the British Mycological Society held at the University of Sussex September 1988

Edited by

J. M. Whipps & R. D. Lumsden

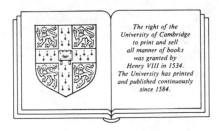

The right of the
University of Cambridge
to print and sell
all manner of books
was granted by
Henry VIII in 1534.
The University has printed
and published continuously
since 1584.

Published for the British Mycological Society by
CAMBRIDGE UNIVERSITY PRESS
Cambridge
New York Port Chester Melbourne Sydney

Published by the Press Syndicate of the University of Cambridge
The Pitt Building, Trumpington Street, Cambridge CB2 1RP
40 West 20th Street, New York, NY10011, USA
10 Stamford Road, Oakleigh, Melbourne 3166, Australia

First published 1989

Designed and produced by **Page Design,** 48 Alan Road, Stockport SK4 4LE

Printed in Great Britain at the University Press, Cambridge

British Library cataloguing in publication data
Biotechnology of fungi for improving plant growth.
　1. Fungi. Biotechnology. 2. Crops. Pests. Biological control. Use of fungi
　I. Whipps, J. M. II. Lumsden, R. D.
　660'.62

Library of Congress cataloguing data available

ISBN 0–521–38236–X

Contents

Contributors

Ralph Baker *Plant Pathology and Weed Science, Colorado State University, Fort Collins, Colorado 80523, USA*

A. K. Charnley *School of Biological Sciences, University of Bath, Claverton Down, Bath BA2 7AY, UK*

R. C. Cooke *Department of Animal and Plant Sciences, University of Sheffield, Sheffield S10 2TN, UK*

C. E. Cordell *Forest Pest Management, USDA Forest Service, Asheville, North Carolina 28802 USA*

J. L. Faull *Biology Department, Birkbeck College, Malet Street, London WC1E 7HX, UK*

S. Gianinazzi *Laboratoire de Phytoparasitologie, Station d'Amélioration des Plantes, INRA, BV 1540, 21034 Dijon Cédex, France*

V. Gianinazzi-Pearson *Laboratoire de Phytoparasitologie, Station d'Amélioration des Plantes, INRA, BV 1540, 21034 Dijon Cédex, France*

A. T. Gillespie *Entomology and Insect Pathology Section, AFRC Institute of Horticultural Research, Littlehampton, West Sussex BN17 6LP, UK*

L. M. Harvey *Department of Bioscience & Biotechnology, University of Strathclyde, Glasgow G1 1XW, UK*

Dana K. Heiny *Department of Plant Pathology, University of Arkansas, Fayetteville, AR 72701, USA*

M. J. Hocart *Microbial Genetics and Manipulation Group, Department of Botany, School of Biological Sciences, University of Nottingham, Nottingham NG7 2RD, UK*

B. R. Kerry *AFRC Institute of Arable Crops Research, Rothamsted Experimental Station, Harpenden, Hertfordshire AL5 2JQ, UK*

B. Kristiansen *Department of Bioscience & Biotechnology, University of Strathclyde, Glasgow G1 1XW, UK*

J. A. Lewis *Biocontrol of Plant Diseases Laboratory, US Department of Agriculture, Agricultural Research Service, Beltsville, Maryland 20705, USA*

Karen Lewis *Department of Animal and Plant Sciences, University of Sheffield, Sheffield S10 2TN, UK*

R. D. Lumsden *Biocontrol of Plant Diseases Laboratory, US Department of Agriculture, Agricultural Research Service, Beltsville, Maryland 20705, USA*

D. H. Marx *Institute for Mycorrhizal Research and Development, USDA Forest Service, Athens, Georgia 30602 USA*

E. R. Moorhouse *Entomology and Insect Pathology Section, AFRC Institute of Horticultural Research, Littlehampton, West Sussex BN17 6LP, UK*

J. Neill *Department of Bioscience & Biotechnology, University of Strathclyde, Glasgow G1 1XW, UK*

J. F. Peberdy *Microbial Genetics and Manipulation Group, Department of Botany, School of Biological Sciences, University of Nottingham, Nottingham NG7 2RD, UK*

N. Poole *ICI Agrochemicals, Jealott's Hill Research Station, Bracknell, Berkshire RG12 6EY, UK*

K. A. Powell *ICI Agrochemicals, Jealott's Hill Research Station, Bracknell, Berkshire RG12 6EY, UK*

Annabel Renwick *ICI Agrochemicals, Jealott's Hill Research Station, Bracknell, Berkshire RG12 6EY, UK*

E. Senior *Department of Bioscience & Biotechnology, University of Strathclyde, Glasgow G1 1XW, UK*

J. E. Smith *Department of Bioscience & Biotechnology, University of Strathclyde, Glasgow G1 1XW, UK*

George E. Templeton *Department of Plant Pathology, University of Arkansas, Fayetteville, AR 72701, USA*

A. Trouvelot *Laboratoire de Phytoparasitologie, Station d'Amélioration des Plantes, INRA, BV 1540, 21034 Dijon Cédex, France*

J. M. Whipps *Microbiology Department, AFRC Institute of Horticultural Research, Littlehampton, West Sussex BN17 6LP, UK*

Preface

The use of fungi for the improvement of plant growth is increasingly being implemented in agriculture, and several fungi have been commercialized for this purpose. For example, ectomycorrhizal fungi are used routinely for inoculation of forest trees, and endomycorrhizas are gaining acceptance in the horticultural industry. Fungi are commercially used to control weeds and they hold promise for the control of plant pathogenic nematodes and fungi, as well as insects. In addition, fungi may stimulate plant growth directly by production of metabolites or growth hormones.

The aim of this book is to describe these diverse uses of fungi to improve plant growth and examine the factors that enhance rapid commercialization. One of the greatest stimulations to research in this area has been the advances in technology associated with the selection, culture and formulation of fungi in relation to specific targets. The Chapters by Marx & Cordell, Harvey *et al.* and Gianinazzi, Gianinazzi-Pearson & Trouvelot explore the use of mutalistically symbiotic mycorrhizal fungi. Ectomycorrhizal fungi can now be grown successfully using large scale solid substrate systems and also liquid fermentation. The biomass obtained is used to inoculate millions of trees. For specific high value crops, problems associated with the inability to culture vesicular-arbuscular mycorrhizas have largely been overcome making inoculum production commercially feasible (Chapter by Gianinazzi *et al.*). Similarly, the control of weeds with fungi is proving to be commercially feasible (Chapter by Templeton & Heiny). Further research with these same basic concepts and approaches will improve the success rate for the control of insects (Chapter by Gillespie & Moorhouse), nematodes (Chapter by Kerry) and soilborne plant pathogens (Chapter by Lumsden & Lewis). Of particular relevance to potential advances are modern molecular approaches to fungal biotechnology such as protoplast fusion and genetic manipulation which provide the opportunity for recombination in a controlled manner of useful traits shown by different strains of fungi. These techniques and possible applications are examined in Chapters by Baker and Hocart & Peberdy. An understanding of the mechanisms of action of fungal biocontrol agents are a prerequisite to molecular studies. Examples of the mechanisms involved in the biocontrol of insects and diseases are described in Chapters by Charnley and Lewis *et al.*, respectively.

Research on fungal biocontrol agents has also been stimulated by problems associated with the use of chemicals for control of weeds, pests and diseases. In addition to cases where no suitable chemical controls are known, concern exists over resistance of pathogens and pests to fungicides and pesticides, environmental damage by excessive chemical usage, and future restrictions on the use of chemicals. These concerns emphasise the need for biocontrol agents that give reproducible results at an economic cost, as explored in the Chapter by Powell & Faull. The selection and release of large numbers of fungi into the environment, particularly if they are genetically manipulated or mutant strains, introduces additional problems and these are considered in relation to disease biocontrol agents in the Chapter by Renwick & Poole.

The book is based on a series of papers presented at a meeting of the British Mycological Society at The University of Sussex, September 19-20, 1988 organized by the Biotechnology Special Interest Committee. We are grateful to all the authors for their contributions to the volume and for accepting suggestions from referees and editors to produce the final balance of the book. We would also like to thank members of the Society for initiating and planning the meeting. The Society wishes to record its thanks to ICI and Shell for donations in support of the meeting. Finally, we would like to thank David Moore for his excellent technical help and efficiency in producing the book from the manuscripts.

J. M. Whipps
AFRC Institute of Horticultural Research, Littlehampton
R. D. Lumsden
Beltsville Agricultural Research Center, USDA-ARS, Beltsville

Chapter 1

The use of specific ectomycorrhizas to improve artificial forestation practices

D. H. Marx[1] & C. E. Cordell[2]

[1]Institute for Mycorrhizal Research and Development, USDA Forest Service, Athens, Georgia 30602 U.S.A. and [2]Forest Pest Management, USDA Forest Service, Asheville, North Carolina 28802 U.S.A.

Introduction

Microorganisms are present in great numbers on and near the feeder roots of trees, and they play vital roles in numerous physiological processes. These dynamic processes are mediated by associations of microorganisms participating in saprotrophic, pathogenic and symbiotic root activities. The major symbiotic associations on tree roots are mycorrhizas. The word mycorrhiza is used to describe a structure that results from a mutually beneficial association between the fine feeder roots of plants and species of highly specialized, root-inhabiting fungi. Mycorrhizal fungi derive most if not all of their organic nutrition from their symbiotic niche in the primary cortical tissues of roots. The mycorrhizal habit probably evolved as a survival mechanism for both partners in the association, allowing each to survive in environments of low soil fertility, drought, disease, and temperature extremes where alone they could not.

Endomycorrhizas are the most widespread and comprise three groups, ericaceous, orchidaceous, and vesicular-arbuscular mycorrhizas (VAM). The VAM are found on more plant species than all other types of mycorrhizas combined; they have been observed in roots of over 1000 genera of plants representing some 200 families. Over 90% of the 300,000 species of vascular plants in the world form VAM. The commercial use of VAM is considered in Chapter 3.

Ectomycorrhizas occur on about 10% of the world flora. Trees belonging to the Pinaceae, Fagaceae, Betulaceae, Salicaceae, Juglandaceae, Myrtaceae, Ericaceae, and a few others form ectomycorrhizas. Numerous fungi have been identified as forming ectomycorrhizas. In North America alone, it has been estimated that more than 2100 species of fungi form ectomycorrhizas with forest trees. Worldwide, there are over 5000 species of fungi that can form ectomycorrhizas on some 2000 species of woody plants.

Importance of ectomycorrhizas to trees

Ectomycorrhizal fungi aid the growth and development of trees. For some trees, such as *Pinus* sp., they are indispensable for growth under natural conditions. The obligate requirement of pine for ectomycorrhizas in a natural environment has been clearly shown by numerous workers in tree regeneration trials in former treeless areas and in countries without native ectomycorrhizal trees (Marx, 1980a). In trees with abundant ectomycorrhizas the combination of fungus and root provides a much larger, and more physiologically active, area for nutrient and water absorption than is present in trees with few or no ectomycorrhizas. This increase in surface area comes both from the multi-branching habit of most ectomycorrhizas and from the extensive vegetative growth of hyphae of the fungal symbionts from the ectomycorrhizas into the soil. These extramatrical hyphae function as additional nutrient and water-absorbing entities and assure maximum nutrient capture from the soil by the host. Ectomycorrhizas are able to absorb and accumulate nitrogen, phosphorus, potassium, and calcium in the fungus mantles more rapidly and for longer periods of time than nonmycorrhizal feeder roots. Ectomycorrhizas also appear to increase the tolerance of trees to drought, high soil temperatures, soil toxins (organic and inorganic), and very low soil acidity. Ectomycorrhizas also deter infection of feeder roots by root pathogens, such as species of *Pythium* or *Phytophthora* (Marx, 1973; Marx & Krupa, 1978) and hormone relationships induced by fungal symbionts cause ectomycorrhizal roots to have greater lengths of physiological activity than nonmycorrhizal roots (Ek, Ljungquist & Stenstrom, 1983). Not all species of fungi form ectomycorrhizas that have equal benefit to their hosts under specific conditions. Some are more effective than others.

Ecology of ectomycorrhizal fungi

Many species of fungi are normally involved in the ectomycorrhizal associations of a forest stand, on a single tree species, on an individual tree, or even on a small segment of lateral root. As many as three species of fungi have been isolated from an individual ectomycorrhiza (Zak & Marx, 1964). A single fungus species can enter into ectomycorrhizal association with numerous tree species on the same site. A fungus can also develop numerous biotypes or clones in a very limited area of a pure stand (Fries, 1987). Some fungi are apparently host-specific; others have broad host ranges and form ectomycorrhizas with members of numerous tree genera in diverse families.

There are distinct early- and late-stage fungi in ectomycorrhizal fungus successions in forests. In aseptic culture (i.e. without competition and other stresses) early- and late-stage fungi form ectomycorrhizas on seedlings equally well. However, only the early-stage fungi are able to rapidly colonize seedlings in natural, nonsterile soil that harbours competitors and other environmental stresses. Early-stage fungi may not totally disappear from mature stands, but they are supplanted by more dominant species (Dighton, Poskitt & Howard, 1986) or are suppressed in reproduction because of changes in canopy and root/soil characteristics.

Increases in fungal species diversity are associated with ectomycorrhizal fungus succession as forest stands age and numbers of host species increase (Last *et al.*, 1984). Fruit body production by these fungi, which is how fungal succession is measured, is also strongly influenced by seasonal changes, rainfall amount and frequency, organic content of soil, root density, and degree of ectomycorrhizal development (Wilkins & Harris, 1946).

Use of specific ectomycorrhizal fungi on seedlings

Ectomycorrhizal fungi have been introduced into deficient soils in various inocula to provide seedlings with adequate ectomycorrhizas to create man-made forests. Most research on inoculation with ectomycorrhizal fungi has been based on two premises. First, any ectomycorrhizal association on roots of tree seedlings is far better than none. Success in correcting deficiencies has contributed greatly to our understanding of the importance of ectomycorrhizas to trees. Second, some species of ectomycorrhizal fungi on certain sites are more beneficial to trees than other fungal species that naturally occur on such sites. Much work in recent years with a few fungal species has been aimed at selecting, propagating, manipulating, and managing the more desirable fungal species to improve tree survival and growth and this has recently been reviewed (Marx & Ruehle, in press). Consequently, the purpose of the present paper is to discuss the research and development that led to present-day ectomycorrhizal technology in the United States.

Selection of candidate fungi

The first and most important step in any inoculation program of tree seedlings is the selection of fungi. Physiological and ecological differences among ectomycorrhizal fungi are great and these differences can be used as criteria for selection. Host specificity is a physiological trait important to consider in the selection process. The consistent association of certain fungi with only a few specific tree hosts is well documented in the literature. Many other fungi are associated with a great number of different tree hosts; these are probably early-stage fungi. Any candidate fungus should exhibit the physiological capacity to form ectomycorrhizas on the desired hosts and the more hosts the better. Isolate variability within any candidate symbiont is another criterion to consider. Several isolates from different tree hosts and geographic regions should be used, at least initially, to determine the amount of variation that exists between isolates.

This point has been stressed by Moser (1958a) and demonstrated with isolates of various fungi. For example, isolates of *Pisolithus tinctorius* from various pines form abundant ectomycorrhizas on pine and oak, but isolates from various oak species form few ectomycorrhizas on pine. Some oak isolates formed no ectomycorrhizas on either host, nor did other pine isolates from Australia and Brazil (Marx, 1979; 1981). Another criterion is the ability of the selected fungus to grow in pure culture and withstand manipulation. A variety of culture media and methods of isolation can be used to obtain pure cultures of the selected fungus (Schenck, 1982; see also chapter 2). Ideally, the fungus should be able to grow rapidly. Once the growth characteristics of a fungus have been confirmed, it is important to determine its capacity to withstand physical, chemical, and biological manipulations. Producing large quantities of vegetative inoculum of a fungus is of little value if the fungus cannot survive the rigours of various manipulations, such as physical processing (blending, leaching, or drying), shipment, and soil incorporation. The fungus inoculum must also be able to survive a minimum of 4 to 6 weeks between nursery soil inoculation and seed sowing, germination, and the production of short roots by the seedlings. During this period, it must survive fluctuations of soil moisture, temperature and microbial competition. Ideally, the fungus should also be able to survive several weeks of storage between inoculum production and use.

Another criterion is the ecological adaptation of the selected fungus to the major type of site on which the seedlings are to be planted. Of equal importance is the ability of the fungus to survive and grow under cultural conditions used in nurseries. In other words, the candidate fungus must be an early-stage fungus in normal succession if it is to be effective on seedlings in the nursery and in the early successional stage of stand establishment. According to Trappe (1977), the ecological adaptability of an ectomycorrhizal fungus hinges on the metabolic pathways it has evolved to contend with environmental variation. Extremes of soil and climate, antagonism from other soil organisms including other ectomycorrhizal fungi, pesticide application, physical disruption of mycelium from nursery cultural practices (undercutting and root pruning), and the abrupt physiological adjustment from a well fertilized and irrigated nursery soil to an uncultivated, low-fertility planting site with all its normal stresses are only a few of the environmental variations to which the selected fungus must adapt.

The effect of temperature on different species and ecotypes of ectomycorrhizal fungi is perhaps the most widely researched environmental factor. Upper and lower temperature limits of the candidate fungi should be determined. Moser (1958a) studied the ability of fungi to survive long periods (up to 4 months) of freezing at -12°C and to grow at 0 to 5°C. He found that high elevation ecotypes of *Suillus variegatus* survived freezing for 2 months, but valley ecotypes were killed after freezing for only 5 days. However, the ability of certain fungi to survive freezing was not correlated with their ability to grow at low temperature. Generally, mountain ecotypes and species had much lower temperature optima than lowland ones. *Pisolithus tinctorius* can grow at temperatures as high as 40 to 42°C (Hung & Chien, 1978) and has a hyphal thermal death point of 45°C (Lamb & Richards, 1971). It not only survives and grows well at high temperatures but it also grows at 7°C (Marx, Bryan & Davey, 1970; Marx & Bryan, 1971) and withstands frozen soil (Marx & Bryan 1975).

Reaction of candidate fungi to soil moisture, organic matter, and pH are also important traits to consider. *Cenococcum geophilum* is not only drought tolerant but forms ectomycorrhizas in natural soils ranging in pH from $3 \cdot 4$ to $7 \cdot 5$ (Trappe, 1977). Unfortunately, the drought tolerant characteristics of *C. geophilum* also make it difficult to establish on pine seedlings in irrigated nurseries where it can be rapidly supplanted by naturally-occurring *Thelephora terrestris* (Marx, Morris & Mexal, 1978). *Pisolithus tinctorius* ectomycorrhizas on pine and oak have been observed in drought-prone coal spoils as acid as pH $2 \cdot 6$. *Suillus bovinus* (Levisohn, 1956) and *Paxillus involutus* (Laiho, 1970) form abundant ectomycorrhizas on seedlings in nurseries with high organic matter, but these ectomycorrhizas are supplanted by others after seedlings are planted out on sites having low organic matter.

The production of hyphal strands and sclerotia are also important traits in candidate fungi. Uptake of nutrients, especially phosphorus (Bowen, 1973), and translocation of carbon compounds (Reid & Woods, 1969) take place through hyphal strands. In Australia, one of the initial criteria for selection of fungi for inoculation programmes is its ability to produce hyphal strands. Although research data are lacking, abundant hyphal strand production by *P. tinctorius* apparently enhances water and nutrient absorption and increases its survival potential under adverse conditions. Yellow-gold hyphal strands of this fungus, easily visible to the naked eye, have been traced through coal spoils as far as 4 m from young seedling roots to fruit bodies. The production of sclerotia by *P. tinctorius* (Marx *et al.*, 1982) and *C. geophilum* (Trappe, 1969) in soil or container rooting media should enhance the abilities of these fungi to survive harsh soil conditions. Therefore, sclerotia production is also a favourable trait in candidate fungi.

All the criteria mentioned are meaningless unless the candidate fungus is aggressive and can form abundant ectomycorrhizas on seedlings as soon as short roots are produced. This is another characteristic common among early-stage fungi. The fungus should be able to maintain superiority over naturally occurring fungi on seedling roots in the nursery. Even though the effect of a fungus on seedlings may only be temporary until it is supplanted by other fungi in the field, this brief advantage may make the difference between survival or death of newly planted seedlings. For tree seedlings to obtain any measurable benefit from a specific ectomycorrhizal association, there is a threshold amount of the specific ectomycorrhizas that must be present on seedling roots.

Types of inocula

The majority of reports on inoculation techniques with ectomycorrhizal fungi involve basidiomycetes on pines, oaks, and eucalypts. The techniques were developed out of necessity to grow tree species requiring ectomycorrhizas in areas of the tropics where ectomycorrhizal fungi were absent. Several types of natural and laboratory-produced inocula and several methods of application have been used through the years. Many of the techniques have proven successful, others have not.

The most widely used natural inoculum, especially in developing countries, is soil or humus collected from established pine plantations (Mikola, 1973; Marx 1980a). In most instances, the original soil inocula came from mature pine plantations on other continents. A major drawback with soil or humus inoculum is the lack of control of species of ectomycorrhizal fungi in the inoculum. There is no assurance that soil inoculum contains the most desirable fungi for the tree species to be produced or for the site on which the seedlings are to be planted out. Soil inoculum from mature pine forests would likely contain more late-stage than early-stage fungi. Also, soil inoculum may contain harmful microorganisms and noxious weeds. Some of these microorganisms may be potentially harmful not only to the tree seedling crop, but also to nearby agricultural crops. Exotic pathogens introduced to the areas in soil inoculum create potential hazards for epiphytotics. The use of natural inoculum, however, does satisfy the premise that any ectomycorrhizas on seedlings used in forest regeneration are better than none.

Spores of various fungi have been used as inoculum to form specific ectomycorrhizas on tree seedlings. The major advantages are that spores require no extended growth phase under aseptic conditions, and spore inoculum is very light. One gram of basidiospores of *Rhizopogon luteolus* or *Pisolithus tinctorius* contains about 1×10^9 potentially infective basidiospores. Another advantage of spores, at least those of certain fungi, is their ability to maintain viability in storage from one season to the next. Such survival is important because spores collected in the summer or early autumn must be stored until the following spring if they are to be used to inoculate spring-sown nursery beds. However, one of the major disadvantages of spore inoculum of most ectomycorrhizal fungi is the lack of appropriate laboratory tests to determine spore viability. Currently, time-consuming ectomycorrhizal synthesis tests are the only reliable means of determining spore viability of many of these fungi. Another disadvantage is that sufficient sporophores of many fungi may not be available every year. One would need ideal storage conditions to maintain a large inventory of spores in order to ensure a constant supply of spore inoculum from year to year.

Formation of ectomycorrhizas by basidiospores usually takes 3 to 4 weeks longer than vegetative inoculum of the same fungus (Theodorou & Bowen, 1970; Marx, Bryan & Cordell, 1976). This delay can be a disadvantage because, during this period, pathogenic fungi or other ectomycorrhizal fungi can colonize the roots and reduce the effectiveness of the introduced spore inoculum. Also, seedlings experiencing a delay in ectomycorrhizal formation lose the physiological advantage that ectomycorrhizas furnish during this period. It should be pointed out, however, that in parts of the world where the natural occurrence of ectomycorrhizal fungi is erratic or deficient, this delay may not have a significant effect on the final results of inoculation. Another significant problem in using spore inoculum is the lack of genetic definition. Genetic diversity in basidiospores can be enormous, particularly if spores are collected from many geographic areas and from different tree hosts and combined into a single inoculum. Basidiospore inoculum of *Pisolithus tinctorius* has been used on an experimental scale, and more recently on an operational scale, in the United States and elsewhere. Recent studies have proved that these spores are effective in various inoculum forms such as: (1) mixed with a physical carrier before soil inoculation, (2) suspended in water and drenched onto soil, (3) dusted or sprayed onto soil, (4) pelleted and broadcast onto soil, (5) encapsulated or coated onto pine seeds, and (6) in hydrocolloid chips (Marx & Ruehle, in press).

Pure mycelial or vegetative inoculum of ectomycorrhizal fungi has been repeatedly recommended as the most biologically sound material for inoculation. Unfortunately, ectomycorrhizal fungi as a group are difficult to grow in the laboratory. Many species have never been isolated and grown in pure culture. Some species grow slowly, others often die after a few months in culture. Most of these fungi require specific growth substances, such as thiamine and biotin, in addition to simple carbohydrates (Schenck, 1982). Several researchers in various parts of the world have developed cultural procedures for producing vegetative inoculum of a variety of fungi for research purposes. Unfortunately, large-scale nursery applications of pure mycelial cultures, even those involving only a few thousand tree seedlings, have been severely hampered by the lack of sufficient quantities of viable inoculum. It is relatively simple to produce a sufficient volume of inoculum, i.e., 30 to 40 litres, for research studies carried out in small con-

tainers, pots, microplots, or even small nursery plots, but it is a completely different matter to produce sufficient quantities of commercial
vegetative inoculum for large-scale nursery inoculation in a practical programme (Marx, 1985). Tens of thousands of litres of vegetative inoculum
would be needed to inoculate just one of the nurseries in the southern US,
where over 1·5 billion pine seedlings are produced annually on some 625
hectares of nursery soil (see also chapter 2).

Propagation of vegetative inoculum

Moser (1958a, b, c) in Austria was one of the first to make a serious attempt to produce vegetative inoculum of ectomycorrhizal fungi. For
production of inoculum, mycelium of *Suillus plorans* was first grown in
liquid culture then in sterile peat moss. Takacs (1961, 1964, 1967) in Ar-
gentina modified Moser's techniques to produce inoculum for new pine
nurseries established in formerly treeless areas lacking native ectomycorrhizal fungi. Inoculum of *Amanita verna, Suillus granulatus, S. luteus,
Hebeloma crustuliniforme*, a *Russula* sp., *Scleroderma verrucosum*, and *S.
vulgare* were produced with this technique. It is now recognized that the
above species of fungi represent both early- and late-stage fungi in normal forest succession. Quantitative data on ectomycorrhizal development
after such inoculations are not available, and it is difficult to evaluate the
success of this inoculation method. Hacskaylo & Vozzo (1967) initiated a
series of inoculation experiments with pure vegetative cultures of various
fungi to correct a deficiency of these fungi on the island of Puerto Rico.
Following Moser's general technique, they grew *Cenococcum geophilum,
Corticium bicolor, Rhizopogon roseolus*, and *Suillus cothuranatus* in poly-
propylene cups containing a 2:1 ratio of sterile peat moss and vermiculite
moistened with nutrient solution. After root inoculation in the nursery, all
fungi except *C. geophilum* formed ectomycorrhizas to varying degrees on
Pinus caribaea seedlings.

Since 1966, the USDA Forest Service Institute for Mycorrhizal Research and Development (IMRD) has been studying the significance of
ectomycorrhizas formed by *Pisolithus tinctorius* and other fungi to survival and growth of tree seedlings on a variety of sites. The interest in *P.
tinctorius* was prompted by its relatively wide geographic and tree host
distribution, range of environmental tolerance, and ease of pure culture
propagation (Schramm, 1966; Marx, 1977, 1980a, 1981). This fungus is
now known to form ectomycorrhizas with over 100 species of woody
plants. Vermiculite and peat moss moistened with modified Melin-Norkrans medium (MMN) was found to be an excellent substrate for the
production of effective vegetative inoculum (Marx, 1969).

Vermiculite-peat moss-MMN inoculum used directly from the container and mixed into fumigated nursery soil is rapidly colonized by

saprophytic microorganisms. Heavy colonization reduces the effectiveness of this inoculum to form ectomycorrhizas on pine. Leaching the inoculum before inoculation to remove most nonassimilated carbohydrates reduces this microbial colonization and increases inoculum effectiveness (Marx, 1980a). Leached inoculum has been used by many workers, but it is difficult to use because it has a very high bulk density ($750 \, g \, l^{-1}$), a high moisture content, and the physical consistency of a sticky paste. These problems were corrected by slowly drying the inoculum to final bulk densities from 320 to 390 $g \, l^{-1}$ and moisture contents from 10 to 35%. Final pH of inoculum ranged from $4 \cdot 4$ to $5 \cdot 2$. The original volume of inoculum was reduced by nearly 60% after leaching and drying. Dried vegetative inoculum mixed more uniformly in soil than the wetter, non-dried formulations, so it was effective when used at lower rates. Also, dried inoculum stored at 5°C for as long as 9 weeks and at room temperature for 5 weeks was still effective in forming ectomycorrhizas (Marx, 1980a).

In 1976, the IMRD and State and Private Forestry, USDA Forest Service, and Abbott Laboratories, North Chicago, Illinois, began a cooperative program of research and nursery evaluation to develop commercial methods of producing vegetative inoculum of *Pisolithus tinctorius*. The effectiveness of different vegetative inoculum formulations was tested under diverse cultural regimes on various tree species in bare-root and container nurseries in the United States and Canada. Inoculum from IMRD and Abbott formed ectomycorrhizas on 13 species of trees in tests with several types of containers and cultural practices at six locations in the United States and one in Canada (Marx *et al.*, 1982). Tests of IMRD and Abbott inoculum formulations in over 75 bare-root seedling nursery tests were completed in 1981 (Marx et al., 1984). In 33 different tree nurseries in 25 states on 14 tree species, research showed that inoculum formulations effective on containerized seedlings were also effective on bare-root seedlings (Fig. 1.1A & 1.1B). Results from over 30 nursery tests with loblolly pine in the southern US showed that abundant *P. tinctorius* ectomycorrhizas reduced seedling losses by over 25% and increased fresh weights of seedlings by over 20%. These were unexpected benefits to the nurserymen. The Abbott inoculum, which was trademarked > MycoRhiz®, was also grown in the vermiculite-peat moss-MMN medium. Liquid cultures (MMN) were grown in shake flasks, 14-litre fermenters, or large industrial fermenters and were used as starter inoculum for the vermiculite-peat moss-MMN substrate. Liquid cultures were continuously agitated and highly aerated at temperatures of 28 to 32°C for 7 to 14 days.

Beds of medium were sterilized by injection of high pressure steam in either deep-tank or rotary drum fermenters. The sterile medium was inoculated with the starter inoculum of *Pisolithus tinctorius*. The level of

Fig. 1.1. Loblolly pine (*Pinus taeda*) seedlings with abundant *Pisolithus tinctorius* ectomycorrhizas following inoculation with commercial vegetative inoculum (A) and nursery-run seedlings with only naturally-occurring ectomycorrhizas (B). Soil was inoculated by machine (C) and, following successful mycorrhizal development, fruit bodies of *P. tinctorius* (D) are produced in abundance.

inoculum varied considerably, but was usually in the range of 5 to 20% of the fermenter volume. The medium was mixed and incubated with or without agitation for one to four weeks, depending on the rapidity of growth and colonization of the substrate.

The colonized substrate was harvested by flotation in water and filtration, a process that also leached the product. The wet product was dried with a fluid bed dryer to a moisture content of 20 to 25%. MycoRhiz® normally had a bulk density of 240 to 310 g l^{-1} and a pH of 5·1 to 5·6. It was packaged in 50 litre units and stored at 5°C. The shelf life under these conditions was 5 to 6 weeks. Using this fermentation technology, Myco-Rhiz® was produced in sufficient volumes for large scale, practical inoculation of tree nurseries.

Inoculum efficacy testing

During the research and development of MycoRhiz®, various laboratory assays were developed to characterize the inoculum in an attempt to predict its eventual effectiveness in forming *Pisolithus* ectomycorrhizas in nursery soils or container substrates. The effectiveness of MycoRhiz® in nursery tests could not be predicted from the results of any laboratory assay, but a pattern did emerge from extensive testing. The most effective formulations in nursery tests were those with: (1) the most hyphae of *P. tinctorius* inside the vermiculite particles (inoculum niche), (2) pH between 4·5 and 6·0 controlled with a 5 to 10% volume of acid peat moss mixed with the pH 7·0 vermiculite, (3) low levels of bacterial and fungal contaminants which apparently colonized inoculum during leaching and drying, and (4) low amounts of residual glucose (< 16 mg g^{-1} inoculum).

MycoRhiz® is no longer being produced by Abbott Laboratories. Between 1978 and 1986, research and nursery evaluations were done with various types of vegetative inoculum of ectomycorrhizal fungi produced by Sylvan Spawn Laboratory, Worthington, Pennsylvania. Results with novel formulations of vegetative inoculum of *Pisolithus tinctorius* produced by this company have been outstanding (Marx *et al.*, 1989a & b). These formulations are better than previous ones because they do not require leaching and drying before use. This procedure is a modification of that used to produce mushroom spawn. A mixture of vermiculite, peat moss, and nutrients is mixed and sterilized in a large rotating vessel. After cooling, starter mycelial inoculum is mixed with the substrate, and 8-9 litre lots of substrate are aseptically dispensed into sterile plastic bags. The ends of the bags are heat sealed and a special breather tape-strip is put in place to allow ventilation during fungal growth. Bags are stacked on carts and incubated at a controlled temperature for specific times (Maul, 1985). After substrate colonization, inoculum in bags is shipped directly to the nursery for use. A newly formed biotechnology company, Mycorr Tech

Inc., University of Pittsburgh Applied Research Center, Pittsburgh, Pennsylvania, is the current producer of vegetative inoculum formulations.

As with the Abbott inoculum, vegetative inoculum produced by Sylvan Spawn could not be successfully characterized for efficacy prior to nursery testing. Inoculum effectiveness could not be predicted on the basis of (1) amount of mycelium visible in inoculum bags, (2) relative amount of mycelium observed microscopically inside vermiculite particles, or (3) number of vermiculite particles that yielded viable cultures after plating on MMN medium. Inoculum efficacy is undoubtedly related to the physiological/morphological status of the mycelium. Unfortunately, procedures to ensure quality prior to inoculum use have not been developed.

Only one vegetative isolate of *Pisolithus tinctorius* was used in the Abbott and Sylvan Spawn inoculum development programs. This so-called 'super-strain' also has been used by many researchers worldwide. This isolate was originally obtained from a fruit body found under pine in northeast Georgia over 25 years ago. After repeated host passage and vegetative reisolation, it is now more aggressive and is capable of forming more ectomycorrhizas more quickly than the original parent culture or isolates obtained from other sources (Marx & Daniel, 1976; Marx, 1981). Currently, the isolate is reisolated each year from either ectomycorrhizas or fruit bodies produced by loblolly pine seedlings in nursery inoculation tests.

Field testing for effects of specific ectomycorrhizas

Improvements in tree seedling performance due to *Pisolithus tinctorius* ectomycorrhizas have been reported on routine reforestation sites in many parts of the world (Marx & Ruehle, in press). Table 1.1 lists published results of studies on reforestation sites in the southern US with pine and oak seedlings. Some studies report only minimal effect on survival while others report differences of over 500% in volume growth. In those studies reporting large differences, control seedlings with only naturally occurring ectomycorrhizas (mostly *Thelephora terrestris*) at planting survived and grew poorly whereas, in those studies reporting small differences, the control seedlings survived and grew considerably better. These findings suggested that seedlings with abundant *P. tinctorius* ectomycorrhizas were tolerating some site or environmental stress factor (e.g. soil water deficits, high temperatures) better than seedlings without *P. tinctorius* ectomycorrhizas.

Two recent studies in Georgia relate to this stress relationship. At planting, seedlings in both studies had the same height, root collar diameter, and total fresh weight regardless of their initial ectomycorrhizal status. The first study was on a good quality site in south Georgia (Marx

Table 1.1. Improvements in survival and growth of tree seedlings due to similar amounts of *Pisolithus tinctorius* ectomycorrhizas on various reforestation sites in the southern United States.

Tree species	No. sites	Years in field	% increase over controls[a,b]		
			Survival	Growth	Growth parameter[c]
Pinus clausa var. *immuginata*	2	2	96-169	270-274	PVI
P. clausa var. *immuginata*	1	7	11	35	PVI
P. echinata	1	4	0	41-89	PVI
P. echinata	2	2	34-39	96-141	PVI
P. elliottii var. *elliottii*	3	2	5-22	6-175	PVI
P. palustris	1	2	6-55	11-99	PVI
P. palustris	1		116	-	-
P. palustris	2	3	7-38	100-180	PVI
P. palustris	1	7	22	58	PVI
P. strobus	1	2	8	500	PVI
P. taeda	4	2	0	21-63	PVI
P. taeda	8	3	5	25	D^2H
P. taeda	1	3	0	28	D^2H
P. taeda	1	2	-	38	height
P. taeda	1	2	20	54	PVI
P. taeda	1	2	0	18	PVI
P. virginiana	2	2	2-4	29-55	PVI
Quercus acutissima	1	2	73	53	D^2H
Q. palustris	1	2	3	39	D^2H

[a]Control seedlings had ectomycorrhizas formed naturally in the nursery.
[b]Ranges of percent increase in survival and growth are caused by different responses to other treatments within the ectomycorrhizal treatments or to different responses on different sites.
[c]PVI = Plot Volume Index: (root collar diameter)^2 × height × number of surviving trees in plot; D^2H = (root collar diameter)^2 × height.

Table 1.2. Survival and growth of loblolly pine (*Pinus taeda*) after 8 years on a good quality forest site in Southwest Georgia with initially different amounts of *Pisolithus tinctorius* (Pt indices) and naturally occurring ectomycorrhizas at planting (Marx, Cordell & Clark, 1988).

Initial ectomycorrhizal condition					Volume (m^3) per		Total wt (kg) per	
Pt index	Total ectomyc	% survival	Height (m)	DBH* (cm)	Tree	ha	Tree	ha ($\times 10^2$)
88	62	72[a]	8·1[a]	14·0[a]	0·054[a]	59[a]	75[a]	812[a]
68	59	65[ab]	8·0[ab]	13·5[b]	0·054[b]	48[b]	69[b]	670[b]
46	46	62[b]	7·8[bc]	13·5[b]	0·048[bc]	45[bc]	66[bc]	622[bc]
27	43	62[b]	7·8[bc]	13·0[c]	0·045[c]	42[bc]	62[c]	585[bc]
0	41	58[b]	7·7[c]	13·0[c]	0·045[c]	38[c]	61[c]	533[c]

Means in a column which are followed by a common letter are not significantly different at $P = 0·05$.

*DBH = diameter at breast height.

& Cordell, 1988). After 4 years, *Pisolithus tinctorius* ectomycorrhizas stimulated increases in per-hectare volume of 43% and increases in green weight of 37% for loblolly pine (*Pinus taeda*), and increases of 22% for both parameters for slash pine (*Pinus elliottii* var. *elliottii*). Trees with only *Thelephora terrestris* ectomycorrhizas grew considerably less during years of low rainfall than trees with *P. tinctorius*. During the driest years, *P. tinctorius* ectomycorrhizas markedly improved diameter growth. In the second study (Marx, Cordell & Clark, 1988), loblolly pine seedlings with initially different amounts of *P. tinctorius* ectomycorrhizas (Pt index 0, 27, 46, 68, or 88) were planted on a good quality old-field site in southwest Georgia. [Pt index is derived from the formula a × (b/c) where a = percent of seedlings with *Pisolithus tinctorius* (Pt) ectomycorrhizas, b = average percent of feeder roots with Pt ectomycorrhizas (including 0 percent for those without Pt), and c = average percent of feeder roots with ectomycorrhizas formed by Pt and other fungi (total ectomycorrhizas)]. After 8 years and crown closure, trees with initial Pt indices of 88 and 68 had significantly better survival and greater heights, diameters, volumes, and green weights per tree and per hectare than did control (Pt index 0) seedlings (Table 1.2). Volume and weight yields per hectare were over 50% greater and volume and weight yields per tree were over 20% greater for trees with Pt index 88 than for controls. Statistical analysis indicated that average volume per hectare was positively correlated with initial Pt index values larger than 58. Tree ring analyses showed that trees with an

Table 1.3. Yearly basal area growth of loblolly pine (*Pinus taeda*) initially with abundant (Pt index 88) *Pisolithus tinctorius* ectomycorrhizas or with only naturally-occurring (Pt index 0) ectomycorrhizas on a forest site in Southwest Georgia (Marx, Cordell & Clark, 1988)[a].

Year	Soil water deficit (mm)[b]	Basal area growth (cm^2)		
		Index 88	Index 0	Difference
1979	71	1·2	0·9	+0·3
1980	328	6·5	4·6	+1·9*
1981	246	18·1	14·3	+3·8*
1982	259	28·4	24·4	+4·0*
1983	218	23·3	21·5	+1·8*
1984	333	25·3	23·9	+1·4
1985	109	30·9	29·9	+1·0
Totals	1564	133·7	119·5	+14·2*

[a]First year diameters (1978) were not obtained in cores.
[b]Derived by Forest Drought Index Model (Zahner & Myers 1986).
*Denotes significantly larger basal area growth.

initial Pt index of 88 had significantly greater annual basal area growth than controls during growing seasons with water deficits between 218 and 328 mm (Table 1.3). Annual growth did not differ when water deficits were greater or less than these amounts. Under conditions of less than 218 mm soil water deficit, all trees regardless of ectomycorrhizal status grew at comparable rates. However, soil water deficits greater than 328 mm apparently were beyond the capacity for *P. tinctorius* ectomycorrhizas to improve tree growth.

For many decades, longleaf pine (*Pinus palustris*) has been a difficult species to regenerate artificially in the southern US. Seedlings that survive planting may stay in the grass stage (no growth in height) for up to 10 years. Longleaf pine seedlings grown under a specific protocol in the nursery, including *P. tinctorius* inoculation, exhibit up to 50% increases in survival and initiate growth in height after only 2 years. This nursery protocol is now in practical use in the southern US.

Not all field tests with *Pisolithus tinctorius* ectomycorrhizas have revealed survival or growth increases. On poor soils improved with fertilizer or sewage sludge, seedlings of various pine species with *Thelephora terre-*

stris ectomycorrhizas have survived and grown as well as those with abundant *P. tinctorius* ectomycorrhizas during growing seasons of sufficient rainfall. Other negative results have been compounded by low Pt indices on test seedlings at planting.

Extensive research has also been done on seedlings with *Pisolithus tinctorius* ectomycorrhizas planted out on disturbed and adverse sites of various types in the eastern US (Marx, 1980b; Cordell *et al.*, 1987). Marx & Artman (1978) reported that bare-root loblolly pine seedlings with abundant *P. tinctorius* ectomycorrhizas had a 400% greater plot volume than control seedlings with *Thelephora terrestris* ectomycorrhizas after 3 years on an acid coal spoil (pH 4·1) in Kentucky. On plots where fertilizer starter tablets were applied to all seedlings, volumes for *P. tinctorius*-infected seedlings were nearly 250% greater. In the same test, shortleaf pine (*Pinus echinata*) seedlings with *P. tinctorius* ectomycorrhizas were over 400% larger without fertilizer tablets and over 100% larger with fertilizer tablets than control seedlings. Seedlings with *P. tinctorius* ectomycorrhizas also contained more foliar N and less foliar S, Fe, Mn, and Al than control seedlings. In this same report, plot volumes for loblolly pine with *P. tinctorius* ectomycorrhizas, after 4 years on an acid coal spoil (pH 3·4) in Virginia, were over 180% greater than those for control seedlings with *Thelephora terrestris* ectomycorrhizas.

Berry (1982) planted out container-grown seedlings of loblolly, pitch (*Pinus rigida*) and loblolly × pitch pine hybrids with *Pisolithus tinctorius* or *Thelephora terrestris* ectomycorrhizas on acid coal spoils in Tennessee and Alabama. After more than two years, on the Tennessee spoil, *P. tinctorius* ectomycorrhizas increased plot volumes of loblolly pine by 585% and increased those of pitch pine by as much as 125%. *P. tinctorius* ectomycorrhizas increased plot volumes of the hybrid pines by 120 to 575%. On the Alabama spoil, loblolly pine plots with *P. tinctorius* ectomycorrhizas had nearly 300% greater volumes than those with *Thelephora terrestris* ectomycorrhizas. Over 150 and 240% differences in growth were found for the pitch pine parents. The response of the hybrids on the Alabama spoil was between 400 and 1400%. Seedlings with *P. tinctorius* ectomycorrhizas had more N and less Mn, Zn, Cu, and Al in foliage than *T. terrestris* seedlings. The largest gains were on the most adverse coal spoils – those having higher air and soil temperatures, lower pH, greater moisture stress, and higher concentrations of Mn.

In these and other field tests on coal spoils, yearly root evaluations confirmed the ecological adaptation of *Pisolithus tinctorius* to these adverse sites. Without exception, seedlings with *P. tinctorius* ectomycorrhizas developed new roots very rapidly and these were rapidly colonized by the fungus. Colonization was so prolific that after the second year hyphal strands were easily detected with the unaided eye in the spoil material as far as 3 m from the nearest *P. tinctorius*-infected seedlings. Production of fruit bodies of *P. tinctorius* in the test plots was also prolific. As many as three fruit bodies m^{-2} per year were produced. In contrast, after the first growing season, roots of seedlings that initially had *Thelephora terrestris* ectomycorrhizas grew little and very few of the new roots were colonized by *T. terrestris*. Also, many of the original *T. terrestris* ectomycorrhizas (i.e. on the original root system) were necrotic and very few appeared to be functional. After the second year, a low incidence (3 to 5%) of *P. tinctorius* ectomycorrhizas from natural sources was detected on newly formed roots of these seedlings. Fruit bodies of *T. terrestris* were only occasionally observed.

Walker *et al*. (1984) planted out loblolly, shortleaf, and Virginia pine (*Pinus virginiana*) seedlings with *Pisolithus tinctorius* or *Thelephora terrestris* ectomycorrhizas in a fertility study on a coal spoil in Tennessee. Their results after 6 years concurred with previous reports by others that *P. tinctorius* ectomycorrhizas improve survival and growth of pines on a stressed site, regardless of fertilization. Loblolly pines with *P. tinctorius* ectomycorrhizas had over 700% greater plot volumes than did *T. terrestris*-infected seedlings in the fertilizer plots. Earlier, Walker, West & McLaughlin (1982) reported that, after 2 years, trees with *P. tinctorius* ectomycorrhizas exhibited an enhanced ability to absorb water during periods of high soil moisture stress as determined by the pressure chamber technique. Their results explain, at least in part, the mechanism by which *P. tinctorius* is able to enhance tolerance and stimulate tree growth on an adverse site.

Ruehle (1980, 1982) planted out container-grown loblolly pines with *Pisolithus tinctorius*, *Thelephora terrestris*, and no ectomycorrhizas on a South Carolina borrow pit (surface 4 m of soil removed) amended with sewage sludge or fertilizer. After 4 years, seedling volumes of *P. tinctorius* + sludge treatments were 75 and 120% greater than seedlings of the other ectomycorrhizal treatments with sludge. In the fertilizer plots, *P. tinctorius* improved seedling growth by 15 and 200 percent, respectively, over the *T. terrestris* and the initially non-mycorrhizal seedlings.

Berry & Marx (1978) planted out loblolly and Virginia pine seedlings with either *Pisolithus tinctorius* or *Thelephora terrestris* ectomycorrhizas on two severely eroded ridgetops in the Copper Basin of Tennessee. Both sites were fertilized and limed to correct nutrient deficiencies. After 2 years, on the worst site (most acid with lowest native fertility), *P. tinctorius* ectomycorrhizas improved growth of both pine species by over 90%. On the less severe site, growth was improved by about 35%.

These field studies on adverse sites show that forestation of such sites can be expedited by using pine seedlings tailored with ectomycorrhizas formed by *Pisolithus tinctorius*. Thus, the planting of seedlings with root systems physiologically and ecologically adapted to the adversities of the planting site can be important in reclamation.

Other ectomycorrhizal fungus candidates

Other species of ectomycorrhizal fungi have been tried on various tree species in experimental inoculation programs in various parts of the world and show promise for eventual practical application. Various types of vegetative inoculum of 22 fungal species and spore inoculum of 18 species have been developed and tested on different tree species under various experimental conditions. Positive results of field testing for the effects of ectomycorrhizas formed by some of these fungi on different tree hosts have also been reported. As with *Pisolithus tinctorius* ectomycorrhizas, however, seedlings with certain of these specific ectomycorrhizas in some field studies have shown no improvements in survival and growth (Marx & Ruehle, in press).

Practical commercial application

Until recently, artificial inoculation with *Pisolithus tinctorius* or any other ectomycorrhizal fungus was limited because procedures, commercial inoculum, and necessary equipment were not readily available to nurserymen and regeneration foresters. As discussed, the USDA Forest Service has been cooperating with several private companies to develop commercial ectomycorrhizal fungus inoculum and the procedures and equipment needed to inoculate bare-root and container-grown seedlings. In addition to *P. tinctorius*, inoculum of *Hebeloma* spp., *Laccaria* spp., and *Scleroderma* spp. have also been developed. However, the significance of those fungi, other than *P. tinctorius*, to seedling performance in the field has not been determined. The types of *P. tinctorius* inoculum that are available include vegetative inoculum from Mycorr Tech, Inc., spore pellets, spore-encapsulated seeds, and bulk spores from either International Forest Tree Seed Co., Odenville, Alabama, or SouthPine, Inc., Birmingham, Alabama. A tractor-drawn nursery seedbed applicator has been developed (Fig. 1.1c) to place *P. tinctorius* vegetative inoculum accurately in seedbeds prior to sowing, in bare-root nurseries (Cordell *et al.*, 1981). Inoculum is applied in bands under seed rows at desired depths. Use of the applicator has reduced the amount of vegetative inoculum needed by 75% and reduced time and labour requirements as compared to broadcast application.

There is a wide range in the cost of commercially available *Pisolithus tinctorius* inoculum (Table 1.4). Operational cost of each inoculum type also varies with such factors as seedbed density, seed size for spore-en-

Table 1.4. Costs of various commercial inoculum types of *Pisolithus tinctorius* in 1987[a]

Inoculum type	Inoculum cost per[c]			
	1000 seedlings		planted hectare	
	$	£	$	£
Vegetative	7·50	4·26	13·45	7·64
Spore-encapsulated seeds	2·22	1·26	3·98	2·26
Spore pellets	2·75	1·56	4·93	2·80
Double-screened[b] bulk spores	0·43	0·24	0·77	0·44

[a]Cost estimates are for loblolly (*Pinus taeda*) and slash (*P. elliottii* var. *elliottii*) pine bare-root nurseries at 270 seedlings m^{-2}, and forest plantings at 1·8×3·0 m spacing for 1794 trees ha^{-1} in the Southern US.
[b]Double screening is required for even flow through spray nozzles with aqueous spore suspensions. Standard bulk spores are screened only once.
[c]£1 sterling = $US 1.76.

capsulated seeds, and spacing of field plantings. In 1987, the *P. tinctorius* vegetative inoculum costs for bare-root nurseries were reduced by increasing efficiency of nursery seedbed inoculations, improving effectiveness of inoculum, and decreasing application rates. The vegetative inoculum is sold on a volume (litre) basis, while the spore inocula are all sold on a weight (kg) basis.

Operational procedures for inoculation vary by inoculum type, but with any inoculum the biological requirements of a second living organism must be considered in addition to those of the seedling. Special precautions are necessary for shipping, storing, and handling the inoculum, as well as for lifting, handling, and field planting of seedlings. For successful inoculation in bare-root seedbeds, populations of pathogenic, saprotrophic, and native ectomycorrhizal fungi that may already be established in the soil must be reduced by spring soil fumigation. Fortunately, soil fumigation is a standard procedure in the majority of nurseries in the US. Prior to spring sowing, vegetative inoculum can be broadcast on the soil surface and incorporated into the fumigated seedbeds or it can be machine-applied with greater effectiveness and efficiency. For container-grown seedlings, vegetative inoculum can be incorporated into the growing medium before filling the containers or placed at selected depths in the growing medium in the container. Bulk spores can be sprayed, drenched, or dusted onto growing medium for containerized seedlings and onto seedbeds in bare-root nurseries. Spore pellets can either be incorporated into the growing medium or soil, or they can be broadcast on the soil surface, lightly covered, and irrigated. Spore pellets have been applied at several nurseries with a standard fertilizer spreader. Spore-encapsulated seeds can be sown by conventional sowing methods.

Fruit bodies of *Pisolithus tinctorius* (Fig. 1.1d) are usually produced in great quantities (3 to 4 m^{-2} of soil) in nursery beds 3 to 4 months after soil inoculation with vegetative inoculum. Spores from these fruit bodies have been used successfully in operational inoculation programmes. These spores appear to maintain the aggressive nature of the parent vegetative culture.

The demand for *P. tinctorius*-tailored nursery seedlings has increased significantly during the past 5 years in the eastern US, despite the added costs. Annual demand for tailored seedlings has increased from 0·5 million in 1984 to over 6 million seedlings in 1988. During the spring of 1988, vegetative or spore inoculum of *P. tinctorius* was applied at 10 bare-root and 2 container nurseries in the southern, central, and northeastern US. Over 6 million seedlings of eight conifer and one hardwood species are being produced. This number includes 2 million seedlings produced at a South Carolina State nursery for operational reforestation at the Savannah River Forest Station operated by the USDA Forest Service and the US Department of Energy. This operation was the largest single, pure vegetative inoculum application of an ectomycorrhizal fungus in a forest tree nursery to date in the world.

Seedlings with *Pisolithus tinctorius* ectomycorrhizas from the above nurseries have been planted on routine reforestation sites and on numerous disturbed sites in the eastern and central US. Currently, field responses of southern pine seedlings to *P. tinctorius* ectomycorrhizas are being monitored in over 75 plantings on diverse reforestation sites. Most of these plantings have been established since 1979. It will take many years, therefore, to determine benefits at stand maturity. It is anticipated, however, that results from many of the older plantings, i.e. those at crown-closure, will be published shortly. Increases of 35 to 50% in tree volume attributable to *P. tinctorius* ectomycorrhizas are still being observed on loblolly, Virginia and eastern white pines after 10 years on a good quality reforestation site in North Carolina which we reported on earlier (Marx, Bryan & Cordell, 1977). Approximately 20 operational plantings of Virginia pine, eastern white pine, pitch-loblolly hybrid pine, and red and black oak seedlings are being monitored on disturbed sites, such as coal spoils. Preliminary observations on these young plantings indicated significant increases in tree survival and early growth resulting from *P. tinctorius* ectomycorrhizas. Planting studies established by the Ohio Division of Mineland Reclamation on coal-mined sites in southern Ohio during 1982 and 1983 showed an average survival increase of 23% for Virginia pine and 24% for eastern white pine seedlings over routine nursery seedlings after 2 years in the field.

Conclusion

Several methods are available to ensure the development of ectomycorrhizas on forest tree seedlings for the establishment of man-made forests. Certain methods have more advantages than others. Pure vegetative inoculum has the greatest biological advantage. There is sufficient information to conclude that pure cultures of certain fungi, such as *Pisolithus tinctorius*, can be used to improve survival and growth of tree seedlings on a variety of sites. This fungus is being used in practical reforestation and reclamation programmes in the United States today. However, these results represent only the beginning of a widespread practical programme. When one considers the millions of hectares of potential exotic forests which must be established on former treeless sites in the Third World, as well as the millions of hectares of former forested lands awaiting artificial regeneration throughout the world, the importance of the selection, propagation, manipulation, and management of superior strains or species of ectomycorrhizal fungi as a forest regeneration tool is paramount. Research so far has only revealed a few of the potential uses of specific ectomycorrhizal fungi in world forestry. A tremendous amount of basic and practical information must be revealed if these fungi are to be fully utilized and integrated into existing forestry programmes.

Since forest trees and ectomycorrhizal fungi have evolved together in naturally stressed environments, a major benefit of ectomycorrhizas is that they aid the host tree in managing stresses. It is logical, therefore, that programmes aimed at using fungi to improve stress management by seedlings directly address these stresses in research and development.

References

Berry, C. R. (1982). Survival and growth of pine hybrid seedlings with *Pisolithus* ectomycorrhizae on coal spoils in Alabama and Tennessee. *Journal of Environmental Quality*, 11, 709-715.

Berry, C. R. & Marx, D. H. (1978). Effects of *Pisolithus tinctorius* ectomycorrhizae on growth of loblolly and Virginia pines in Tennessee Copper Basin. *US Department of Agriculture, Forest Service Research Note*, SE-264. US Department of Agricultures: Asheville, North Carolina.

Bowen, G. D. (1973). Mineral nutrition of ectomycorrhizae. In *Ectomycorrhizae: Their Ecology and Physiology*, ed. G. C. Marks & T. T. Kozlowski, pp. 151-205. Academic Press: New York.

Cordell, C. E., Marx, D. H., Lott, J. R. & Kenney, D. S. (1981). The practical application of *Pisolithus tinctorius* ectomycorrhizal inoculum in forest tree nurseries.

In *Forest Regeneration*, Proceedings of Symposium on Engineering Systems for Forest Regeneration, pp. 38-42. American Society of Agricultural Engineers: Raleigh, North Carolina.

Cordell, C. E., Owen, J. E., Marx, D. H. & Farley, M. E. (1987). Ectomycorrhizal fungi—beneficial for mineland reclamation. In *Proceedings 1987 National Symposium on Mining, Hydrology, Sedimentology, and Reclamation*, pp. 321-326. University of Kentucky Press: Lexington, Kentucky.

Dighton, J., Poskitt, J. M. & Howard, D. M. (1986). Changes in occurrence of basidiomycete fruit bodies during forest stand development: with specific reference to mycorrhizal species. *Transactions of the British Mycological Society*, **87**, 163-171.

Ek, M., Ljungquist, P. O. & Stenström, E. (1983). Indole-3-acetic acid production by mycorrhizal fungi determined by gas chromatography mass spectrometry. *New Phytologist*, **94**, 401-407.

Fries, N. (1987). Somatic incompatibility and field distribution of the ectomycorrhizal fungus *Suillus luteus* (Boletaceae). *New Phytologist*, **107**, 735-739.

Hacskaylo, E. & Vozzo, J. A. (1967). Inoculation of *Pinus caribaea* with pure cultures of mycorrhizal fungi in Puerto Rico. In *Proceedings of the 14th International Union of Forestry Research Organizations*, vol. 5, pp. 139-148. IUFRO: Munich.

Hung, L. L. & Chien, C. Y. (1978). Physiological studies on two ectomycorrhizal fungi, *Pisolithus tinctorius* and *Suillus bovinus*. *Transactions of the Mycological Society of Japan*, **19**, 121-127.

Laiho, O. (1970). *Paxillus involutus* as a mycorrhizal symbiont of forest trees. *Acta Forestalia Fennica*, **106**, 1-73.

Lamb, R. J. & Richards, B. N. (1971). Effect of mycorrhizal fungi on the growth and nutrient status of slash and radiata pine seedlings. *Australian Forestry*, **35**, 1-7.

Last, F. T., Mason, P. A., Ingleby, K. & Fleming, L. V. (1984). Succession of fruitbodies of sheathing mycorrhizal fungi associated with *Betula pendula*. *Forest Ecology & Management*, **9**, 229-234.

Levisohn, I. (1956). Growth stimulation of forest tree seedlings by the activity of freeliving mycorrhizal mycelia. *Forestry*, **29**, 53-59.

Marx, D. H. (1969). The influence of ectotrophic mycorrhizal fungi on the resistance of pine roots to pathogenic infections. I. Antagonism of mycorrhizal fungi to root pathogenic fungi and soil bacteria. *Phytopathology*, **59**, 153-163.

Marx, D. H. (1973). Mycorrhizae and feeder root diseases. In *Ectomycorrhizae: Their Ecology and Physiology*, ed. G. C. Marks & T. T. Kozlowski, pp. 351-382. Academic Press: London.

Marx, D. H. (1977). Tree host range and world distribution of the ectomycorrhizal fungus *Pisolithus tinctorius*. *Canadian Journal of Microbiology*, **23**, 217-223.

Marx, D. H. (1979). Synthesis of ectomycorrhizae by different fungi on northern red oak seedlings. *US Department of Agriculture, Forest Service Research Note*, SE-282. US Department of Agricultures: Asheville, North Carolina

Marx, D. H. (1980a). Ectomycorrhizal fungus inoculations: a tool for improving forestation practices. In *Tropical Mycorrhiza Research*, ed. P. Mikola, pp. 13-71. Oxford University Press: London.

Marx, D. H. (1980b). Role of mycorrhizae in forestation of surface mines. In *Proceedings of a Symposium on Trees for Reclamation in the Eastern United States,*

Forest Service General Technical Report, NE-61, pp. 109-116. US Department of Agriculture: Lexington, Kentucky.

Marx, D. H. (1981). Variability in ectomycorrhizal development and growth among isolates of *Pisolithus tinctorius* as affected by source, age, and reisolation. *Canadian Journal of Forest Research*, 11, 168-174.

Marx, D. H. (1985). Trials and tribulations of an ectomycorrhizal fungus inoculation program. In *Proceedings of the 6th North American Conference on Mycorrhizae*, ed. R. Molina, pp. 62-63. Forestry Research Laboratory: Corvallis, Oregon.

Marx, D. H. & Artman, J. D. (1978). Growth and ectomycorrhizal development of loblolly pine seedlings in nursery soil infested with *Pisolithus tinctorius* and *Thelephora terrestris* in Virginia. Forest Service Research Note, SE-256. US Department of Agriculture: Asheville, North Carolina.

Marx, D. H. & Bryan, W. C. (1971). Influence of ectomycorrhizae on survival and growth of aseptic seedlings of loblolly pine at high temperature. *Forest Science*, 17, 37-41.

Marx, D. H. & Bryan, W. C. (1975). Growth and ectomycorrhizal development of loblolly pine seedlings in fumigated soil infested with the fungal symbiont *Pisolithus tinctorius*. *Forest Science*, 21, 245-254.

Marx, D. H., Bryan, W. C. & Cordell, C. E. (1976). Growth and ectomycorrhizal development of pine seedlings in nursery soils infested with the fungal symbiont *Pisolithus tinctorius*. *Forest Science*, 22, 91-100.

Marx, D. H., Bryan, W. C. & Cordell, C. E. (1977). Survival and growth of pine seedlings with *Pisolithus* ectomycorrhizae after two years on reforestation sites in North Carolina and Florida. *Forest Science*, 23, 363-373.

Marx, D. H., Bryan, W. C. & Davey, C. B. (1970). Influence of temperature on aseptic synthesis of ectomycorrhizae by *Thelephora terrestris* and *Pisolithus tinctorius* on loblolly pine. *Forest Science*, 16, 424-431.

Marx, D. H. & Cordell, C. E. (1988). *Pisolithus* ectomycorrhizae improve 4-year performance of loblolly and slash pines in South Georgia. Georgia Forestry Research Report No. 4, Macon, Georgia.

Marx, D. H., Cordell, C. E. & Clark, A., III (1988). Eight-year performance of loblolly pine with *Pisolithus* ectomycorrhizae on a good-quality forest site. *Southern Journal of Applied Forestry* 12, 275-280.

Marx, D. H., Cordell, C. E., Kenney, D. S., Mexal, J. G., Artman, J. D., Riffle, J. W. & Molina, R. J. (1984). Commercial vegetative inoculum of *Pisolithus tinctorius* and inoculation techniques for development of ectomycorrhizae on bare-root seedlings. *Forest Science Monograph*, 25.

Marx, D. H., Cordell, C. E., Maul, S. B. & Ruehle, J. L (1989a). Ectomycorrhizal development on pine by *Pisolithus tinctorius* in bare-root and container seedling nurseries. I. Efficacy of various vegetative inoculum formulations. *New Forests*, 3, 45-56.

Marx, D. H., Cordell, C. E., Maul, S. B. & Ruehle, J. L (1989b). Ectomycorrhizal development on pine by *Pisolithus tinctorius* in bare-root and container seedling nurseries. II. Efficacy of various vegetative and spore inocula. *New Forests*, 3, 57-66.

Marx, D. H. & Daniel, W. J. (1976). Maintaining cultures of ectomycorrhizal and plant pathogenic fungi in sterile water cold storage. *Canadian Journal of Microbiology*, 22, 338-341.

Marx, D. H. & Krupa, S. V. (1978). Mycorrhizae. A. Ectomycorrhizae. In *Interactions Between Nonpathogenic Soil Microorganisms and Plants*, ed. Y. R. Dommergues & S. V. Krupa, pp. 373-400. Elsevier Scientific Publishing Company: Amsterdam.

Marx, D. H., Morris, W. G. & Mexal, J. G. (1978). Growth and ectomycorrhizal development of loblolly pine seedlings in fumigated and nonfumigated soil infested with different fungal symbionts. *Forest Science*, 24, 193-203.

Marx, D. H. & Ruehle, J.L. (in press). Ectomycorrhizae as biological tools in reclamation and revegetation of waste lands. In *Proceedings of the Asian Conference on Mycorrhizae, January 1988, Madras, India*.

Marx, D. H., Ruehle, J. L., Kenney, D. S., Cordell, C. E., Riffle, J. W., Molina, R. J., Pawuk, W. H., Navratil, S., Tinus, R. W. & Goodwin, O. C. (1982). Commercial vegetative inoculum of *Pisolithus tinctorius* and inoculation techniques for development of ectomycorrhizae on container-grown tree seedlings. *Forest Science*, 28, 373-400.

Maul, S. B. (1985). Production of ectomycorrhizal fungus inoculum by Sylvan Spawn Laboratory. In *Proceedings of the 6th North American Conference on Mycorrhizae*, ed. R. Molina, pp. 64-65. Forestry Research Laboratory: Corvallis, Oregon.

Mikola, P. (1973). Application of mycorrhizal symbiosis in forestry practice. In *Ectomycorrhizae: Their Ecology and Physiology*, eds. G. C. Marks & T. T. Kozlowski, pp. 383-411. Academic Press: New York.

Moser, M. (1958a). Die kunstliche Mykorrhizaimpfung an Forstpflanzen. I. Erfahrungen bei der Reinkultur von Mykorrhizapilzen. *Forstwissenschaftliches Centralblatt*, 77, 32-40.

Moser, M. (1958b). Die kunstliche Mykorrhizaimpfung und Forstpflanzen. II. Die Torfstreukultur von Mykorrhizapilzen. *Forstwissenschaftliches Centralblatt*, 77, 273-278.

Moser, M. (1958c). Die Einfluss tiefer Temperaturen auf das Washstum und die Lebenstatigkeit hoherer Pilze mit spezieller Berucksichtigung von Mykorrhizapilzen. *Sydowia*, 12, 386-399.

Reid, C. P. P. & Woods, F. W. (1969). Translocation of ^{14}C-labeled compounds in mycorrhizae and its implications in interplant nutrient cycling. *Ecology*, 50, 179-187.

Ruehle, J. L. (1980). Growth of containerized loblolly pine with specific ectomycorrhizae after 2 years on an amended borrow pit. *Reclamation Review*, 3, 95-101.

Ruehle, J. L. (1982). Mycorrhizal inoculation improves performance of container-grown pines planted on adverse sites. In *Proceedings of the Southern Containerized Forest Tree Seedling Conference*, Forest Service General Technical Report, SO-37, pp. 133-35. US Department of Agriculture: Savannah, Georgia.

Schenck, N. C. (1982). *Methods and Principles of Mycorrhizal Research*. The American Phytopathological Society: St. Paul, Minnesota.

Schramm, J. R. (1966). Plant colonization studies on black wastes from anthracite mining in Pennsylvania. *Transactions of the American Philosophical Society*, 56, 1-194.

Takacs, E. A. (1961). Inoculacion de especies de pinos con hongos formadores de micorrizas. *Silvicultura*, 15, 5-17.

Takacs, E. A. (1964). Inoculacion artificial de pinos de regiones subtropicales con hongos formadores de micorrizas. *India Supplemento Forestal*, 12, 41-44.

Takacs, E. A. (1967). Produccion de cultivos puros de hongos micorrizogenos en el Centro Nacional de Investigaciones Agropecuarias, Castelar. *India Supplemento Forestal*, **4**, 83-87.

Theodorou, C. & Bowen, G. D. (1970). Mycorrhizal responses of radiata pine in experiments with different fungi. *Australian Forestry*, **34**, 183-191.

Trappe, J. M. (1959). Studies on *Cenococcum graniforme*. I. An efficient method for isolation of sclerotia. *Canadian Journal of Botany*, **47**, 1389-1390.

Trappe, J. M. (1977). Selection of fungi for ectomycorrhizal inoculation in nurseries. *Annual Review of Phytopathology*, **15**, 203-222.

Walker, R. F., West, D. C. & McLaughlin, S. B. (1982). *Pisolithus tinctorius* ectomycorrhizae reduce moisture stress of Virginia pine on a Southern Appalachian coal spoil. In *Proceedings of the 7th North American Forest Biology Workshop*, ed. B. A. Thielges, pp. 374-383. University of Kentucky Press: Lexington, Kentucky.

Walker, R. F., West, D. G., McLaughlin, S. B. & Amundsen, C. C. (1984). The performance of loblolly, Virginia, and shortleaf pine on a reclaimed surface mine as affected by *Pisolithus tinctorius* ectomycorrhizae and fertilization. In *Proceedings of the 3rd Biennial Southern Silvicultural Research Conference*, Forest Service General Technical Report, SO-54, pp. 410-416. U.S. Department of Agriculture: Atlanta, Georgia.

Wilkins, W. H. & Harris, G. C. M. (1946). The ecology of the larger fungi. V. An investigation into the influence of rainfall and temperature on the seasonal production of fungi in a beechwood and a pinewood. *Annals of Applied Biology*, **33**, 179-188.

Zahner, R. & Myers, R. K. (1986). Assessing the impact of drought on forest health. In *Proceedings of the Society of American Foresters 1986 National Convention, Birmingham, AL*, pp. 227-234. Society of American Foresters: Bethesda, Maryland.

Zak, B. & Marx, D. H. (1964). Isolation of mycorrhizal fungi from roots of individual slash pines. *Forest Science*, **10**, 214-222.

Chapter 2

The cultivation of ectomycorrhizal fungi

L. M. Harvey, J. E. Smith, B. Kristiansen, J. Neill & E. Senior

Department of Bioscience & Biotechnology, University of Strathclyde, Glasgow G1 1XW, UK

Introduction

Mycorrhizal fungi are ubiquitous, occurring in natural ecosystems in most climatic zones throughout the world. Most agricultural plants of economic importance form mycorrhizal associations, including field crops such as maize and wheat and tropical plantation crops such as coffee, tea and rubber. The most common associations are those formed between the host plant and vesicular-arbuscular mycorrhizal fungi. Ectomycorrhizal associations are formed by as few as 5% of vascular plants, but they predominate in several economically important families such as Pinaceae (Pine, Fir, Larch), Betulaceae (Alder, Birch) and Fagaceae (Oak, Chestnut, Beech) (Harley, 1969; Trappe 1977). Plants with mycorrhizal associations are, in general, healthier, sturdier and more disease resistant than their counterparts which lack the association. Indeed, many forest trees would not survive without mycorrhizas. There are many reports in the literature of tree plantations where host species fail to grow until mycorrhizal fungi are introduced into the ecosystem (Brisco, 1959; Madu, 1967; Mikola, 1970).

The inoculation of agricultural crops with microbial species that benefit the plant has long been practiced in agriculture. With increased public awareness of certain adverse effects of fertilizers, pesticides and other routinely used agrochemicals on both animals (including man) and the environment, the deliberate application of beneficial microorganisms is becoming extremely attractive (Paau, 1988).

Advances in biotechnology leading to the improved performance of microbial inocula has stimulated interest in the use of microorganisms to increase plant productivity. The use of mycorrhizal inoculants can increase plant productivity by:

- improving the survival rate of transplanted seedlings;
- improving plant growth rate;
- increasing disease resistance;

- improving the survival rate of plants in adverse conditions such as drought or unfertile soils.

Mycorrhizal fungi have been introduced to nursery or forest soils in a number of different forms. For example, by transferring natural soil containing established mycorrhizal fungi (Mikola, 1973); by incorporating sporophore material – the sporophore being chopped, dried and mixed with a carrier before being broadcast onto the soil or mixed directly with the growth medium in containerised systems (Marx *et al.*, 1982); and by inoculating with pure cultures of vegetative mycelia. This final approach has been recommended as the most reliable method of inoculation (Mikola, 1973; Trappe, 1977). However, a major drawback has been the difficulty of large scale axenic cultivation of ectomycorrhizal fungi. This chapter will consider the methods of laboratory scale propagation of ectomycorrhizal fungi with special emphasis on production of sufficient quantities for use as commercial inocula.

Applications of ectomycorrhizal inocula

Applications of ectomycorrhizal inocula are found mainly in the forestry industry. The presence of mycorrhizas on pine seedlings planted on previously forested land can greatly influence performance of the plant. At the nursery level, fumigation procedures used to eliminate pathogenic organisms can also kill beneficial mycorrhizal fungi. In the past, nurseries have depended on the slow and unreliable process of recolonization by the few surviving organisms or by air-borne spores (Mexal, 1980). Introduction of a laboratory-produced inoculum in the seed bed would greatly accelerate the recolonization process and also ensure that the fungal species present was the one best suited to the plant being cultivated.

Mycorrhizal fungi may also be of use in land reclamation. The most common method of waste disposal in many countries is by controlled landfill (Gulley, 1981). Restoration of completed landfill sites must become a priority if landfill is to remain popular, not only for aesthetic reasons but also to regenerate the land. On many waste sites the most effective means of restoration is by introduction of continuous vegetation cover (Johnson, Bradshaw & Harwood, 1976). Given time, any landfill site revegetates naturally (Bradshaw, 1984), but plant growth can be limited by several factors including the presence of toxic leachates (Clouston, 1984) and gases (Arthur, Leone & Flower, 1981). In common with many other waste sites, growth of trees on landfill sites may be further limited by the absence of suitable mycorrhizal fungi in the soil (Good, Wilson & Last, 1980). Ectomycorrhizal fungi may be of immense value in the establishment of trees on industrial waste such as anthracite waste (Schramm, 1966), metallic mine tailings (Harris & Jurgensen, 1977) and coal spoils (Marx & Artman-1979; Ingleby, Last & Mason, 1985). On such sites the presence of ectomycorrhizal fungi appears to increase the host plant tolerance to drought, high soil temperatures, soil toxins (both inorganic and organic) and extremes of soil pH (Marx, 1975, 1976, 1977, 1980a).

Isolation and production of axenic cultures

The first stage in the cultivation of any microorganism is isolation from the natural environment and the production of a pure culture. Mycorrhizal fungi are, in general, difficult to isolate and culture in the laboratory. Some species are obligate symbionts and have not yet been cultured in the absence of host cells. Such species are therefore not suitable for bulk production of ectomycorrhizal inoculants.

Jackson & Mason (1984) indicate that although mycorrhizas can be successfully isolated at any time of the year, best results are obtained from isolates collected in summer and autumn. For ease of identification and to minimise contamination, the fungus should be isolated from the fruiting body as soon as possible after it has been picked (Moser, 1958), as well as from the mycorrhiza. Isolation from mycorrhizas involves washing procedures followed by treatment with hydrogen peroxide or other chemical sterilants to reduce the number of contaminating organisms. Addition of antibiotics, such as streptomycin and penicillin, to growth media may be required to suppress bacterial growth. Once the fungus has been successfully isolated and cultivated, the host plant should be inoculated with the fungus to demonstrate the mycorrhizal association. Once these steps have been carried out and proven, one can confidently identify the fungal isolate. Emergence of the hyphae can take between 3 and 20 days.

In many cases, attempts to isolate and grow mycorrhizal fungi on agar media have failed. Routine preliminary isolation steps should therefore employ more than one isolation medium in order to increase the chances of isolating specific fungi. Once isolated, care should be taken to preserve and maintain isolates of fungi. To this end a variety of preservation techniques should be employed to ensure that at least one method is successful.

Selection and maintenance of fungal species

Selection of the most appropriate fungus or mixture of fungal species is of the utmost importance for commercial success. Fungi are selected to meet the following criteria:

- They must be able to form mycorrhizas with the host. Strict host specificity of ectomycorrhizal fungi is rare (Trappe, 1962; Molina, 1979) but one plant may commonly form mycorrhizas with several species of fungi at the same time (Dominik, 1959). For example, Trappe (1977) estimated that some 2000 species of ectomycorrhizal fungi are potential partners of Douglas fir (*Pseudotsuga men-ziesii*), few being host specific.

- They must be suited to the soil for which they are destined. Often the soil may be very acidic, may contain high levels of specific metal ions, be lacking in essential nutrients or subject to drought.
- They should be readily propagated and maintained in the laboratory. Because mycorrhizal fungi, as with many Basidiomycete fungi, grow poorly in culture, it is important to chose a fungus that is easy to grow and maintain in a viable condition.

Opinions vary regarding the maintenance of these cultures. Takacs (1967) suggested that stock cultures should be subcultured every 60 days to keep the fungi viable. Moser (1958) recommended that the cultures be transferred frequently, that is, every 2 to 5 weeks depending on the fungal species. Marx (1980b) found that continuous culturing of ectomycorrhizal fungi could lead to reduced growth rates and loss of ability to form mycorrhizas. For culture security, the researcher should therefore adopt all procedures in order to ensure continued production of healthy cultures.

Cultivation of ectomycorrhizal fungi

The cultivation of certain microorganisms in the laboratory remains an exacting task. This is particularly true of ectomycorrhizal fungi because, removed from their symbiont partners, they are naturally slow growers. In the field, ectomycorrhizal fungi depend on their host plant for their supply of carbohydrates. Generally speaking the carbohydrates take the form of simple sugars such as glucose, sucrose and fructose. Melin (1925) carried out the first experiments to determine which carbon compounds are used by mycorrhizal fungi. Since then various researchers have studied the growth of such fungi on a range of simple and complex carbon sources (Norkrans, 1950; Palmer & Hacskaylo, 1970). The response of the ectomycorrhizal fungi *in vitro* tends to be genus-specific. The majority of ectomycorrhizal fungi use simple sugars but some species can use disaccharides, sugar alcohols and polysaccharides. For example, Norkrans (1950) showed that *Tricholoma fumosum* produces cellulase and can utilise cellulose.

During *in vitro* cultivation, a small amount of glucose $(0 \cdot 1 \text{ g l}^{-1})$ can greatly enhance the hydrolysis of more complex sugars by stimulating the formation of adaptive enzymes like cellulases and amylases (Jackson & Mason, 1984). These enzymes are normally suppressed in nature since the fungus readily obtains simple carbohydrates from the host plant.

Nitrogen nutrition of ectomycorrhizal fungi appears to be relatively simplistic and is similar to other soil fungi. Defined nitrogen sources such as ammonium salts, are usually the best source of nitrogen to use. Ectomycorrhizal fungi are also able to use complex nitrogen sources such as organic nitrogen compounds, amino acids, peptones, yeast and plant extracts. Some species will grow better if an amino acid is incorporated into the medium. Many species are stimulated by the presence of group B vitamins, particularly thiamin and biotin. For example, *Boletus* species are heterotrophic for thiamin. Jackson & Mason (1984) recommend that both thiamin and biotin be incorporated into synthetic culture media. Ectomycorrhizal fungi can also use a wide range of phosphate compounds including organic phosphates (Bowen, 1973). Other essential minerals required for good growth of ectomycorrhizal fungi include K, Mn, Zn, sometimes required only as trace elements. Most species of ectomycorrhizal fungi grow best in acidic environments with a pH range between 4 and 7 and at temperatures between 18 and 27 °C (Jackson & Mason, 1984).

Cultivation of ectomycorrhizal fungi on solid media

A wide range of agar media have been employed in the isolation and cultivation of ectomycorrhizal fungi. Suitable media for cultivating mycorrhizal fungi from sporophores or 'sterilised' mycorrhizas are documented in the literature (Jackson & Mason, 1984). By far the most popular agar is Modified Melin Norkrans (MMN) which appears to suit a wide range of fungi and is often a first choice medium for the isolation of mycorrhizal fungi from nature. Though this is a suitable general cultivation medium, improvement in culture performance and maintenance is often achieved by developing a medium more suited to the specific needs of individual strains. In practice a good rate of growth on solid media produces a colony 6 to 8 cm diameter after 15 to 20 days culture (Marx & Kenney, 1982).

Agar plugs have been used to inoculate seedlings such as pine (Grenville & Peterson, 1985). The plugs were transferred into growth pouches 5 to 10 mm from the actively growing root and modified Melin-Norkrans nutrient solution added. In addition, this method has been used in the laboratory to induce sclerotium formation in the fungus. This method is not routinely used as an inoculation method for nursery trials, but can be used for screening candidate fungi on different hosts.

Agar media are also useful for storing and maintaining ectomycorrhizal fungi. Cultures can be stored on slants or plates under refrigerated conditions (4°C) for up to 6 months, or mycelial agar discs can be stored in sterile distilled water at 5°C and held for long periods without loss of viability. Marx & Daniel (1976), using the sterile water cold storage technique, have held discs for periods up to three years without loss of viability. This method also appears the most suitable for retaining infectivity characteristics of ectomycorrhizal fungi.

Static flask culture

Static flask cultivation of ectomycorrhizal fungi is a popular method of cultivation for routine laboratory use and for small-scale field/nursery trials. Both solid and liquid culture methods are employed.

Liquid static flask culture is useful for production of small quantities of pure cultures of a fungal species which can subsequently be used for the inoculation of solid substrates such as vermiculite-peat. The advantage of liquid culture techniques over solid substrates are that:

- contamination of the culture by other microbial species is easy to detect;
- determination of the amount of biomass present is very simple;
- changes in culture morphology can be seen easily so deterioration of the culture can be detected readily. This is in stark contrast to solid-substrate cultures where laborious removal of the soil and microscopic examination of the organism is necessary.

Oxygen transfer in a static liquid flask culture is negligible and is solely dependent on the surface transfer of oxygen at the air-liquid interface. It has been suggested that ectomycorrhizal fungi are obligate aerobes (Jackson & Mason, 1984); consequently, the maximum surface area possible should be available for oxygen transfer. Typically, a 500 ml conical flask will contain 100 ml of medium. To compensate for losses by evaporation, sterile distilled water can be added during cultivation, though this procedure can be a source of contamination.

Many ectomycorrhizal fungi form large, dense pellets when cultivated in this way. In large pellets only the periphery has access to nutrients and oxygen. Consequently, much of the mycelial biomass is non-viable. To overcome this problem, many researchers add broken glass to the nutrient solution before sterilization. Regular shaking of the culture breaks up large clumps or pellets and helps to produce a more dispersed (and more viable) fungal mass. However, there are many disadvantages to this system. In order to produce sufficient vegetative inocula to use for nursery or field trials, many flasks must be prepared and inoculated. Also, although it is possible to grow the organisms at room temperature, a constant temperature room is advisable for consistent production, particularly if the process is intended for a commercial production facility.

Solid substrate culture of ectomycorrhizal fungi in the laboratory is mainly carried out in flasks or bottles containing vermiculite-peat (9:1, v/v) or peat moss. The substrate is moistened with MMN in the ratio of 180 ml solution to 250 ml vermiculite-peat (Marx 1969). The mixture is then sterilised and inoculated either with cultures from agar plates or with liquid cultures produced as above. The cultures must then be incubated for periods of up to 3-4 months to allow the organism to grow through the substrate (Mason, 1980).

Shake flask culture

Shake flask culture, introduced by Kluyver & Perquin (1933), is widely used in the fermentation industry and in other research areas for the production of inocula, and empirical medium design experiments, but has not been practiced in mycorrhizal studies. It is hard to say why this is the case, particularly considering the obligately aerobic nature of ectomycorrhizal fungi. The tendency of ectomycorrhizal fungi to form large clumps could be one reason. A simple homogenization step, thus providing a large number of growing points from which the culture develops can get around this problem to some extent. Since some species may be sensitive to homogenization, mild sonication might be used to fragment hyphae.

Once inoculated, flasks can be incubated in rotary incubators where temperature is readily controlled. One group, at the Centre de Recherches Forestières de Nancy have used this method to cultivate a strain of *Hebeloma crustuliniforme* on a solution of diluted malt extract at 25°C (Mauperin *et al.*, 1987).

Production of ectomycorrhizal fungi by submerged liquid fermentation

Few reports on the successful cultivation of ectomycorrhizal fungi by submerged liquid fermentation have been published. The basic factors influencing growth of microorganisms, including ectomycorrhizas, can be divided into two groups (Pirt, 1975):

(1) Intracellular factors, which depend on the structure and internal organization of the cell. These include, metabolic mechanisms and genetic organization.

(2) Extracellular factors, such as nutritional and environmental factors (Table 2.1).

Table 2.1. Extracellular factors influencing the growth of microorganisms

Nutritional parameters	Environmental parameters
Carbon	Temperature
Nitrogen	pH
Phosphate	Agitation
Minerals	Dissolved oxygen tension
Trace elements	
Growth factors	

It is not possible to control directly the intracellular factors influencing the growth of microorganisms. However, by employing sophisticated fermentation technology in conjunction with nutritional control it is now possible to influence intracellular factors by the control of parameters external to the cell.

Nutritional parameters and temperature can be controlled only to a limited extent using static or shake flask culture and consequently such methods are not ideal for producing a standard reproducible inoculum. This is particularly important when cultivating ectomycorrhizal fungi since small changes in growth conditions can result not only in a loss of biomass production but also loss of infectivity. Thus, fermenters are potentially advantageous in solving these problems.

Many types of fermenters are available for the cultivation of microorganism and cell cultures. The most popular and versatile fermenter used for cultivating fungi is the stirred tank reactor which can be modified easily to suit the needs of the organism. When dealing with filamentous fungi (as is the case with ectomycorrhizal fungi) certain factors should be considered. For example, good mixing and mass transfer conditions are required, sample ports should be suitable for fungal fermentation broths, and the area available for attachment to surfaces within the vessel should be minimal as fungi have an affinity to adhere to surfaces.

Little information has been published about the growth of ectomycorrhizal fungi in submerged liquid culture. It is possible that some strains may be shear sensitive, in which case the standard six blade Rushton turbine, routinely used in stirred tank reactors, may prove to be the wrong type of agitation system to employ. If this is the case, a gentler agitation system, such as the Anne propeller which has proved useful in cultivation of shear-sensitive animal cells can be used. In extreme cases, air lift fermenters which employ no mechanical agitation system are a possibility. Given the obligately aerobic nature of ectomycorrhizal fungi and the fact that oxygen is sparingly soluble in water and culture media, the stirred tank reactor is likely to be the best type of fermenter to use to ensure a sufficient supply of oxygen for microbial growth.

The slow growing nature of ectomycorrhizal fungi and the relatively low concentrations of biomass achieved by liquid fermentation (Le Tacon *et al.*, 1985; Litchfield & Lawhon, 1982) require that large sample volumes be taken in order to minimise sample errors. This, together with the length of the fermentation time, implies that small volume fermenters (less than 5 l) are unsuitable for research purposes.

The rate of growth of most ectomycorrhizal fungi is so slow that it is unlikely that continuous culture techniques will be employed to study the physiology of these organisms. Consequently, only when an increased understanding of the growth requirements of the organisms is obtained and growth rates are improved, may it be possible to adopt continuous culture techniques. This would be a great advantage, as continuous culture allows a more accurate assessment of the effects of nutritional parameters on the growth of an organism since in batch cultivation the nutritional status of the medium is constantly changing.

Solid substrate fermentation

To date, commercial inocula of ectomycorrhizal fungi preparations have been produced using solid substrate fermentation techniques. Abbott produced an inoculum using a vertical deep tank solid substrate fermenter with vermiculite (Marx & Kenney, 1982). The inoculum used for the solid substrate fermenter, however, was produced by submerged liquid culture using the same liquid medium.

Solid substrate fermentations have presented a number of practical disadvantages in the past:

- heat and mass transfer problems;
- difficulty in monitoring changes in fermentation;
- non-uniformity of conditions within the reactor;
- difficulty in scaling-up the process;
- difficulty in achieving and maintaining sterile conditions;
- slow process.

Nevertheless, successful production of infective inocula using solid substrate techniques has been demonstrated by Marx & Cordell (Chapter 1). Solid substrate cultivation also employs simple media, low aeration requirements and has lower energy expenditure compared with stirred tank bioreactors (Smith, 1985).

Infectivity testing and carrier systems

Production of large quantities of ectomycorrhizal fungi by large scale fermentation could provide sufficient quantities of inocula for field trials and, ultimately, for the production of a commercially available product. It is essential however, that the biomass produced is both infective and viable when it is introduced into the ecosystem of the host plant.

In order to determine whether the mycelium is suitable for field testing, the mycelium must first be tested in the laboratory to determine whether it is infective. At the research level, each time a fermentation is carried out with a changed parameter, the infectivity of the fungus must be tested to determine whether this change in the process has influenced infectivity. Routine quality control should be carried out, even on inocula-produced from a fermentation which is known to produce infective inocula. A suitable bioassay technique must, therefore, be used for that organism-host plant system. Unfortunately, bioassays can be very time consuming, taking up to 14 weeks to determine the quality of the inoculum. A more rapid assay method is, therefore, required. Marx *et al.* (1982) has described various fast assay techniques for testing for ectomycorrhiza efficiency. One such technique is the inoculum slurry system which reduces the assay time to 25 days.

Once produced, the inoculum must be retained in a viable condition until it can be used in the field. Once in the soil, the organism must also remain viable and infective for several weeks, until the mycorrhizal association has developed. The inoculum produced should be in a form which can be incorporated readily into existing forestry or nursery practices. Production of commercial inocula has been carried out almost exclusively by Marx and colleagues in conjunction with Abbott Laboratories (Marx *et al.*, 1982). They found that the most suitable carrier was vermiculite (expanded mica), an inert compound which allows the hyphae to grow between the layers, affording some degree of protection to the organism. Peat is normally added to provide pH buffering.

Abbott marketed the product as MycoRhiz®. The mixture did not, however, perform well in trials carried out in 1977-78. Marx & Rowan (1981) produced a viable inoculum by drying down the vermiculite-peat mixture to 12 to 20% moisture content. The 'dry' product was found to be infective after storage at room temperature for 5 weeks and 5°C for 9 weeks. In both cases the above inocula were produced by solid substrate fermentation methods.

Le Tacon *et al.* (1985) compared inocula produced on vermiculite- peat with inoculum produced by liquid fermentation followed by entrapment in polymeric gels and clay, for infectivity on Douglas fir and Norwegian spruce seedlings. The inoculum produced in the fermenter significantly improved the growth of both types of seedlings compared with that of the vermiculite-peat inoculum. They postulated that the inoculum produced in the fermenter and entrapped in gel was significantly better than that produced in vermiculite-peat because the mycelium had a much higher metabolic activity. Gel formulations are, however, more expensive than a vermiculite-peat.

Litchfield & Lawhorn (1982) grew cultures of *Pisolithus tinctorius* and *Thelephora terrestris* by submerged liquid fermentation. The ectomycorrhizal fungi were grown for 30 days until large fungal colonies were produced. Addition of vermiculite to the fermentation broth served as a site for nucleating fungal growth. They suggested that the carrier and fungus could be freeze-dried and readily dispersed in soil. Thus, it appears that a range of carriers affording some degree of protection to the organism can be used. Selection of the method employed will therefore come down to a matter of cost and preference by forestry nursery management.

Conclusion

The future of ectomycorrhizal fungal inoculants in nursery and forestry applications looks bright. Seven to eight million forest tree seedlings are being 'tailored' with superior symbionts for improved growth each year in the USA alone. Sufficient quantities of viable infective inocula must be produced at relatively low costs in order to ensure the commercial success of inoculation programmes. The most promising method of achieving this may well be to employ submerged liquid fermentation techniques which are quicker and more controlled than the solid substrate fermentation systems currently employed. Lack of published data on the use of submerged liquid fermentation techniques is a hindrance but because the science is fairly new to mycorrhizal research it is recognised that many advances are being made but are cloaked in industrial secrecy. The decision on whether to employ solid or liquid cultivation techniques will ultimately come down to process economics.

References

Arthur, J. J., Leone, I. & Flower, F. B. (1981). Floating and landfill gas effects on red and sugar maples. *Journal of Environmental Quality*, 10, 431-435.

Bowen, G. P. (1973). Mineral nutrition of Ectomycorrhizae In *Ectomycorrhizae, their Ecology and Physiology*, ed. G. C. Marks & T. T. Kozlowski, pp. 151-205. Academic Press: London.

Bradshaw, A. D. (1984). Ecological principles and land reclamation practice. *Landscape Planning*, 11, 35-48.

Briscoe, C. B. (1959). Early results of mycorrhizal inoculation of pine in Puerto Rico. *Caribbean Forester*, 20, 73-77.

Clouston, B. (1984). Landscape project. Reclamation and landform design. The Liverpool International Festival. *Landscape Planning*, 11, 327-335.

Dominik, T. (1959). Development dynamics of mycorrhizae formed by *Pinus sylvestris* and *Boletus luteus* in arable soils. *Prace Szczecinskiego Towarzystwa Naukowe*, 1, 1-30.

Good, J. E. G., Wilson, J. & Last, F. T. (1980). Clonal selection of amenity trees. In *Research for Practical Arboriculture*, Proceedings of the Forestry Commission/Arboriculture Association Seminar, pp. 109-115. Forestry Commission: Edinburgh.

Grenville, D. J. & Peterson, R. L. (1985). The development, structure and histochemistry of sclerotia of ectomycorrhizal fungi. *Canadian Journal of Botany*, 63, 1402-1411.

Gulley, B. W. (1981). Setting the scene. Paper 1, Landfill Growth Symposium Papers, Harwell.

Harley, J. L. (1969). *The Biology of Mycorrhiza*. Leonard Hill: London.

Harris, M. M. & Jurgensen, M. F. (1977). Development of *Salix* and *Populus* mycorrhizae in metallic mine tailings. *Plant and Soil*, 47, 509-517.

Ingleby, K., Last, F.T. & Mason, P. A. (1985). Vertical distribution and temperature relations of sheathing mycorrhizas of *Borila* spp., growing on coal spoils. *Forest Ecology and Management*, **12**, 279-285.

Jackson, R. M. & Mason, P. A. (1984). *Mycorrhiza*, Studies in Biology No. 159. Edward Arnold Ltd: London.

Johnson, M. S., Bradshaw, A. D. & Harwood, J. H. (1976). Plant growth problems of non-ferrous metal mine waste. In *Land Reclamation Conference*, ed. J. Essex & I. Higgins, pp. 481-499. Thurrock Borough Council: Essex.

Kluyver, A. J. & Perquin, L. H. C. (1933). Zur Methodik der Schimmelstoffwechseluntersuchung. *Biochemische Zeitschrift*, **266**, 68.

Le Tacon, F., Jung, G., Mugnier, J., Michelot, P. & Mauperin, C. (1985). Efficiency in a forest nursery of an ectomycorrhizal fungus inoculum produced in a fermentor and entrapped in polymeric gels. *Canadian Journal of Botany*, **63**, 1664-1668.

Litchfield, J. H. & Lawhon, W. T. (1982). Aerobic submerged fermentation of sporulating ectomycorrhizal fungi. US Patent 4,327,181.

Madu, M. (1967). The biology of ectotrophic mycorrhiza with reference to the growth of pines in Nigeria. *Obeche , Journal of the Tree club, University of Ibadan*, **1**, 9-16.

Marx, D. H. (1969). The influence of ectotrophic mycorrhizal fungi on the resistance of pine roots to pathogenic infections. 1. Antagonism of mycorrhizal fungi to root pathogenic fungi and soil bacteria. *Phytopathology*, **59**, 153-163.

Marx, D. H. (1975). Mycorrhizae and the establishment of trees on strip mineral land. *Ohio Journal of Science*, **75**, 288-297.

Marx, D. H. (1976). Use of specific mycorrhizal fungi on tree roots for forestation of disturbed lands. In *Proceedings, Conference on the Forestation of Disturbed Surface Areas*, ed. K. A. Utz, pp. 47-65. USDA Forest Service: Atlanta.

Marx, D. H. (1977). The role of mycorrhizae in forest production. In *TAPPI Conference Papers*, Annual Meeting, 151-161.

Marx, D. H. (1980a). Role of mycorrhizae in forestation of surface mines. *USDA Forest Service General Technical Report* NE **61**, 109-116.

Marx, D. H. (1980b). Ectomycorrhizal fungus inoculations: a tool for improving forestation practices. In *Tropical Mycorrhizal Research*, ed. P. Mikola, pp. 13-71. Clarendon Press: Oxford.

Marx, D. H. & Artman, J. D. (1979). *Pisolithus tinctorius* ectomycorrhizae improve survival and growth of pure seedlings on acid coal spoils in Kentucky and Virginia. *Reclamation Review*, **2**, 23-31.

Marx, D. H. & Daniel, W. J. (1976). Maintaining cultures of ectomycorrhizal and plant pathogenic fungi in sterile water cold storage. *Canadian Journal of Microbiology*, **22**, 338-341.

Marx, D. H. & Kenney, D. S. (1982). Production of ectomycorrhizal fungus inoculum. In *Methods and Principles of Mycorrhizal Research*, ed. N. C. Schenck, pp. 131-146. The American Phytopathology Society: St Paul, Minnesota.

Marx, D. H. & Rowan, S. J. (1981). Fungicides influence growth development of specific ectomycorrhizae on loblolly pine seedlings. *Forest Science*, **27**, 167-176.

Marx, D. H., Ruehle, J. L., Kenney, D. S., Cordell, C. E., Riffle, J. W., Molina, R. J., Pawuk, W. H., Navratil, S., Tinus, R. W. & Goodwin, O. C. (1982). Commercial vegetative inoculum of *Pisolithus tinctorius* and inoculation techniques for development of ectomycorrhizae on container grown tree seedlings. *Forest Science*, **28**, 373-400.

Mason, P. A. (1980). Aseptic synthesis of sheathing ectomycorrhizas. In *Tissue Culture Methods for Plant Pathologists*, ed. D. S. Ingram & J. P. Heleson, pp. 173-178. Blackwell Scientific Publications: Oxford.

Mauperin, C., Mortier, F., Garbaye, J., Le Tacon, F. & Carr, G. (1987). Viability of an ectomycorrhizal inoculum produced in a liquid medium and entrapped in a calcium alginate gel. *Canadian Journal of Botany*, **65**, 2326-2329.

Melin, E. (1925). *Untersuchungen über die Bedeutung der Baummykorhiza. Eine okologische physiologische Studie*. Fischer: Jena.

Mexal, J. G. (1980). Aspects of mycorrhizal inoculation in relation to reforestation. *New Zealand Journal of Forest Science*, **10**, 208-217.

Mikola, P. (1970). Mycorrhizal inoculation in afforestation. *International Review of Forestry Research*, **3**, 123-196.

Mikola, P. (1973). Application of mycorrhizal symbiosis in forestry practice. In *Ectomycorrhizae: their Ecology and Physiology*, ed. G. C. Marks & T. T. Kozlowski, pp. 383-411. Academic Press: New York.

Molina, R. (1979). Pure culture synthesis and host specificity of red alder mycorrhizae. *Canadian Journal of Botany*, **57**, 1223-1228.

Moser, M. (1958). Di kunstliche Mykorrhizaimpfung an Forstpflanzen. 1. Erfahrungen bei der Reinkultur von Mykorrhizapilzen. *Forstwissenschaftliches Zentralblatt*, **77**, 32-40.

Norkrans, B. (1950). Studies in growth and cellulytic enzymes of *Tricholoma*. *Symbolae Botanicae Upsaliensis*, **11**, 1-126.

Paau, A.S. (1988). Formulations useful in applying beneficial microorganisms to seeds. *Trends in Biotechnology*, **66**, 276-279.

Palmer, G. G. & Hacskaylo, E. (1970). Ectomycorrhizal fungi in pure culture. 1. Growth on single carbon sources. *Physiologia Plantarum*, **23**, 1187-1197.

Pirt, S. J. (1975). *Principles of Microbe and Cell Cultivation*. Blackwell Scientific Publications: London.

Schramm, J. R. (1966). Plant colonization studies on black waste from anthracite mining in Pennsylvania. *Transactions of the American Philosophical Society*, **56**, 1-194.

Smith, J. E. (1985). Fermentation technology. In *Biotechnology Principles*, ed. J. E. Smith, pp. 39-74. Van Nostrand Reinhold: London.

Takacs, E. A. (1967). Produccion de cultivos puros de hongos micorrizogenos en el Centro Nacional de Investigaciones Agropecuanas Castelar. *Idia, Suplemento Forestal*, **4**, 83-87.

Trappe, J. M. (1962). Fungus associates of ectotrophic mycorrhizae. *Botanical Reviews*, **28**, 538-606.

Trappe, J. M. (1977). Selection of fungi for ectomycorrhizal inoculation in nurseries. *Annual Review of Phytopathology*, **15**, 203-222.

Chapter 3

Potentialities and procedures for the use of endomycorrhizas with special emphasis on high value crops

S. Gianinazzi, V. Gianinazzi-Pearson & A. Trouvelot

Laboratoire de Phytoparasitologie, Station d'Amélioration des Plantes, INRA, BV 1540, 21034 Dijon Cédex, France

Introduction

Vesicular-arbuscular endomycorrhizas (VAM), the most common type of mycorrhizal association, are formed by nearly all cultivated plants, whether they are agricultural, horticultural, or fruit crops. The importance of this type of symbiotic fungal infection for plant mineral nutrition, and more generally for plant health (see Gianinazzi-Pearson & Gianinazzi, 1986; Smith & Gianinazzi-Pearson, 1988), makes it one of the potentially more useful biological means of assuring plant production with a minimum input of chemicals such as fertilizers or pesticides. The sort of impact that endomycorrhizas can have on plant development is illustrated by the nursery trials in Fig. 3.1. Until recently, agricultural techniques have not taken into consideration endomycorrhizas, but more and more growers are now becoming aware of them as a new means for ensuring plant production, and several private enterprises are at present interested in producing VAM inoculum. This situation, together with the present concern about man's negative impact on the environment, is creating the need and the conditions for introducing and rationally using VAM in plant production systems.

As with other soil microorganisms, the successful use of VAM fungi can only be achieved under certain conditions and it is therefore necessary not only to develop techniques for inoculum production but also to define strategies for successful inoculation. Failure to do this has no doubt contributed to the inconclusive results obtained in many field experiments in the past. It is not, however, our intention to dwell on this aspect here. The interested reader can find further details and references in reviews by Powell (1984) and Hall (1988). The aim of the present paper is rather to outline recent research, mainly from our own laboratory, on methods of inoculum production, application and actual areas of use, and to specu-

Fig. 3.1. VAM inoculated (I) and non inoculated (NI) fumigated nursery beds of (a) *Chamaecyparis lawsoniana* and (b) lilac (*Syringa vulgaris*). Inoculation with (a) a soil-based inoculum containing *Glomus mosseae* and *G. intraradices*, and (b) an agricultural waste inoculum of *Glomus mosseae* (Levavasseur & Sébir Nurseries, Ussy, 14420 Potigny, France).

late on the potential openings for the introduction of VAM fungi in agriculture in the future.

Inoculum production

VAM fungi cannot be grown in pure culture and consequently the techniques used for inoculum production are different from those generally employed for other biotechnologically interesting fungi (Gianinazzi & Gianinazzi-Pearson, 1986). VAM inoculum has to be produced on living roots which is usually considered a major disadvantage. However, this process can have the advantage of minimising the risk of maintaining fungi that have lost their symbiotic properties. On the other hand, producing VAM fungi on living plants is problematic because of the precautions that have to be taken in order to obtain 'clean' inoculum. It is possible to produce VAM on excised roots or whole plants under axenic conditions using disinfected spores as inoculum (see Hepper, 1984), but for the moment this procedure is not practically viable for mass production.

Procedures that are being used for inoculum production can be grouped into three main areas (Fig. 3.2). In all cases, fungi are maintained as mycorrhizal roots of stock culture plants (Ferguson & Woodhead, 1982) and multiplied up on roots of species having preferably no diseases in common with the crop(s) for which the inoculum is intended. The roots of these plants can be used to produce inoculum in large quantities either as soil-based inoculum, in artificial substrates or directly as surface-disinfected VAM. These different steps are carried out using disinfected soil or substrates under controlled conditions to reduce contamination by airborne pathogens or pests, and cultures are quality controlled regularly.

In Dijon, the method we have developed for producing a soil-based type of inoculum in large quantities involves inoculating disinfected soil in large containers (4 m^3) with mycorrhizal roots, and sowing with clover (Fig. 3.3). After 3 to 5 months the soil containing roots, spores and hyphae is recovered, air dried and sieved for use. This procedure can potentially yield up to $3 \times 10^3 \text{ m}^3$ of inoculum ha^{-1} of glasshouse per year, and it has produced highly infective inocula ($5–10 \times 10^3$ propagules kg^{-1}) for several VAM fungal species, which have been used successfully in both field and nursery trials (Gianinazzi & Gianinazzi-Pearson, 1986; Fig. 3.1a). Warner et al. (1985) patented a system for glasshouse production of a peat based inoculum which involved growing mycorrhizal plants in peat blocks standing in a shallow nutrient-flow culture. The resulting inoculum consisted of spores, infected roots and infested peat ground up together, and it was estimated that $1.05 \times 10^4 \text{ m}^3$ of inoculum ha^{-1} of glasshouse could be produced per year.

In Germany, Dehne & Backhaus (1986) are producing inocula in pots using expanded clay as a substratum. After plant growth, the substratum containing hyphae and spores can be separated easily from the roots to serve directly as inoculum. Native Plants Inc. (NPI) have recently put an inoculum incorporated into an inorganic carrier (Nutri-link®) (Wood, 1987) on the American market. It is based on a patented technique involving the large scale production of spores from VAM cultures which are bound to a clay support and sold in a dried form. Other techniques have been developed for large scale inoculum production in suitable binding agents, some of which have been patented, but these have not led to commercial applications (Hall & Kelson, 1981; Hayman, Morris & Page, 1981; Ganry, Diem & Dommergues, 1982). Another form of inoculum we are developing consists of surface disinfected VAM prepared by treating infected root pieces with a mild disinfectant like Chloramine T. This provides an excellent inoculum for certain special crops where it is essential to reduce microbial contamination to a minimum. This is the case, for example, with micropropagated plants which are planted out from axenic

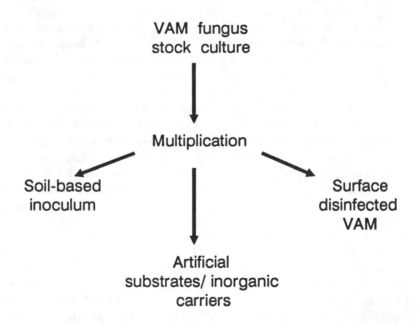

Fig. 3.2. Procedures used for VAM production.

Fig. 3.3. General (a) and detailed (b) views of large-scale soil-based inoculum production of VAM fungi on clover under glasshouse conditions.

conditions and where only small quantities of inoculum are required to inoculate a large number of plants (see below).

Each method has its own advantages and limitations. The system of inoculum production needs to be chosen in relation to the techniques used in plant production and economic criteria. For example, this year in France we have prepared a 'soil-based inoculum' in an agricultural waste soil produced by a sugarbeet company (S.R.D.-Sucre Union), as a means of commercially utilising this side product. This type of inoculum has given excellent results in nursery trials with woody ornamentals like *Ampelopsis-quinquefolia*, *Liquidambar styraciflua* and lilac (*Syringa vulgaris*) (Fig. 3.1b) (Sébir & Levavasseur, 1988).

Little attention has been given to the problem of storing VAM inoculum over long periods. Long-term preservation of VAM fungi would provide a practical means of maintaining cultures, of ensuring against their contamination or loss, and of having continuously available supplies of inoculum with defined characteristics and high viability. Fungi in colonised roots or spores can survive for extended periods in dry soil, although their viability tends to decline (Tommerup & Abbott, 1981; Tommerup, 1985) and cultures in dried soil inoculum (containing no free water), sealed in plastic bags and stored at 5°C, have been kept for several years (Ferguson & Woodhead, 1982). Attempts to simply freeze dry VAM inoculum as a method of storage have not been successful (Crush & Pattison, 1975) but it can be preserved using L-drying or cryopreservation techniques. L-drying consists of drying under vacuum over a desiccant at a temperature above zero; spores and hyphae in dry mycorrhizal roots have been preserved for up to six years with 100% survival using this technique (Tommerup & Kidby, 1979). The success of cryopreservation of mycorrhizal roots, which involves freezing them in the presence of cryoprotective compounds and storage at low temperatures, depends very much on the cryoprotectant used and the freezing rate (Tommerup & Bett, 1985).

Inoculum application

In order to use any of these different types of inoculum rationally and successfully, it is necessary to define a strategy of inoculation. That is, to determine whether in a given situation it is advantageous to inoculate and which fungi to use. Simple biological tests can be used to determine the endomycorrhizal potential (number of VAM fungal propagules) of a soil or substrate, the effectiveness of indigenous VAM fungi for plant production and the receptivity of a soil or substrate to VAM fungi that may be introduced into the system (Gianinazzi, Trouvelot & Gianinazzi-Pearson, 1989). The information obtained from these biotests, together with current knowledge about the endomycorrhizal dependency of different plant genera and species, provide a basis for developing a reliable strategy of inoculation.

The VAM potential of inert potting mixtures or disinfected soil is usually zero. In this case inoculation should be beneficial, especially for highly mycorrhiza-dependent plant species. The amount of VAM fungal propagules in non-disinfected soils can vary greatly and a distinction is often made between 'high' and 'low' levels. However, this terminology depends very much on the researcher as well as on the infectiveness of indigenous fungi and the susceptibility of the plant species involved. For example, in a field experiment to determine minimum propagule numbers required for optimum plant growth of different crops, we found that using a rapidly infecting efficient fungus, 167 propagules kg^{-1} of soil were adequate for leeks whilst this inoculum potential was not sufficient to give maximum responses in highly mycorrhiza-dependent *Liquidambar styraciflua* seedlings. In situations where the VAM potential of soils is extremely low and the fungi are inefficient, it has been shown that controlled inoculation with suitable VAM fungi can benefit crop growth (Black & Tinker, 1977; Azcon-G. de Aguilar, Barea & Azcon, 1979; Hall, 1980). Where indigenous VAM fungal populations are effective, it is possible to envisage management practices (e.g. crop rotations including a highly infective plant or avoidance of harmful pesticides) which increase or maintain the endomycorrhizal potential of the soils. It is not at present possible to predict the outcome of inoculation in soils containing high populations of ineffective VAM fungi, as this will depend very much on the competitive ability of the introduced fungal species. For the moment, very little is known about competition between VAM fungi or between them and other soil microorganisms (Abbott, Robson & Hall, 1983; Lopez-Aguillon & Mosse, 1987) and more research is needed. However, in spite of this, inoculation has been reported to improve plant production in non-disinfected field situations producing relatively high levels of VAM infection (Bagyaraj & Sreeramulu, 1982; Sreeramulu & Bagyaraj, 1986; Baltruschat, 1987).

Inoculation trials usually employ inoculum at excessively high and impractical rates in order to guarantee rapid infection by the introduced fungal species (see Powell, 1984; Hall, 1988 for values). Application rates are expressed in terms of weight of inoculum without any attempt to define or standardize the mycorrhizal potential of the product used. This aspect requires more attention if inoculation is to be carried out efficiently using quantities of inoculum that are practically and economically compatible with plant production systems. Furthermore, since VAM fungi can show widely different edaphic requirements, it is necessary to have a collection of species which are adapted to different physical and chemical soil properties, in particular pH, available P levels, intended fertilizer treatments and possibly soil texture. The possibility of using 'cocktail' inocula of known propagule titre and containing more than one VAM fungus is being tested this year in nursery trials in France (see below) to ensure inoculation responses under different soil conditions.

Areas for inoculum use

The most probable areas for immediate inoculum use are those where indigenous fungi are completely absent. In France, we are concentrating our efforts on the production and management of endomycorrhizal plant species in situations such as nurseries. Here, soil is disinfected before planting, artificial (inert) potting mixtures or substrates are used, or micropropagation is a routine technique for plant production.

Soil disinfection, a current practice in commercial nurseries in France, results not only in the elimination of pathogens but also of indigenous mycorrhizal fungi. This often leads to patchy growth of plants due to lack of mycorrhizal development, with subsequent problems of plant production (Fig. 3.1). Consequently, certain nurseries have begun VAM inoculation trials. After testing soils for their compatibility to selected VAM fungal strains using a fungal effectiveness/soil receptivity test (Gianinazzi, Trouvelot & Gianinazzi-Pearson, 1989), nursery beds have been inoculated by broadcasting and raking-in inoculum (10^4 propagules m^{-2}) before seeding. Usual nursery practices have otherwise been used for plant production. Better growth has been observed with all the tested plant species (Sébir & Levavasseur, 1988). However, the different fungi used were not always equally effective for all plants, underlining the importance of plant-fungus interactions. This year several plants have been inoculated on a large scale with a 'cocktail' inoculum containing more than one fungus. This controlled inoculation has again been successful (Fig. 3.1a), and the introduction of the technique into commercial nursery production systems is now envisaged.

The most important limiting factor for VAM inoculations is the bulk of inoculum required. Soil inoculum rates as high as 17×10^4 kg ha^{-1} have been used in field trials (Powell, 1984), though lower rates can be used for high density crops (e.g. Menge (1985) mentions the use of up to $4\cdot2 \times 10^3$ kg ha^{-1} of soil inoculum). The practical and economical use of high quantities of inoculum will depend on the type of plant and production system. For example, our trials in nurseries where soil is disinfected and high value crops are produced in high densities (500 plants m^{-2}), have shown that the use of inoculum with high VAM potential becomes practically and economically viable.

Soil is rarely incorporated into artificial or inert potting mixtures used in nurseries for growing seeds, cuttings or micropropagated plants, so that mycorrhizal fungi are usually absent from these. In a programme on the

Table 3.1. Effect of two inoculation techniques on the growth and mycorrhizal infection of grapevine root-stocks (S04102), 8 weeks after transplanting into a peat/soil/gravel mix (1:2:1 v/v) (Ravolanirina, unpublished data).

Inoculation technique	Fresh weight (g)*		Level of infection*		
	Shoot	Root	F%	M%	A%
In vitro					
Non inoculated	$0 \cdot 92^b$	$1 \cdot 04^b$	$0 \cdot 0^b$	$0 \cdot 0^b$	$0 \cdot 0^b$
Gigaspora margarita	$1 \cdot 56^b$	$1 \cdot 35^b$	$51 \cdot 2^a$	$29 \cdot 6^a$	$20 \cdot 9^a$
Post vitro					
Non inoculated	$4 \cdot 23^b$	$4 \cdot 29^b$	$0 \cdot 0^b$	$0 \cdot 0^b$	$0 \cdot 0^b$
Gigaspora margarita	$13 \cdot 86^a$	$11 \cdot 73^a$	$75 \cdot 2^a$	$34 \cdot 1^a$	$26 \cdot 7^a$

*Means of six replicates; values in columns followed by the same superscript letter do not differ significantly ($P = 0 \cdot 01$; Duncan's test). F% = frequency of root infection, M% = intensity of cortical infection, A% = arbuscule frequency in roots (Trouvelot, Kough & Gianinazzi-Pearson, 1985).

---

introduction of VAM into the production cycle of micropropagated temperate and tropical fruit crops, a number of fertilized substrates or potting mixtures showed different relative abilities to ensure mycorrhiza development. Infection by three different VAM fungi was always improved when sterile soil was included in the potting mixtures, but only one substrate, a peat/clay soil/gravel (1:2:1 v/v) mix, proved suitable for rapid VAM development by all the fungi. For fungi with acid pH requirements, this mix can be modified using a low pH clay or sandy soil. Micropropagated plants can be inoculated *in vitro* and the efficiency of this procedure has been compared with that of *post vitro* inoculation with surface disinfected VAM at planting out into the above mix in pots (Pons *et al.*, 1983; Gianinazzi, Gianinazzi-Pearson & Trouvelot, 1986). In the case of a grapevine root-stock, for example, infection intensities were comparable using the two inoculation techniques but plant growth after transplanting into pots was considerably better when plants were *post vitro* inoculated (Table 3.1).

Table 3.2. Effect of pre-inoculation with two VAM fungi on the growth of a micropropagated grapevine root-stock (S04102), 12 weeks after planting out into disinfected field plots (Ravolanirina & Gianinazzi, unpublished data).

Treatment	Shoot fresh weight (g)	
	Expt. 1	Expt. 2
Non inoculated controls	150[b]	238[b]
Plants pre-inoculated with:		
Glomus mosseae	508[a]	–
Glomus fasciculatum	474[a]	546[a]

Values in columns followed by the same superscript letter do not differ significantly ($P = 0·05$; Duncan's test).

Thus, it appears that, at least for grapevine, *in vitro* inoculation is not of practical interest.

In order to introduce VAM as early as possible into micropropagation systems, the efficiency of coupling inoculation with a *post vitro* rooting procedure has also been investigated. This involves transplanting micropropagated cuttings shortly after the induction of root initials into a mycorrhizal substrate for root development before planting out. Such a combination of *post vitro* rooting and VAM inoculation with rational fertilization has given promising results with micropropagated grapevine root-stock and oil palm (Ravolanirina *et al.*, 1989)(Table 3.2). This technique has the advantage of requiring relatively small amounts of inoculum (100 g surface disinfected VAM per 500 plants). After transplanting into mycorrhiza-deficient field plots, pre-inoculated plants continue to show large growth increases in comparison to uninoculated plants.

Future developments

Plant production systems where VAM exploitation is practically conceivable are still limited. However, certain approaches can be envisaged which may open new possibilities and encourage industrial interest in their use. On a short term basis, the development of inoculum 'cocktails', containing mixtures of VAM fungi adapted to different edaphic conditions, could constitute a product ensuring successful wide scale inoculation of a variety of plant species in nurseries. An interesting potential also lies in the use of agricultural chemicals that specifically increase VAM infection whilst controlling the development of plant diseases (Jabaji-Hare & Kendrick, 1985; Schönbeck & Dehne, 1987; Morandi,

1989 and unpublished data). This introduces the possibility of improving use and management of efficient indigenous VAM fungal populations on an agricultural scale.

Genetic variability in fungal efficiency and host response exists naturally. This can and has been used to select VAM fungal isolates that are able to improve plant production even at high levels of soil fertility (Gianinazzi & Gianinazzi-Pearson, 1986; Trouvelot, Gianinazzi & Gianinazzi-Pearson, 1987). Variations in the extent and effect of VAM infection has also been linked to the genotype of the host plant (Bertheau *et al.*, 1980; Krishna *et al.*, 1985; Lackie *et al.*, 1988), and the recent isolation of *myc⁻* plant mutants (Duc *et al.*, 1989) represents a first step towards the identification and manipulation of host genes regulating the formation and expression of VAM symbiosis. On a long term basis, this could open the way to developing plants that will form an endomycorrhiza specifically with certain VAM fungi. If plants could be so selective then it would be possible to guarantee the efficiency of high quality inoculum introduced into field soils, even in the presence of high populations of non-effective indigenous fungi.

References

Abbott, L. K., Robson A. D. & Hall, I. R. (1983). Introduction of vesicular-arbuscular mycorrhizal fungi into agricultural soils. *Australian Journal of Agricultural Research*, **34**, 741-749.

Azcon-G de Aguilar, C., Azcon, R. & Barea, J. M. (1979). Endomycorrhizal fungi and *Rhizobium* as biological fertilisers for *Medicago sativa* in normal cultivation. *Nature*, **279**, 325-327.

Bagyaraj, D. J. & Sreeramulu, K. R. (1982). Preinoculation with VA mycorrhiza improves growth and yield of Chili transplanted in the field and saves phosphatic fertilizer. *Plant and Soil*, **69**, 375-381.

Baltruschat, H. (1987). Field inoculation of maize with vesicular arbuscular mycorrhizal fungi by using expanded clay as carrier material for mycorrhiza. *Journal of Plant Diseases and Protection*, **94**, 419-430.

Bertheau, Y., Gianinazzi-Pearson, V. & Gianinazzi, S. (1980). Développement et expression de l'association endomycorhizienne chez le blé. I. Mise en évidence d'un effet variétal. *Annales d'Amélioration des Plantes*, **30**, 67-78.

Black, R. L. B. & Tinker, P. B. (1977). Interactions between effects of vesicular-arbuscular mycorrhiza and fertilizer phosphorus on yield of potatoes in the field. *Nature*, **267**, 510-511.

Crush, J. R. & Pattison, A. C. (1975). Preliminary results on the production of vesicular-arbuscular mycorrhizal inoculum by freeze drying. In *Endomycorrhizas*, eds F. E. Sanders, B. Mosse & P. B. Tinker, pp. 485-493. Academic Press: London, New York.

Dehne, H. W. & Backhaus, G. F. (1986). The use of vesicular-arbuscular mycorrhizal fungi in plant production. I. Inoculum production. *Zeitschrift für Pflanzenkrankheiten und Pflanzenschutz*, **93**, 415-424.

Duc, G., Trouvelot, A., Gianinazzi-Pearson, V. & Gianinazzi, S. (1989). First report of non-mycorrhizal plant mutants (*Myc-*) obtained in pea (*Pisum sativum* L.) and fababean (*Vicia faba* L.). *Plant Science*, **60**, 215-222.

Ferguson, J. L. & Woodhead, S. H. (1982). Production of endomycorrhizal inoculum A. Increase and maintenance of vesicular-arbuscular mycorrhizal fungi. In *Methods and Principles of Mycorrhiza Research*, ed. N. C. Schenck, pp. 47-54. American Phytopathological Society: St Paul, Minnesota.

Ganry, F., Diem, H. G. & Dommergues, Y. R. (1982). Effect of inoculation with *Glomus mosseae* on nitrogen fixation by field grown soybeans. *Plant and Soil*, **68**, 321-329.

Gianinazzi, S. & Gianinazzi-Pearson, V. (1986). Progress and headaches in endomycorrhiza biotechnology. *Symbiosis*, **2**, 139-149.

Gianinazzi, S., Gianinazzi-Pearson, V. & Trouvelot, A. (1986). Que peut-on attendre des mycorhizes dans la production des arbres fruitiers? *Fruits*, **41**, 553-556.

Gianinazzi, S., Trouvelot, A. & Gianinazzi-Pearson, V. (1989). Conceptual approaches for the rational use of VA endomycorrhizae in agriculture: possibilities and limitations. *Agriculture, Ecosystems and Environment*, in press.

Gianinazzi-Pearson, V. & Gianinazzi, S., eds. (1986). *Physiological and genetical aspects of mycorrhizae*. INRA Press: Paris.

Hall, I. R. (1980). Growth of *Lotus pedunculatus* Cav. in an eroded soil containing soil pellets infested with endomycorrhizal fungi. *New Zealand Journal of Agricultural Research*, **23**, 103-105.

Hall, I. R. & Kelson, A. (1981). An improved technique for the production of endomycorrhizal infested pellets. *New Zealand Journal of Agricultural Research*, **24**, 221-222.

Hall, I. R. (1988). Potential for exploiting vesicular-arbuscular mycorrhizas in agriculture. *Advances in Biotechnological Processes*, **9**, 141-174.

Hayman, D. S., Morris, E. T. & Page, R. J. (1981). Methods for inoculating field crops with mycorrhizal fungi. *Annals of Applied Biology*, **99**, 247-253.

Hepper, C. M. (1984). Isolation and culture of VA mycorrhizal (VAM) fungi. In *VA Mycorrhiza*, ed. C. Ll. Powell & D. J. Bagyaraj, pp. 95-112. CRC Press Inc.: Boca Raton, Florida.

Jabaji-Hare, S. H. & Kendrick, W. B. (1985). Effects of Fosetyl-Al on root exudation and on composition of extracts of mycorrhizal and non mycorrhizal leek roots. *Canadian Journal Plant Pathology*, **7**, 118-126.

Krishna, K. R., Shetty, K. G., Dart, P. J. & Andrews, D. J. (1985). Genotype dependent variation in mycorrhizal colonization and response to inoculation of pearl millet. *Plant and Soil*, **86**, 113-125.

Lackie, S. M., Bowley, S. R. & Peterson, R. L. (1988). Comparison of colonization among half-sib families of *Medicago sativa* L. by *Glomus versiforme* (Daniels & Trappe) Berch. *New Phytologist*, **108**, 477-482.

Lopez-Aguillon, R. & Mosse, B. (1987). Experiments on competitiveness of three endomycorrhizal fungi. *Plant and Soil*, **77**, 155-170.

Menge, J. A. (1985). Developing widescale VA mycorrhizal inoculations: is it practical or necessary? In *Proceedings of the 6th North American Conference on Mycorrhizae*, ed R. Molina, pp. 80-82. Forest Research Laboratory: Corvallis, Oregon.

Morandi, D. (1989). Effect of endomycorrhizal infection and biocides on phytoalexin accumulation in soybean roots. *Agriculture, Ecosystems and Environment*, in press.

Pons, F., Gianinazzi-Pearson, V., Gianinazzi, S. & Navatel, J. C. (1983). Studies of VA mycorrhizae *in vitro*: mycorrhizal synthesis of axenically propagated wild cherry (*Prunus avium* L.) plants. *Plant and Soil*, 71, 217-221.

Powell, C. Ll. (1984). Field inoculation with VA mycorrhizal fungi. In *VA Mycorrhiza*, ed. C. Ll. Powell & D. J. Bagyaraj, pp. 205-222. CRC Press Inc.: Boca Raton, Florida.

Ravolanirina, F., Blal, B., Gianinazzi, S. & Gianinazzi-Pearson, V. (1989). Mise au point d'une méthode rapide d'endomycorhization de *vitro* plants. *Fruits*, 44, 165-170.

Schönbeck, F. & Dehne, H. W. (1987). Improvement of the VA mycorrhiza status in plant production. In *Mycorrhizae in the Next Decade: Practical Applications and Research Priorities*, eds D. M. Sylvia, L. L. Hung & J. H. Graham, p. 37. University of Florida: Gainesville, Florida.

Sébir, P. & Levavasseur, J. (1988). L'endomycorhization VA: une réponse à certains problèmes. *L'Horticulture Française*, 205, 4-6.

Smith, S. E. & Gianinazzi-Pearson, V. (1988). Physiological interactions between symbionts in vesicular-arbuscular mycorrhizal plants. *Annual Review of Plant Physiology and Plant Molecular Biology*, 39, 221-244.

Sreeramulu, K. R. & Bagyaraj, D. J. (1986). Field responses of chili to VA mycorrhiza on black clayey soil. *Plant and Soil*, 92, 299-304.

Tommerup, I. C. (1985). Strategies for long-term preservation of VA mycorrhizal fungi. In *Proceedings of the 6th North American Conference on Mycorrhizae*, ed R. Molina, pp. 87-88. Forest Research Laboratory: Corvallis, Oregon.

Tommerup, I. C. & Abbott, L. K. (1981). Prolonged survival and viability of VA mycorrhizal hyphae after root death. *Soil Biology and Biochemistry*, 13, 431-433.

Tommerup, I. C. & Bett, K. B. (1985). Cryopreservation of genotypes of VA mycorrhizal fungi. In *Proceedings of the 6th North American Conference on Mycorrhizae*, ed R. Molina, p. 235. Forest Research Laboratory: Corvallis, Oregon.

Tommerup, I. C. & Kidby, D. K. (1979). Preservation of spores of vesicular arbuscular endophytes by L-drying. *Applied Environmental Microbiology*, 37, 831-835.

Trouvelot, A., Gianinazzi, S. & Gianinazzi-Pearson, V. (1987). Screening of VAM fungi for phosphate tolerance under simulated field conditions. In *Mycorrhizae in the Next Decade: Practical Applications and Research Priorities*, eds D. M. Sylvia, L. L. Hung & J. H. Graham, p. 39. University of Florida: Gainesville, Florida.

Trouvelot, A., Kough, J. L. & Gianinazzi-Pearson, V. (1986). Mesure du taux de mycorhization VA d'un système radiculaire. Recherches de méthodes d'estimation ayant une signification fonctionnelle. In *Physiological and Genetical Aspects of Mycorrhizae*, eds V. Gianinazzi-Pearson & S. Gianinazzi, pp. 217-221. INRA Press: Paris.

Warner, A., Mosse, B. & Dingemann, L. (1985). The nutrient film technique for inoculum production. In *Proceedings of the 6th North American Conference on Mycorrhizae*, ed R. Molina, p. 85. Forest Research Laboratory: Corvallis, Oregon.

Wood, T. (1987). Commercial production of VA mycorrhiza inoculum: axenic versus non axenic techniques. In *Mycorrhizae in the Next Decade: Practical Applica-*

tions and Research Priorities, eds D. M. Sylvia, L. L. Hung & J. H. Graham, p. 274. University of Florida: Gainesville, Florida.

Chapter 4

The use of fungi to control pests of agricultural and horticultural importance

A. T. Gillespie & E. R. Moorhouse

Entomology and Insect Pathology Section, AFRC Institute of Horticultural Research, Littlehampton, West Sussex BN17 6LP, UK

Introduction

The importance of entomogenous fungi as natural regulators of insect populations has been recognised for many years. Over two thousand years ago, the Chinese were aware of the infection (by species of *Cordyceps* and *Isaria*) that caused mortality in populations of silkworms and cicadas (cited in Roberts & Humber, 1981). The possibilities of using entomogenous fungi as biological control agents were first explored by Metchnikoff (1879) and Krassilstschik (1888) in the Ukraine. These two pioneers mass-produced *Metarhizium anisopliae* and tested it against wheat cockchafer, *Anisoplia austriaca* and sugarbeet curculionid *Cleonus punctiventris*.

During the late 19th and early 20th centuries many entomogenous fungi were examined as possible control agents for various insect pests. In France, field experiments were carried out by Le Moult after World War One (cited by Ferron, 1985) using the Deuteromycete *Beauveria brongniartii* (*B. tenella*) against the European Cockchafer, *Melolontha melolontha*. In the United States, *Aschersonia* was developed as a control agent of whitefly by Berger (1909) and remained on the market for several years, until it was replaced by more effective chemical insecticides in the 1920's. The sequence of development, commercialization and later withdrawal has been repeated subsequently with a number of other fungal products such as Vertalec® and Mycotal® (*Verticillium lecanii*) and Mycar® (*Hirsutella thompsonii*). Currently, there are no fungal products commercially available for insect control in the Western world although products are used in other parts of the world, e.g. Russia, China and Brazil.

During recent years there has been a resurgence of interest in entomogenous fungi caused by factors such as increasing insecticide resistance and environmental concerns over pesticide use. This has led to the isolation of many new strains of fungi from a wide range of hosts, and further emphasised the potential that exists for the use of fungi in insect control.

It has also led to a greater understanding of the life cycles of entomogenous fungi, and ecological factors important to their use as biological control agents and these areas are included in this review.

Taxonomy and host range of entomogenous fungi

Over 100 different fungal genera (representing all the main fungal subdivisions) have been shown to parasitise living insects (Roberts & Humber, 1981; McCoy, Sampson & Boucias, 1988). Both saprotrophs and primary pathogens have been isolated from dead or diseased insects in most ecological niches (Zacharuk, 1981). They tend to be more common in tropical areas where temperature and humidity favour their growth, but they are also important in the regulation of insect numbers in temperate areas. Entomogenous fungi can be found in most ecosystems ranging from natural aquatic systems to high technology horticultural systems of temperate areas. Examples of some of the fungal pathogens that have been isolated from a range of agricultural and horticultural pest species are given in Table 4.1. The host range of individual genera is variable, with some such as *Massaspora* (Entomophthoraceae) being isolated only from cicadas whilst others such as *Metarhizium* have a broad host range (Veen, 1968).

Biology of entomogenous fungi

Effects of relative humidity on germination and growth

The entomogenous fungi, unlike other microbial control agents, infect their insect hosts by active penetration of the insect cuticle rather than via the host's digestive tract. The processes of conidial germination and growth on the cuticle are therefore highly dependent on both available moisture and temperature. Gillespie & Crawford (1986) examined conidial germination of four Deuteromycete genera (*Beauveria bassiana*, *Metarhizium anisopliae*, *Paecilomyces* spp. and *Verticillium lecanii*) over a range of relative humidities (r.h.) between 100 and 92%. The germination rates were highest from 100 to 97% r.h., there was a significant decrease in the rate at 96% r.h. and a further decrease at 94% r.h., and there was no germination at 92% r.h. The mycelial growth profile of a typical strain of *M. anisopliae* is shown in Fig. 4.1. Interestingly, a slightly reduced water potential increased the growth rate. Thus, the rate of extension at 99% was greater than that at 100% r.h. This occurred with all the *M. anisopliae* var. *anisopliae* strains examined but not with *M. anisopliae* var. *major*.

The importance of relative humidity for *in vivo* germination of *Nomuraea rileyi* was demonstrated by Getzin (1961). Larvae of *Trichoplusia ni* were treated and maintained at r.h. values of 42, 80, 90 and 100%, with resultant mortalities of 4, 3, 49 and 100% respectively. The presence of free water did not inhibit germination and may even enhance disease spread

Table 4.1. Examples of the range of pest species that are infected by entomogenous fungi.

Order	Host species	Pathogen	Reference
Orthoptera	*Schistocerca gregaria*	*Metarhizium anisopliae*	Gunnarson, 1988
Dermaptera	*Forficula* spp.	*Entomophthora forficula*	Deacon, 1983
Hemiptera sub-order Heteroptera	*Scotinophara coarctata*	*Beauveria bassiana*	Rombach *et al.*, 1986
Hemiptera sub-order Homoptera	*Trialeurodes vaporariorum*	*Aschersonia aleyrodis*	Fransen, 1987
Thysanoptera	*Thrips tabaci*	*Verticillium lecanii*	Gillespie, 1984
Coleoptera	*Melolontha melolontha*	*Beauveria brongniartii*	Keller, 1978
Diptera	*Delia antiqua*	*Entomophthora muscae*	Carruthers *et al.*, 1985a
Lepidoptera	*Anticarsia gemmatalis*	*Nomuraea rileyi*	Kish & Allen, 1978
Hymenoptera	*Diprion similis*	*Entomophthora tenthredinis*	Klein & Coppel, 1973
Non insect pests	*Tetranychus urticae*	*Hirsutella thompsonii*	Gardner *et al.*, 1982

provided the rainfall is not too heavy (Kish & Allen, 1978). There are occasional reports of fungi infecting insects at low r.h. For example, *Aschersonia aleyrodis* was able to infect whitefly larvae even when the r.h. was maintained at a constant 50% (Fransen, 1987). This unexpected observation could be due to the increased humidity around the leaf surface and insect cuticle resulting from transpiration. The importance of the microclimate was also emphasised by Hall & Burges (1979). Control of two different aphid species, *Myzus persicae* and *Macrosiphoniella sanborni* was similar under optimal conditions for *Verticillium lecanii*, yet in the glasshouse there was a significant difference in mortality. This was probably due to humidity differences in the preferred feeding sites of the two species; *Myzus persicae* generally feeds on the undersides of leaves

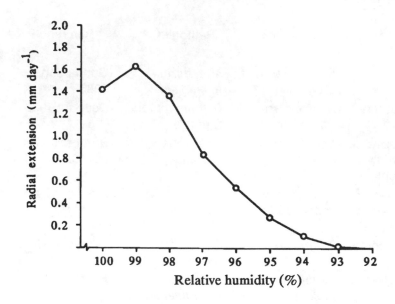

Fig. 4.1. Radial extension of *Metarhizium anisopliae* strain 275 at a range of relative humidities.

where the humidity would be high, compared to the more exposed parts of the stem preferred by *Macrosiphoniella sanborni*. The spread of *V. lecanii* through other aphid populations was also highly dependent on r.h. (Milner & Lutton, 1986). Disease spread was maximal when free water was available, the rate of transmission was reduced at 100% r.h., further reduced at 97% r.h. and completely prevented at 93% r.h.

Effects of temperature on germination and growth

The other major environmental factor that affects the infection processes of all entomogenous fungi is temperature. The effect of temperature on mycelial growth of a range of *Metarhizium anisopliae* strains is shown in Fig. 4.2. Optimal temperatures for radial growth were strain dependent and ranged from 20°C to greater than 30°C. Similarly, Walstad, Anderson & Stambaugh (1970) showed that optimum germination and growth occurred at 25-30°C, with both germination and growth of *Beauveria bassiana* and *M. anisopliae* markedly reduced below 15°C and above 35°C.

Temperature also affected infection of *Heliothis zea* by *Nomuraea rileyi* (Mohamed, Sikorowski & Bell, 1977). Mortalities were greatest at

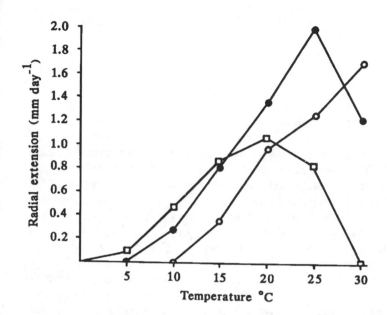

Fig. 4.2. Effect of temperature on radial extension of *Metarhizium anisopliae* isolates on Sabouraud dextrose agar. (open squares, strain 108; open circles, strain 171; closed circles, strain 275.

20 and 25°C and were reduced at 15 and 30°C. Similar reductions in pathogenicity at high temperatures were reported by Gardner (1985). The reduced virulence at elevated temperatures is probably due to reduced germination and eventual spore death.

The infection process

Only a brief account of the infection process is given here, as a more detailed account is provided by Charnley in Chapter 5. The main steps leading to the infection and subsequent death of insects are shown in Fig. 4.3.

The Deuteromycetes only form primary conidia (the infective propagule), whereas the Entomophthoraceae may form primary, secondary, tertiary or even quaternary conidia. Conidial germination is dependent on a number of factors besides temperature and moisture availability, including stimulation by epicuticular lipids on the integument (Veen, 1968) and inhibition from other microorganisms on the cuticle (Schabel, 1976).

Under suitable conditions, conidia imbibe water and produce germ tubes which may penetrate the cuticle directly as in *Verticillium lecanii* (Hughes & Gillespie, 1985) or form appressoria, as in *Metarhizium anisopliae* (Zacharuk, 1970a-c). Penetration results primarily from enzymic degradation and to a lesser extent, mechanical pressure (Zacharuk, 1970c). The importance of enzyme systems was demonstrated by the *in vivo* production of proteases, lipases and chitinases by *M. anisopliae* (St. Leger, Cooper & Charnley, 1987).

Once through the cuticle, the fungus alters its growth morphology and produces hyphal bodies which circulate in the haemolymph and prolife-rate by budding. Insect death usually occurs 3 to 14 days after application of conidia. The LT_{50} (the time taken to achieve 50% mortality) is depend-ent on the temperature and the initial spore dose (Carruthers *et al*.; 1985a; Moorhouse & Gillespie, unpublished). Death results from a combination of factors: mechanical damage resulting from tissue invasion, depletion of nutrient resources and toxicosis. Following death, fungal growth reverts back to the typical hyphal form (the saprotrophic stage) and the fungus colonises the cadaver. Antibiotic substances, such as beauvericin pro-duced by *Beauveria bassiana* (Zacharuk, 1981) may reduce competition for nutrients during the saprotrophic stage. Finally, the hyphae again penetrate the cuticle (provided there is adequate moisture) and sporula-tion occurs on the insect cadaver.

Spore formation, dispersal and survival

Sporulation is dependent on relative humidity and temperature. Thus, *Beauveria bassiana* and *Metarhizium anisopliae* sporulated on artificial media within 4 days at 100% r.h., but took over 5 days at 98-92·5% r.h. Below 92·5% r.h. there was no sporulation (Walstad *et al*., 1970). *Hirsutella thompsonii* required even higher moisture levels, sporulation from mite cadavers only occurring at relative humidities above 98% (Gerson, Kenneth & Muttath, 1979). *B. bassiana* and *M. anisopliae* started sporulating within 5 days at 25-30°C, but at 10°C no sporulation was ob-served for at least 15 days (Walstad *et al*., 1970).

Spores from the cuticles of infected insects spread to other host insects by a variety of routes. Garcia & Ignoffo (1977) consider that wind trans-port is the most important transfer mechanism for *Nomuraea rileyi*. Heavy rainfall may inhibit the transfer of inoculum within a population because spores are washed off the cadavers into the soil rather than being blown onto fresh hosts (Garcia & Ignoffo, 1977; Kish & Allen, 1978). The Ento-mophthoraceae are characterised by their ability to forcibly discharge conidia. Carruthers *et al*. (1985b) noted that *Delia antiqua* infected with *Entomophthora muscae* moved to an elevated position during the late afternoon and actively discharged conidia overnight. In this way, conidia

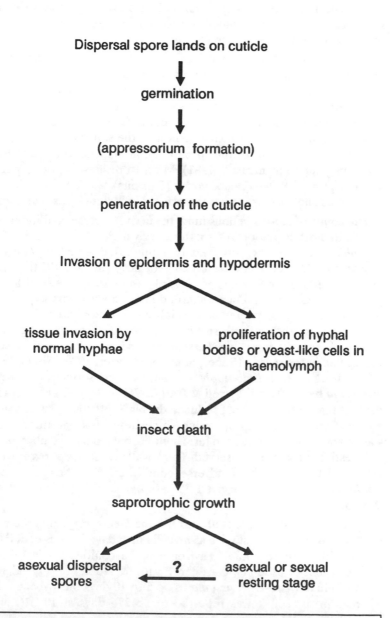

Fig. 4.3. A simplified diagram of the infection process. Brackets indicate that stages may be absent in some species. From Deacon, 1983.

are transferred to fresh hosts when conditions are optimal for conidial survival and germination. Flies that produce resting spores die on the soil surface and release spores directly into the soil.

The movement of conidia within a soil profile is poorly understood, although the most important transport agent is likely to be soil water (Storey & Gardner, 1987). It is also probable that soil organisms contribute to both lateral and vertical movement within the soil. Zimmermann & Bode (1983) demonstrated that Acari, Collembola, Diptera and Coleoptera can passively transport *M. anisopliae* conidia in the soil. Non-host organisms may also be important for the spread of conidia from the soil to foliage and from plant to plant (Hall, 1981). Mites were shown to be responsible for transport of *B. bassiana* conidia (Samsinakova & Samsinak, 1970). Parasitoids and predators may also spread inoculum by passive transfer.

The ability of entomogenous fungi to survive adverse conditions inside the dead host is important to the survival of some species. *Erynia neoaphidis* (*E. aphidis*) survived in mummified *Acyrthosiphon pisum* cadavers for 32 weeks at 0°C and 20% or 50% r.h.; at 20°C the survival time was reduced to 8 weeks at 20% r.h. and under 1 week at 80% r.h. (Wilding, 1973). Survival in mummified insect hosts is less important for species that can overwinter as specialised resting spores. Some Entomophthorales produce both resting spores and conidia throughout the season whereas others such as *Massospora levispora* only produce resting spores late in the season (Soper, Smith & Delyzer, 1976). Similarly, Carl (1975) working with *Entomophthora* sp. (later described as *Neozygites parvispora* by MacLeod, Tyrrell & Carl, 1976) in Switzerland, found that the first resting spores were produced during September. There was then a transitional phase, during which some individuals produced resting spores and others continued to form conidia, but from November onwards only resting spores were observed. Thick walled resting spores are able to withstand long periods of adverse conditions. For example, those of *Conidiobolus obscurus* survived 12 months at 6°C (Perry, Lateur & Wilding, 1982).

Specialised resting spores are rare in the Deuteromycete entomogenous fungi as these fungi survive as mycelia in cadavers, or as conidia. The soil provides a more suitable environment for conidial survival than the leaf surface. For example, Gardner, Sutton & Noblet (1977) calculated the half life of *Beauveria bassiana* to be 5 to 10 days on soybean foliage, whereas in the soil it was reported to be 276 days at 10°C (Lingg & Donaldson, 1981). The main factor contributing to conidial mortality is sunlight. In an experiment in Brazil, Daoust & Pereira (1986) found that a one week exposure to sunlight killed all the conidia of *Metarhizium anisopliae* and *Beauveria bassiana* whereas conidia protected from sunlight remained viable for over three weeks. They found that other factors, such as high temperature, were less detrimental to conidial survival. Glass effectively filters out the damaging UV light from sunlight, so under glasshouse conditions conidia on leaves may remain viable longer than under field conditions.

Use of fungi to control arthropod pests

Aschersonia aleyrodis

The glasshouse whitefly, *Trialeurodes vaporariorum* is a serious pest of many glasshouse crops and is commonly controlled by the parasitoid, *Encarsia formosa*. Sometimes the level of control is insufficient and so recent attention has focussed on *Aschersonia aleyrodis* as a possible complementary control agent for *T. vaporariorum*. Work in the Netherlands by Ramakers & Samson (1984) demonstrated that consistent control of *T. vaporariorum* could be obtained on cucumber crops over 4 weeks old, but that control was reduced on younger crops, possibly due to the effects of low relative humidities. Fransen, Winkelmman & Van Lenteren (1987) obtained 94% control after treatment of young eggs of *T. vaporariorum* (the first instar larvae are infected at emergence) compared with only 28% control of fourth instar larvae. They suggested that several applications should be made to maximize the control of the more vulnerable first instars. Earlier work by Fransen (1986) showed that conidia remained viable on foliage for 22 days at 20°C and 70% r.h. and that adding nutrients to the formulation could increase the speed of infection. However, nutrient addition may decrease persistence.

The natural spread of *A. aleyrodis* through glasshouse whitefly populations is very limited as only first instar larvae are mobile. Full leaf cover with conidial sprays will be required to maximize infection rates. The development of *A. aleyrodis* infections under conditions of high humidity is very rapid (Rombach & Gillespie, 1988) and it is possible that one night of high humidity in a glasshouse would be sufficient to initiate widespread infection. If rapid infection can be achieved, the future of *A. aleyrodis* as a biocontrol agent for whitefly may be promising.

Beauveria bassiana

The use of *Beauveria bassiana* as a microbial control agent for a range of insect pests has been investigated in many parts of the world. As early as 1962, attempts were made by Nutrilite Products in the USA to obtain registration of *B. bassiana* as Biotrol®, but these were unsuccessful (Ignoffo et al., 1979).

Much of the subsequent development work necessary for the production of *B. bassiana* as a microbial insecticide has been carried out in the Soviet Union. Roberts & Humber (1981) reported that in 1977, twenty-two tonnes of *B. bassiana* were produced in the USSR, by submerged or submerged-surface fermentation (Ignoffo *et al.*, 1979). The *B. bassiana* preparation, in the form of a wettable powder, is distributed under the name Boverin and is primarily used for control of Colorado potato beetle, *Leptinotarsa decemlineata*, and of the codling moth, *Cydia pomonella* (Ferron, 1981). The area treated with Boverin in 1977 was calculated as 7900 ha (Lipa, 1985).

The natural mortality from disease of over-wintering *L. decemlineata* beetles caused by fungal disease was occasionally as high as 80-100% (Sikura & Sikura, 1983). From this evidence, it is clear that there is a real possibility of using *B. bassiana* to control *L. decemlineata*. After a detailed examination of the efficacy of Boverin in field trials in Russia it was found that *B. bassiana* was more active against weakened insects and in practice it is recommended that Boverin should be used in conjunction with reduced doses of insecticides (Ferron, 1981). The current recommendation is for 2 kg ha^{-1} of Boverin to be used with a quarter dose of insecticide. The reduced rate of insecticide weakens the insects to such an extent that the fungus can easily infect and kill them (Lipa, 1985). Without insecticides, 3 or 20 kg ha^{-1} of Boverin were required to kill 90% of first and fourth instar larvae respectively, but such high doses are probably uneconomic.

Attention has also been focussed in the USA on the possibility of using *B. bassiana* for control of *L. decemlineata*. In a 3-year pilot test, the control of the beetle was found to be highly variable. In 8 of the 24 trials, potato yields from the *B. bassiana* treated plots were significantly greater than the control plots, but only two of these gave yields that were not significantly different from the insecticide treatment (Hajek *et al.*, 1987). One of the possible reasons for the poor results may be the high pest numbers present at the time of fungal application.

Currently, preparations of *B. bassiana* spores are used on several crops in China to control the European corn borer, *Ostrinia nubilalis*, pine caterpillars, *Dendrolimus* spp., and green leafhoppers, *Nephottix* spp. At least 400,000 ha are treated annually (Franz & Krieg, 1980; Hussey & Tinsley, 1981). *B. bassiana* is particularly suitable for use in the communal agricultural system of China because individual communes can produce their own supplies of inoculum. Spores are produced in covered pits in the ground using boiled rice. The rice is inoculated with a liquid culture of *B. bassiana* (Pinnock, personal communication) and although conditions are not sterile, contamination is rarely a problem due to antibiotics produced by *B. bassiana*. Spores of *B. bassiana* have also been used in France to control *O. nubilalis* giving control equal to standard insecticides in 17 out of 18 field trials conducted in 1985 (in Roberts & Wraight, 1986). In Brazil,

Daoust & Pereira (1986) suggested the use of *B. bassiana*-treated *Ceratosanthes hilariana* tubers, to attract chrysomelid beetles.

Rice pests have recently received much attention as possible target pests for entomogenous fungi; including *B. bassiana* and *Metarhizium anisopliae*. The rice ecosystem is particularly suitable for fungi because of the high humidity in paddy fields and the relatively high ambient temperatures in rice growing areas. Natural epizootics have been frequently observed and attempts to control rice pests in the field by applications of fungal inoculum have been relatively successful. For example, populations of the rice black bug, *Scotinophara coarctata* were significantly reduced after application of *B. bassiana* and other fungi (Rombach *et al.*, 1986).

Beauveria brongniartii

This pathogen has been used almost exclusively against soil inhabiting beetles, especially the European cockchafer, *Melolontha melolontha*. In France the potential of *B. brongniartii* against *M. melolontha* was demonstrated by Ferran (1978). Blastospores were incorporated into the soil at 2×10^{13} and 2×10^{14} spores ha^{-1} and the lower rate caused 14·5 to 56·6% mortality whereas the higher rate increased mortality to 38 to 80% (the control mortality was 0 to 32%). Treatment of adult *M. melolontha* females with *B. brongniartii* spores resulted in subsequent infection of their progeny (Keller, 1978) and this might be very important to the spread of the fungus when spores are applied to localised swarming areas. Keller (1986) applied $2 \cdot 6 \times 10^{14}$ blastospores ha^{-1} to 14 swarming sites of *M. melolontha* (89 ha in total) and obtained 85% larval mortality. This successful trial demonstrates the possibility of controlling *M. melolontha*. Although the initial spore dose was high, it provided insect control over a large area adjoining the swarming sites and the effect persisted for several years.

Recently, Tillemans & Coremans-Pelseneer (1987) demonstrated that *B. brongniartii* may be suitable for control of the black vine weevil, *Otiorhyncus sulcatus*. Under glasshouse conditions they achieved up to 84% control of introduced larvae after an application of 2×10^{8} conidia l^{-1} of peat. Two strains of *B. brongniartii* were also compared with other entomopathogens for control of *O. sulcatus* under glasshouse conditions by Gillespie (unpublished). The two strains (both isolated from *M. melolontha*) reduced larval numbers on *Begonia* plants by 46 and 48%, but this was less than the reduction obtained using some *Metarhizium anisopliae* strains (the best of which reduced larval numbers by 100%). It is clear that the efficiency of control is highly strain dependent and highlights the importance of strain evaluation and selection.

Entomophthoraceae

Members of the Entomophthoraceae have been isolated from a wide range of insect species, the most common hosts being aphids, muscoid flies, lepidopteran larvae and grasshoppers (Roberts & Humber, 1981). Some species such as *Zoophthora radicans* (*Entomophthora sphaerasperma*) have a wide host range (Gustafsson, 1965) whereas others are very specific. For example, *Erynia neoaphidis* is primarily associated with aphids (Deacon, 1983). In some years, when weather conditions are favourable, naturally occurring epizootics of entomophthoraceous fungi can significantly reduce pest populations.

A study of the parasites, predators and pathogens of potato aphids by Shands, Hall & Simpson (1962) revealed that *E. neoaphidis*, *Conidiobolus obscurus* and *Z. radicans* were the most important natural regulators of aphid numbers. *Entomophthora muscae* and *Strongwellsea castrans* were found to be the only natural agents to cause mortality frequently in adult *Leptohylemyia coarctata*. Annual variation in the number of infected adults was relatively high. In 1970, twenty-nine percent of the adults captured were infected, whereas in the previous year the level was only 1% (Wilding & Lauckner, 1974). There was similar variability in the natural infection of thrips by *Neozygites parvispora* in onion and leek fields (Carl, 1975). The first infections were found in the field in July but epizootics did not develop until September, by which time the crop had already been damaged.

Probably one of the most successful applications of the Entomophthoraceae is in the control of the lucerne aphid, *Therioaphis trifolii*, in Australia. The control of this pest represents one of the few uses of an entomopathogen as a classical biological control agent. *T. trifolii* was introduced to Australia with none of its natural enemies and it rapidly became a major pest. In the USA several Entomophthorales are considered to be important natural control agents of *T. trifolii* (Hall & Dunn, 1958). A search for natural control agents was initiated and several pathogenic isolates of *Zoophthora radicans* were identified in Israel. After a period of testing against *T. trifolii* and native Australian pests one of these isolates from Israel was introduced; it spread rapidly throughout E. Australia and now helps to maintain the pest population at a low level (Milner, Soper & Lutton, 1982).

Early attempts to use the Entomophthoraceae as microbial control agents involved providing infected insects as foci for the development of epizootics. Dunstan (1927) succeeded in initiating epizootics of *Z. radicans* on *Psylla mali* early in the season by introducing diseased insects. More recently, Carl (1975) demonstrated that introducing *Neozygites parvispora* to a population of *Thrips tabaci*, under conditions similar to those in commercial glasshouses, could infect 97% of the population after 30 days. It must be emphasised that in this example the conditions were ideal for the fungus in terms of temperature, humidity and a very dense host population. In a repeat of the experiment in the glasshouse; the results were inconclusive, possibly due to a sparse host population restricting fungal spread. Limited success has been achieved by applying species of Entomophthoraceae to field crops. Wilding (1982) obtained 48% infection of apterous aphids on a crop of field beans (*Vicia faba*) and doubled the yields by introducing *Aphis fabae* infected with *Erynia neoaphidis* and *Neozygites fresenii*. The control achieved in the field beans, however, was inconsistent and compared poorly with a chemical pesticide comparison. The production and introduction of diseased insects into crops is very time consuming and the costs of such a control system would be prohibitively high. Even then, introductions of diseased insects do not guarantee the initiation of an epizootic because the factors that control epizootics are complex and varied, with environmental conditions being perhaps the most important.

There is an urgent need to develop a fungal formulation that is economic to produce, easy to apply, not too dependent on precise environmental conditions and that can consistently control the target pest. The forcibly-discharged conidia are generally considered unsuitable as an inoculum because of their fragile nature and short life span. In many cases infected larvae have to be used as initial inocula but the costs incurred by this process are high. However, spraying homogenised larvae onto the crop can reduce expense (Kelsey, 1965). The use of resting spores has also been considered. Latgé, Soper & Madore (1977) produced approximately 10^9 resting spores of *Conidiobolus obscurus* l^{-1} of fermenter medium. However, it was calculated that an uneconomic 1000 to 2000 l of medium ha^{-1} would be required to give a field LD_{50} dose (Latgé, 1982). Another major restriction on the use of resting spores lies in their dormancy, but this may be overcome in some species (Soper *et al.*, 1975). Recent attention has focussed on the possibility of using hyphal bodies and mycelium as a source of inoculum (McCabe & Soper, 1985). The use of hyphal bodies would not be ideal because, like resting spores, they are not the infective propagule.

Hirsutella spp.

There are about 40 species in the genus *Hirsutella*, of which the most important is the mite pathogen, *H. thompsonii* (McCoy, 1981). *H. thompsonii* has been observed frequently in the citrus groves of Florida as a natural regulator of the citrus rust mite, *Phyllocoptruta oleivora*. The mites feed on the fruit surface producing cosmetic damage which leads to downgrading and a reduced value product. Natural epizootics of *H.*

thompsonii do not develop until late in the season by which time the damage threshold has been exceeded (McCoy, 1981).

Many attempts have been made to understand the pathogenicity of *H. thompsonii* to *P. oleivora* and to apply the knowledge gained to devise an effective control strategy. Optimal environmental conditions can allow rapid infection and death of the host within 72 hours (McCoy & Selhime, 1974). Under suboptimal conditions (the relative humidity was not maintained at 100% for 24 hours) mortality rates fell dramatically (Gerson *et al.*, 1979). In the field, the environmental conditions are highly variable and consequently the control of *P. oleivora* in field experiments has been inconsistent. The fungus was applied to citrus trees as a mycelial preparation. Moist conditions were found to be needed not only for infection but also to permit mycelial sporulation and production of infective conidia (McCoy *et al.*, 1971). In the field, prolonged moist periods cannot be guaranteed, so various attempts have been made to improve mycelial preparations and encourage conidia production and survival.

Abbott Laboratories recognised the potential of *H. thompsonii* in the early 1970's and after some initial problems, a wettable powder formulation was developed. The wettable powder contained fragments of mycelium and also nutrients, to encourage sporulation on plant surfaces, thereby improving efficacy. The product was registered as Mycar® with the Environmental Protection Agency in 1981. In trials, the formulation significantly reduced mite numbers on leaves by a mean of 63% compared to the control (McCoy, 1981). Production ceased in 1985 for several reasons: inconsistent pest control; the need for cold storage to maintain product viability; problems in producing infective spores; and quality control problems associated with the absence of a satisfactory bioassay.

The need for high humidity is a major problem with *H. thompsonii* but high moisture levels could possibly be achieved in the field by irrigation. However, it is easier and more cost effective to regulate conditions in glasshouses and this could provide a better environment for the use of this fungus (Gardner, Oetting & Story, 1982).

Metarhizium spp.

There are three species of *Metarhizium*: *M. flavoride*, *M. album* and *M. anisopliae*, the latter being the most important (Rombach, Humber & Evans, 1987). *M. anisopliae* is separated into two varieties, var. *anisopliae* isolated from a wide range of insect species and var. *major* only isolated from the rhinoceros beetle, (*Oryctes rhinoceros*) (Tulloch, 1976). *M. anisopliae* has attracted a good deal of interest from researchers all over the world as a result of its wide geographic and species distribution. Latch (1964) calculated that 166 species of insects are naturally attacked by *M. anisopliae*.

M. anisopliae is currently used commercially in Brazil for the control of various spittlebug species (such as *Mahanarva posticata*) on sugar cane. The fungus is produced on sterilized rice in glass bottles or autoclavable bags (Aquino *et al.*, 1975). Once adequate sporulation has occurred, the grains and conidia are dried, ground and marketed under a variety of trade names (for example Metaquino and Metabiol). As the process is very simple and does not require complex fermentation or processing facilities, it is highly suited to production on a localised basis by grower co-operatives. The average spore yield is 10^{12} spores kg^{-1} rice and this makes the product very cost effective with 1 kg of rice producing enough conidia to treat 1 ha. Ferron (1981) reported that an aerial spray of 6×10^{11} to $1 \cdot 2 \times 10^{12}$ conidia ha^{-1} significantly increased cane sugar content, even though the maximum mortality recorded was only 65%. Control of the spittlebugs (and corresponding increase in sugar content) could perhaps be improved by increasing the dose rate without greatly increasing costs of production or application. Spittlebug control is one of the few major success stories resulting from research into the entomopathogenic fungi. In 1978, fifty thousand hectares were treated in the state of Pernambuco alone.

M. anisopliae also has great potential for the control of various soil pests. The soil is a good environment for the fungus in terms of moisture (which often limits the use of fungi to control aerial pests) and steady temperatures, although low temperatures may inhibit growth of the fungus in some field soils (Latch, 1964). More recently in Australia, Coles & Pinnock (1984) used conidia produced on bran as a bait for the pasture cockchafer, *Aphodius tasmaniae*. Spores were applied at a number of sites at $1 \cdot 4 \times 10^{14}$ ha^{-1} or $1 \cdot 3 - 3 \times 10^{15}$ ha^{-1}. After one month, 55% of the larvae at one site were dead and after four months the mortality at each site exceeded 80%. These results were encouraging, but the high rates used would most probably be uneconomic.

A good deal of attention has been focussed recently on the possibility of utilising *M. anisopliae* to control an increasingly serious horticultural pest, the black vine weevil, *Otiorhynchus sulcatus*. Soil pests, such as *O. sulcatus*, are best controlled using preventative rather than curative treatments, as it is difficult to assess larval numbers in a growing crop and consequently hard to time treatments. Prado (1979) obtained complete control in one trial using 10^8 *M. anisopliae* conidia l^{-1} soil. In general, higher spore doses of *M. anisopliae* increase *O. sulcatus* mortality (Zimmermann, 1981; Soares, Marchal & Ferron, 1983).

Zimmermann (1981) compared two preventative application strategies for *M. anisopliae* conidia and reported mortalities of 81 and 16% for soil drench and pre-incorporation respectively. The efficacy of the fungus was also reduced by delaying application for four weeks after egg application.

Zimmermann's work and studies in our laboratory suggest that a *Metarhizium anisopliae* product may be available for glasshouse growers in the near future. There is also an urgent need for control of *O. sulcatus* on nursery stock and soft fruit, but it is not yet certain whether an isolate of *M. anisopliae* can be found which will provide effective weevil control at the reduced temperatures occurring in field crops. In small-scale, outdoor trials a prophylactic treatment of *M. anisopliae* controlled 30 to 50% of weevil larvae (Zimmermann & Simons, 1986, 1986). It was also observed that *M. anisopliae* may have a long term effect on weevil control.

M. anisopliae is also used to control coconut pests, for example some control of *O. rhinoceros* was achieved using a baculovirus combined with *M. anisopliae* var. *major* to provide short term control whilst the baculovirus spreads. Conidia are produced on oat grain (Latch, 1976) or rice (Bedford, 1980) and distributed to the breeding sites. Marschall (1980) treated artificial breeding sites with spores and obtained high mortality. Most infected adults died at the treated breeding site and hence there was only limited spread of the fungus to natural breeding sites. Treatment of palms with *M. anisopliae* var. *anisopliae* resulted in 32% mortality of the indigenous rhinoceros beetle, *Sapanes australis*, compared to a natural mortality of 1%, over 48 weeks. This treatment also infected another important pest, the black palm weevil, *Rhynchophorus bilineatus*. Hänel (1982) demonstrated that the termite, *Nasutitermes exitiosus*, was susceptible to *M. anisopliae* under laboratory conditions. However, field experiments using *M. anisopliae* against termites gave variable results. Some colonies were eradicated whilst others declined gradually, possibly due to the natural behaviour of certain species which eject diseased individuals from their colonies prior to the formation of infective propagules.

Nomuraea spp.

The genus *Nomuraea* consists of two species: *N. rileyi*, a pathogen of Lepidoptera, and *N. atypicola*, a pathogen of spiders. A recent survey by Sikarowski in Mississippi (in Mohamed *et al.*, 1977) concluded that *N. rileyi* was one of the major pathogens regulating *Heliothis* spp. in field crops. Natural epizootics of *N. rileyi* occur annually, sometimes destroying the entire larval population of *Heliothis zea* and *Trichoplusia ni* (Mohamed *et al.*, 1977). Unfortunately, the caterpillar population often exceeds the damage threshold before *N. rileyi* is able to reduce pest numbers.

Good control of *Plathypena scabra* and *H. zea* in the field was achieved by application of *N. rileyi*-infected *Heliothis* cadavers that were cut into 3 mm pieces (Sprenkel & Brooks, 1975). Although the control achieved was significant, in practice the production and application of inoculum by this means would be uneconomic. Ignoffo *et al.* (1976a) induced an epizootic in soybean caterpillars by applications of *N. rileyi* conidia. Whilst this approach again produced acceptable results, the costs incurred in the production and application of large doses of conidia would be high. Further examination of the soybean caterpillar-*N. rileyi* interaction revealed differences in the susceptibility of different caterpillar species (Ignoffo, 1981) and in the pathogenicity of different *N. rileyi* strains (Ignoffo *et al.*, 1976b). An assessment of the pathogenicity of different fungal strains may lead to the selection of species specific strains with greatly reduced LC_{50} values.

The results obtained from several field experiments have been disappointing. Getzin (1961) applied 3×10^{12} conidia ha^{-1} of cabbages and although the resulting larval mortality (of *T. ni*) was 67%, this was too low to prevent significant damage. Even when Bell (in Ignoffo, 1981) increased the application frequency and rate to seven applications of $1 \cdot 4 \times 10^{14}$ conidia ha^{-1} at weekly intervals the level of damage was still unacceptable. Mohamed, Bell & Sikorowski (1978) similarly found that they could kill a significant number of *H. zea* larvae on sweet corn (62-90%) without reducing the damage by a similar amount. The maximum damage reduction was only 36%.

The future of *N. rileyi* as a microbial insecticide is dependent on the availability of strains that can kill their target pests with enough consistency and speed to maintain damage at a low level. The production of adequate spore numbers is also hindering its development. *N. rileyi* can now be grown in liquid culture with yields $1 \cdot 5$ times greater than those obtained using agar. However, the blastospores produced are not pathogenic to larvae (Bell, 1975). Ignoffo *et al.* (1975) concluded that *N. rileyi* could possibly be commercialised as a prophylactic treatment to induce early epizootics and thereby reduce damage. The treatment could be based on blastospores applied together with nutrients (resulting in conidial production on the plants). However, such a formulation would increase the dependence on prolonged moisture availability for efficient control. Holdom & Van De Klashorst (1986) estimated that the hyphal bodies take 2 to 4 days at 25°C to sporulate under ideal conditions and point out during this time many of the spores could be killed by sunlight. In addition, growth of unwanted microorganisms on the nutrients in the formulation would occur. The use of blastospores formulated with nutrients obviously justifies further research, but many problems will need to be solved if blastospores are to be used in practice.

Verticillium spp.

The most important entomopathogenic species of this genus is *V. lecanii* (*Cephalosporium lecanii*). The fungus has been isolated from a wide range of insects, mites and spiders both in temperate and tropical areas (Rombach & Gillespie, 1988). Interest in the use of *V. lecanii* has been stimulated by the demands of growers for an efficient control agent for aphids and whitefly in glasshouse crops. The use of *V. lecanii* for aphid control on chrysanthemum was investigated by Hall & Burges (1979) in the UK and by Gardner, Oetting & Storey (1984) in the USA. Hall & Burges (1979) maintained the population of *Myzus persicae* below its damage threshold by a single application of *V. lecanii*. The control of two other minor aphid species, *Macrosiphoniella sanborni* and *Brachycaudus helichrysi*, was less satisfactory. The control of *M. sanborni* and *Aphis gossypii* were improved by combining *V. lecanii* with the aphid alarm pheromone ε-β-farnesene (Hockland *et al.*, 1986) and by frequent applications of low doses of the pathogen (L. R. Wardlow, personal communication). It should, however, be emphasised that increasing the application frequency also increases the costs of control. Hall (1982) demonstrated that formulation of *V. lecanii* with carbohydrate carriers improved the establishment of the fungus on cucumbers and suggested that control of *A. gossypii* could be prolonged without resorting to frequent applications. Recent work by Sopp, Palmer & Gillespie (1989) described improved *A. gossypii* control on chrysanthemum by applying *V. lecanii* using an electrostatic sprayer. The charged spray droplets penetrated the crop more efficiently and increased deposition on the undersides of leaves (the feeding size of *A. gossypii*).

Two commercial formulations of *V. lecanii*, Mycotal® and Vertalac®, were developed and initially marketed in Britain by Tate & Lyle and later by Microbial Resources. Vertalac was recommended for aphid control in glasshouse crops and Mycotal was recommended for control of glasshouse whitefly. Grower resistance to microbial insecticides and inconsistent pest control resulted in both formulations being withdrawn from sale in 1986.

The Mycotal formulation was based on a strain of *V. lecanii* isolated from whitefly. The pathogenicity of this strain to whitefly was superior to the aphid strain (Hall, 1982). On cucumbers, the whitefly strain spread from sprayed leaves to scales on unsprayed leaves. A single application of Mycotal controlled a whitefly population for 2 to 3 months (Hall, 1982). The Mycotal formulation was also effective against thrips and significantly reduced the *Thrips tabaci* population on glasshouse cucumbers (Gillespie, 1984).

There has been limited interest in the use of *V. lecanii* in field environments. *V. lecanii* has been used against soft green scale, *Coccus viridis* on coffee plants in India (Easwaramoorthy & Jarayaj, 1977, 1978; Easwaramoorthy *et al.*, 1978). Two applications (two weeks apart) of $1 \cdot 6 \times 10^7$ ml^{-1} caused a maximum mortality of 73·1% under field conditions. The mortality could be increased to 97·6% by adding 0·05% Tween 20. The inclusion of nutrients in the formulation also increased fungal pathogenicity.

V. lecanii may have a potential future as a microbial pesticide with both insecticidal and fungicidal properties, For example, *V. lecanii* can reduce the number of carnation rust (*Uromyces dianthi*) pustules on carnation leaves under experimental conditions (Deacon, 1983). Interestingly, three out of the eight strains of *V. lecanii* examined were also highly pathogenic to aphids. There is clearly potential for further work in this area.

Future research directions

The lack of commercially available mycoinsecticides in the USA and W. Europe indicates the difficulties encountered in development of fungi for pest control. Indeed, products have been launched and then withdrawn, not least because they failed to provide reliable pest control. Despite these problems, commercial interest has increased over the last few years and it is widely accepted that fungi have the potential to control noxious insects and mites. Before considering the research strategies needed to realise the potential of fungi, the possible advantages of mycoinsecticides compared to conventional chemical insecticides should be examined.

Fungi infect most genera of insects and mites but pathogenic species or strains can be quite specific and may only infect one type of host. Thus, fungi can be used to control important pests without affecting beneficial predators and parasitoids. For example, the use of *V. lecanii* for whitefly control on glasshouse crops can be integrated with use of the parasitoid, *Encarsia formosa* (Gillespie, unpublished). Small numbers of parasitoids do become infected with *V. lecanii*, but the fungus is not able significantly to reduce populations. Entomogenous fungi cannot infect mammals and thus reduce the hazards normally encountered with insecticide applications. There is no evidence to date of fungal resistance occurring in insect populations, although this could be due to their limited application. Some of the entomogenous fungi can provide prolonged pest control without polluting the environment (a problem encountered with certain persistent insecticides). Thus, widespread mycoinsecticide use could provide great benefits to both man and his environment. The problem facing research scientists is how this goal might be achieved.

At present, we believe that research on entomogenous fungi should be concentrated on insects whose habit provides a favourable environment for fungal growth. Such research is already taking place, increasingly concentrated on soil insects together with rice pests and certain stem borers. Soil is a particularly suitable environment for exploiting fungi as moisture levels rarely limit conidial germination and growth and soil pests are increasingly difficult to control with available insecticides. The ability of entomogenous fungi to parasitise insects probably provides an important ecological advantage over other soil microorganisms and allows them to successfully compete. In the soil, fungal spores are also protected from UV light which often causes spore mortality on leaf surfaces. The use of fungicides to control plant disease also poses problems for mycoinsecticide use. One possible strategy to overcome this problem is to produce strains of entomogenous fungi that are resistant to certain fungicides. These can then perhaps be used in integrated pest and disease control programmes.

The importance of strain selection cannot be over-emphasised when considering fungi for pest control as many characters can vary greatly between strains of a given species. Thus strains should be considered as individual entities and generalizations about species characteristics avoided. Important characters to be considered during strain selection are given in Table 4.2.

In addition, intra specific selection can be used to optimize biological characteristics as multi-spore isolates may comprise a number of genotypes which can vary in pathogenicity (Samsinakova & Kalalova, 1983; Jimenez & Gillespie, unpublished). Thus it is recommended that selection should be made of both strains and single spore isolates. Single spore isolates of *Beauveria bassiana* were not stable, however, and lost pathoge-

---

Table 4.2. Characters for consideration when selecting fungal isolates for control of pests.

a) Pathogenicity

b) Spore germination and growth rate

c) Relative humidity required for spore germination

d) Effect of temperature on fungal growth

e) Sporulation on dead insects

f) Disease spread

g) Feasibility of production

h) Inoculum survival during storage

i) Fungicide resistance

nicity with repeated agar subculture, in some cases more rapidly than the parent multi-spore isolate (Jimenez & Gillespie, unpublished).

For commercialization, a fungus must be amenable to large-scale production. The most attractive production method is liquid fermentation using stirred tank reactors. Some fungi, such as *Verticillium lecanii*, produce many hyphal bodies in submerged culture (Hall, 1981) and these can be harvested by centrifugation, dried and formulated. Unfortunately, because hyphal bodies are not adapted for survival and have thin walls, low temperature storage is necessary. Other fungi, including *Metarhizium anisopliae*, mainly produce mycelia in liquid culture with relatively few hyphal bodies (Adamek, 1963; Gillespie, 1984). Mycelia and hyphal bodies can be dried and formulated using the method of McCabe & Soper (1985). Storage problems and delays in pest control caused by the mycelium having to rehydrate and sporulate before insect infection pose problems for this approach. *M. anisopliae* is best produced in semi-solid media using cereal grains as practiced on a large scale in Brazil (Aquino *et al.*, 1975). Certain isolates of *B. bassiana* and *Hirsutella thompsonii* conidiate in submerged culture (Van Winklehoff & McCoy, 1984; Thomas, Khachatourians & Ingledew, 1987, Hilber & Gillespie, unpublished). Furthermore, spores produced in liquid can be more pathogenic than those produced on solid or semi-solid media (Jimenez & Gillespie, unpublished).

Many of the Entomophthoraceae would possibly be good biological control agents but production limitations impose a major constraint to their use. Some such as *Neozygites parvispora* are obligate parasites (Carl, 1975) whilst others require very complex media for production. The forcibly discharged conidia of the Entomophthorales are very fragile and survive poorly. It is possible to produce resting spores of some species such as *Conidiobolus obscurus* by liquid fermentation (Latgé *et al.*, 1978). However, field control obtained with these spores was poor. Production and application of ground, dried mycelium may make it possible to utilise some of the Entomophthoraceae hitherto considered unsuitable (McCabe & Soper, 1985).

Storage of spore preparations is a very important consideration both for commercial producers and users. On-farm storage facilities are normally not designed to maintain products at precise temperatures and humidities. Fermenter-produced spores thus require processing for storage and field use. Blachère *et al.* (1973) used silica powder to dry *Beauveria brongniartii* spores and clay coating was used by Fargues *et al.* (1983) to prolong spore survival. Conidia of some *Metarhizium anisopliae* strains can be stored for at least 16 weeks in 0·05% Triton X–100 at 5°, 15° and 25°C without significantly affecting viability (Moorhouse &

Gillespie, unpublished). Prior, Jollands & La Patourel (1988) demonstrated that survival of *B. bassiana* conidia can be improved with formulation. Conidial suspensions in 0·01% Tween-80 were unable to germinate after 3 days at room temperature whereas an oil-based formulation increased survival time to 12 days at room temperature and over 40 days in the refrigerator. Couch & Ignoffo (1981) stated that a storage life of 18 months was necessary for a commercial product.

It is likely that in many cases the success or failure of a mycoinsecticide application is determined by the temperatures and relative humidities occurring after spore application. There is an urgent need for improved formulation technologies to preserve spore viability during storage and to allow spore germination at suboptimal r.h. Published data on formulation is rare, but recently Philipp & Hellstern (1986) reported a formulation based on liquid paraffin that allowed the fungus *Ampelomyces quisqualis* to parasitise powdery mildew mycelium at 80% r.h. Studies at Littlehampton (Jimenez & Gillespie, unpublished) have shown that conidia from one *B. bassiana* isolate can be stored for 9 months at 20°C without significant loss in viability or pathogenicity. It is likely that formulation technology will advance rapidly in the next decade, provided an increased research effort can be made in this area. It should be remembered, however, that formulation can only assist initial infection, subsequent disease spread will still be reliant on favourable humidities.

Molecular biology provides exciting opportunities for improving fungi for pest control, though in the short term the technology is likely to be more valuable in elucidating mechanisms of pathogenicity. Transformation of *Metarhizium anisopliae* has just been reported using the *Aspergillus nidulans PBEN A3* gene (Goettel, personal communication). Progress in understanding mechanisms of pathogenicity is now being made, particularly in the area of cuticular penetration where the key enzyme is probably an endoprotease (St. Leger, Cooper & Charnley, 1986; Charnley, Chapter 5). In the future it may be possible to insert toxin genes from *Bacillus thuringiensis* into fungi to increase the speed of insect kill. However, any improvement in strains achieved by genetic manipulation could still be overshadowed by the requirement for high r.h. levels. Future research should concentrate on formulation development and the targetting of pests which live in environments with high r.h.

If these strategies are pursued, the understanding of entomogenous fungi will increase and mycoinsecticides may once again become commercially available. If this aim is achieved, agriculture will benefit from prolonged pest control, reduced risk of resistance and a high degree of safety to non-target organisms with associated environmental benefits.

Acknowledgements. The authors would like to thank Dr A. K. Charnley for securing funds for E. R. M. from Bath University, Dr. C. C. Payne for critically reading the manuscript and Sue Bewsey for her excellent typing. Vertalec and Mycotal are Trade Marks of Koppert B.V.

References

Adamek, L. (1963). Submerged cultivation of the fungus *Metarhizium anisopliae* (Metsch.). *Folia Microbiologia (Praha)*, **10**, 255-257.

Aquino, de M. L., Cavalanti, V. A., Sena, R. C. & Queiroz, G. F. (1975). Nova technologia de multicacao do fungo *Metarhizium anisopliae*. *Boletim Tecnico da Comissao Executiva de Defesa Fitossanitaria da Lavoura Canavieira de Pernambuco*, **4**, 1-31.

Bedford, G. O. (1980). Biology, ecology and control of palm rhinoceros beetles. *Annual Review of Entomology*, **25**, 309-339.

Bell, J. V. (1975). Production and pathogenicity of the fungus *Spicaria rileyi* from solid and liquid media. *Journal of Invertebrate Pathology*, **26**, 129-130.

Berger, E. W. (1909). Whitefly studies in 1908. *Bulletin of the Florida Agricultural Experimental Station*, **97**, 42-59.

Blachère, H., Calves, J., Ferron, P., Corrieu, G. & Peringer, P. (1973). Étude de la formulation et de la conservation d'une preparation entompathogène a base de blastospores de *Beauveria tenella* (Delacr. Siemasko). *Annales du Zoologie et Ecologie Animale*, **11**, 247-257.

Carl, K. F. (1975). An *Entomophthora* sp. (Entomophthorales, Entomophthoraceae) pathogenic to *Thrips* spp. (Thysan.: Thripidae) and its potential as a biological control agent in glasshouses. *Entomophaga*, **20**, 381-388.

Carruthers, R. I., Feng, Z., Robson, D. S. & Roberts, D. W. (1985a). *In vivo* temperature-dependent development of *Beauveria bassiana* mycosis of the European Cornborer, *Ostrinia nubialis*. *Journal of Invertebrate Pathology*, **46**, 305-311.

Carruthers, R. I., Haynes, D. L. & MacLeod, D. M. (1985b). *Entomophthora muscae* (Entomophthorales: Entomophthoraceae) in the onion fly, *Delia antiqua* (Diptera: Anthomyiidae). *Journal of Invertebrate Pathology*, **45**, 81-93.

Coles, R. & Pinnock, D. E. (1984). Current status of the production and use of *Metarhizium anisopliae* for control of *Aphodius tasmaniae* in South Australia. In *Proceedings of the 4th Australian Applied Entomological Research Conference*, ed. P. Bailey & D. Swincer, pp 357-361.

Couch, T. L. & Ignoffo, C. M. (1981). Formulation of insect pathogens. In *Microbial Control of Pests and Plant Diseases 1970-1980*, ed. H. D. Burges, pp. 621-634. Academic Press: London.

Daoust, R. A. & Pereira, R. M. (1986). Stability of entomopathogenic fungi *Beauveria bassiana* and *Metarhizium anisopliae* on beetle attracting tubers and cowpea foliage in Brazil. *Environmental Entomology*, **15**, 1237-1243.

Deacon, J. W. (1983). *Microbial Control of Plant Pests and Diseases*, pp. 31-42, Aspects of Microbiology 7. Van Nostrand Reinhold Co. Ltd: Wokingham, England.

Dunstan, A. G. (1927). The artificial culture and dissemination of *Entomophthora sphaerosperma* Fres., a fungous parasite for the control of European apple sucker (*Psylla mali* Schmidb.). *Journal of Economic Entomology*, **20**, 68-75.

Easwaramoorthy, S. & Jayaraj, S. (1977). The effect of temperature, pH and media on the growth of the fungus *Cephalosporium lecanii*. *Journal of Invertebrate Pathology*, **29**, 399-400.

Easwaramoorthy, S. & Jayaraj, S. (1978). Effectiveness of the white halo fungus *Cephalosporium lecanii* against field populations of the coffee green bug, *Coccus viridis*. *Journal of Invertebrate Pathology*, **32**, 88-96.

Easwaramoorthy, S., Regupathy, A., Santharam, G. & Jayaraj, S. (1978). The effect of subnormal concentrations of insecticides in combination with the fungal pathogen *Cephalosporium lecanii* Zimm. in the control of the coffee green scale, *Coccus viridis* Green., *Journal of Applied Entomology*, **86**, 161-166.

Fargues, J., Reisinger, O., Robert, P. H. & Aubart, C. (1983). Biodegradation of entomopathogenic hyphomycetes: influence of clay coating on *Beauveria bassiana* blastospore survival in soil. *Journal of Invertebrate Pathology*, **41**, 131-142.

Ferron, P. (1978). Etiologie et epidemiologie des muscardines. Thèse Doctorat d'État, Université Pierre et Marie Curie, Paris.

Ferron, P. (1981). Pest control by the fungi *Beauveria* and *Metarhizium*. In *Microbial Control of Pests and Plant Diseases 1970-1980*, ed. H. D. Burges, pp. 465-482. Academic Press: London.

Ferron, P. (1985). Fungal control. In *Comprehensive Insect Physiology, Biochemistry and Pharmacology*, vol. 12, ed. G. A. Kerkut & L. I. Gilbert, pp. 313-346. Pergamon Press: Oxford.

Fransen, J. J. (1986). The survival of spores of *Aschersonia aleyrodis*, an entomopathogenic fungus of greenhouse whitefly, *Trialeurodes vaporariorum*. In *Fundamental and Applied Aspects of Invertebrate Pathology*, ed. R. A. Samson, J. M. Vlak & D. Peters, p. 226. Foundation of the 4th International Colloquium of Invertebrate Pathology: Wageningen, The Netherlands.

Fransen, J. J. (1987). *Aschersonia aleyrodis* as a microbiological control agent of greenhouse whitefly. Ph.D. Thesis, University of Wageningen, The Netherlands.

Fransen, J. J., Winkleman, K. & Van Lenteren, J. C. (1987). The differential mortality at various life stages of the greenhouse whitefly, *Trialeurodes vaporariorum* (Homoptera: Aleyrodidae) by infection with the fungus *Aschersonia aleyrodis* (Deuteromycotina: Coelomycetes). *Journal of Invertebrate Pathology*, **50**, 158-165.

Franz. J. M. & Krieg, A. (1980). Mikrobiologische Schadlingsbekampfung in China. *Ein Riesebert Forum Mikrobiologie*, **3**, 173-176.

Garcia, C. & Ignoffo, C. M. (1977). Dislodgement of conidia of *Nomuraea rileyi* from cadavers of cabbage looper, *Trichoplusia ni*. *Journal of Invertebrate Pathology*, **30**, 114-116.

Gardner, W. A. (1985), Effects of temperature on the susceptibility of *Heliothis zea* larvae to *Nomuraea rileyi*. *Journal of Invertebrate Pathology*, **46**, 348-349.

Gardner, W. A., Oetting, R. D. & Storey, G. K. (1982). Susceptibility of the two spotted spider mite, *Tetranychus urticae* Koch, to the fungal pathogen *Hirsutella thompsonii* Fischer. *Florida Entomologist*, **65**, 458-465.

Gardner, W. A., Oetting, R. D. & Storer, G. K. (1984). Scheduling of *Verticillium lecanii* and benomyl applications to maintain aphid (Homoptera; Aphidae) control on chrysanthemums in greenhouses. *Journal of Economic Entomology*, **77**, 514-518.

Gardner, W. A., Sutton, R. M. & Noblet, R. (1977). Persistence of *Beauveria bassiana*, *Nomuraea rileyi* and *Nosema necatrix* on soybean foliage. *Environmental Entomology*, 6, 616-618.

Gerson, U., Kenneth, R. & Muttath, T. I. (1979). *Hirsutella thompsonii*, a fungal pathogen of mites. II. Host-pathogen interactions. *Annals of Applied Biology*, 91, 29-40.

Getzin, L. W. (1961). *Spicaria rileyi* (Farlow) Charles, an entomogenous fungus of *Trichoplusia ni* (Hubner). *Journal of Insect Pathology*, 3, 2-10.

Gillespie, A. T. (1984). The potential of entomogenous fungi to control glasshouse pests and brown planthoppers of rice. Ph.D. Thesis, University of Southampton.

Gillespie, A. T. & Crawford, E. (1986). Effect of water activity on conidial germination and mycelial growth of *Beauveria bassiana*, *Metarhizium anisopliae*, *Paecilomyces* spp. and *Verticillium lecanii*. In *Fundamental and Applied Aspects of Invertebrate Pathology*, eds R. A. Samson, J. M. Vlak & D. Peters, p. 254. Foundation of the 4th International Colloquium of Invertebrate Pathology: Wageningen, The Netherlands.

Gunnarson, S. G. S. (1988). Infection of *Schistocerca gregaria* by the fungus *Metarhizium anisopliae*: cellular reactions in the integument studied by scanning electron and light microscopy. *Journal of Invertebrate Pathology*, 52, 9-17.

Gustafsson, M. (1965). On the species of the genus *Entomophthora* in Sweden. II. Cultivation and physiology. *Lantbrukshögskolons Annaler*, 31, 405-457.

Hajek, A. E., Soper, R. S., Roberts, D. W., Anderson, T. E., Biever, K. D., Ferro, D. N. LeBrun, R. A. & Storch, R. H. (1987). Foliar applications of *Beauveria bassiana* (Balsamo) Vuillemin for control of the Colorado potato beetle, *Leptinotarsa decemlineata* (Say) (Coleoptera: Chrysomelidae): An overview of pilot test results from the Northern United States. *Canadian Entomologist*, 119, 959-974.

Hall, I. M. & Dunn, P. H. (1958). Artificial dissemination of entomophthorous fungi pathogenic to the spotted alfalfa aphid in California. *Journal of Economic Entomology*, 51, 341-344.

Hall, R. A. (1981). The fungus *Verticillium lecanii* as a microbial insecticide against aphids and scales. In *Microbial Control of Pests and Plant Diseases 1970-1980*, ed. H. D. Burges, pp. 483-498. Academic Press: London.

Hall, R. A. (1982). Control of whitefly, *Trialeurodes vaporariorum* and cotton aphid, *Aphis gossypii* in glasshouses by two isolates of the fungus *Verticillium lecanii*. *Annals of Applied Biology*, 101, 1-11.

Hall, R. A. & Burges, H. D. (1979), Control of aphids in glasshouses with the fungus *Verticillium lecanii*. *Annals of Applied Biology*, 102, 455-466.

Hänel, H. (1982). Propagation of *Metarhizium anisopliae* infection in termite colonies in the laboratory and in the field. In *Proceedings of the 3rd International Colloquium of Invertebrate Pathology*, p. 107.

Hockland, S. H., Dawson, G. H., Griffiths, D. C., Marples, B., Pickett, J. A. & Woodcock, C. M. (1986). The use of aphid alarm pheromone (ε-β-farnesene) to increase effectiveness of entomophilic fungus *Verticillium lecanii* in controlling aphids on chrysanthemums under glass. In *Fundamental and Applied Aspects of Invertebrate Pathology*, ed. R. A. Samson, J. M. Vlak & D. Peters, p. 252. Foundation of the 4th International Colloquium of Invertebrate Pathology: Wageningen, The Netherlands.

Holdom, D. G. & Van de Klashorst, G. (1986). Sporulation by hyphal bodies of *Nomuraea rileyi* and subsequent infection of *Heliothis* spp. *Journal of Invertebrate Pathology*, **48**, 232-245.

Hughes, J. G. & Gillespie, A. T. (1985). Germination and penetration of the aphid *Macrosiphoniella sanborni* by two strains of *Verticillium lecanii*. In Programme and Abstracts, XVIII Annual Meeting of the Society for Invertebrate Pathology, p. 28.

Hussey, N. W. & Tinsley, T. W. (1981). Impressions of insect pathology in the People's Republic of China. In *Microbial Control of Pests and Plant Diseases 1970-1980*, ed. H. D. Burges, pp. 785-795. Academic Press: London.

Ignoffo, C. M. (1981). The fungus *Nomuraea rileyi* as a microbial insecticide. In *Microbial Control of Pests and Plant Diseases 1970-1980*, ed. H. D. Burges, pp. 513-538. Academic Press: London.

Ignoffo, C. M., Puttler, B., Marston, N. L., Hostetter, D. L. & Dickerson, W. A. (1975). Seasonal incidence of the entomopathogenic fungus *Spicaria rileyi* associated with noctuid pests of soybeans. *Journal of Invertebrate Pathology*, **25**, 135-137.

Ignoffo, C. M., Marston, N. L., Hostetter, D. L. & Puttler, B. (1976a). Natural and induced epizootics of *Nomuraea rileyi* in soybean caterpillars. *Journal of Invertebrate Pathology*, **27**, 191-198.

Ignoffo, C. M., Puttler, B., Hostetter, D. L. & Dickerson, W. A. (1976b). Susceptibility of the cabbage looper, *Trichoplusia ni*, and the velvetbean caterpillar, *Anticarsia gemmatalis*, to several isolates of the entomopathogenic fungus *Nomuraea rileyi*. *Journal of Invertebrate Pathology*, **28**, 259-262.

Ignoffo, C. M., Garcia, C., Alyoshina, O. A. & Lappa, N. V. (1979). Laboratory and field studies with Boverin: a mycoinsecticidal preparation of *Beauveria bassiana* produced in the Soviet Union. *Journal of Economic Entomology*, **72**, 562-565.

Keller, S. (1978). Infektionsversuche mit dem Pilz *Beauveria tenella* an Maikafers (*Melolontha melolontha* L.). *Mitteilungen der Schweizerischen Entomologischen Gesellschaft*, **51**, 13-19.

Keller, S. (1986), Control of May beetle grubs (*Melolontha melolontha* L.) with the fungus *Beauveria brongniartii* (Sacc.) Petch. In *Fundamental and Applied Aspects of Invertebrate Pathology*, ed. R. A. Samson, J. M. Vlak & D. Peters, pp. 525-528. Foundation of the 4th International Colloquium of Invertebrate Pathology: Wageningen, The Netherlands.

Kelsey, J. M. (1965). *Entomophthora sphaerosperma* (Fres.) and *Plutella maculipennis* (Curtis) control. *New Zealand Entomologist*, **3**, 47-49.

Kish, L. P. & Allen, G. E. (1978). The biology and ecology of *Nomuraea rileyi* and a program for predicting its incidence on *Anticarsia gemmatalis* in soybean. *Bulletin of the Florida Agricultural Experimental Station*, **795**, 1-48.

Klein, M. G. & Coppel, H. C. (1973). *Entomophthora tenthredinis*, a fungal pathogen of the introduced pine sawfly in northwestern Wisconsin. *Annals of the Entomological Society of America*, **66**, 1178-1180.

Krassilstschik, J. (1888). La production industrielle des parasites vegeteux pour la destruction des insectes nuisible. *Bulletin Science France Belgique*, **19**, 461-472.

Latch, G. C. M. (1964). *Metarhizium anisopliae* (Metschnikoff) Sorokin strains in New Zealand and their possible use for controlling pasture-inhabiting insects. *New Zealand Journal of Agricultural Research*, **8**, 384-396.

Latch, G. C. M. (1976). Studies on the susceptibility of *Oryctes rhinoceros* to some entomogenous fungi. *Entomophaga*, **21**, 31-38.

Latgé, J.-P. (1982). Production of Entomophthorales. In *Proceedings of the 3rd International Colloquium on Invertebrate Pathology*, pp. 164-169.

Latgé, J.-P., Soper, R. S. & Madore, C. D. (1977). Media suitable for industrial production of *Entomophthora virulenta* zygospores. *Biotechnology & Bioengineering*, **19**, 1269-1284.

Latgé, J.-P., Remaudiere, G., Soper, R. S., Madore, C. D. & Diaquin, M. (1978). Growth and sporulation of *Entomophthora virulenta* on semidefined media in liquid culture. *Journal of Invertebrate Pathology*, **31**, 225-233.

Lingg, A. J. & Donaldson, M. D. (1981). Biotic and abiotic factors affecting the stability of *Beauveria bassiana* in soil. *Journal of Invertebrate Pathology*, **38**, 191-200.

Lipa, J. J. (1985). Progress in biological control of the Colorado beetle (*Leptinotarsa decemlineata*) in Eastern Europe. *European and Mediterranean Plant Protection Organization Bulletin*, **15**, 207-211.

MacCleod, D. M., Tyrrell, D. & Carl, K. P. (1976). *Entomophthora parvisora* sp. nov., a pathogen of *Thrips tabaci*. *Entomophaga*, **21**, 307-312.

Marschall, K. J. (1980), Biological control of rhinoceros beetles: experiences from Samoa. In *Proceedinqs of the International Conference on Cocoa and Coconuts, 1978,* paper 44. Incorporated Society of Planters: Kuala Lumpur.

McCabe, D. & Soper, R. S. (1985). Preparation of an entomopathogenic fungal insect control agent. United States Patent, 4,530,834.

McCoy, C. W. (1981). Pest control by the fungus *Hirsutella thompsonii*. In *Microbial Control of Pests and Plant Diseases 1970-1980*, ed. H. D. Burges, pp. 499-512. Academic Press: London.

McCoy, C. W. & Selhime, A. G. (1974). The fungus pathogen, *Hirsutella thompsonii* and its potential use for control of the citrus rust mite in Florida. *Proceedings of the International Citrus Congress*, **2**, 521-527.

McCoy, C. W., Sampson, R. A. & Boucias, D. G. (1988). Entomogenous fungi. In *Handbook of Natural Pesticides*, vol. V, *Microbial Insecticides, Part A, Entomogenous Protozoa and Fungi*, ed. C. M. Ignoffo & N. B. Mandore. CRC Press Inc.: Boca Raton, Florida.

McCoy, C. W. & Selhime, A. G., Kanavel, R. F. & Hill, A. J. (1971). Supression of citrus rust mite populatíons with application of fragmented mycelia of *Hirsutella thompsonii*. *Journal of Invertebrate Pathology*, **17**, 270-276.

Metchnikoff, E. (1879). Maladies des hannetons duble. *Zapiski imperatorskogo obshcestaa sel'skago Khlzyaistra yuzhnoi rossii*, 17-50.

Milner, R. J. & Lutton, G. G. (1986). Dependence of *Verticillium lecanii* (fungi: Hyphomycetes) on high humidities for infection and sporulation using *Myzus persicae* (Homoptera: Aphididae) as host. *Environmental Entomology*, **15**, 380-382.

Milner, R. J, Soper, R. S. & Lutton, G. G. (1982). Field release of an Israeli strain of the fungus *Zoophthora radicans* (Brefeld) Batko for biological control of *Therioaphis trifolii* (Monell) f. *maculata*. *Journal of the Australian Entomological Society*, **21**, 113-118.

Mohamed, A. K. A., Sikorowski, P. P. & Bell, J. V. (1977). Susceptibility of *Heliothis zea* larvae to *Nomuraea rileyi* at various temperatures. *Journal of Invertebrate Pathology*, **30**, 414-417.

Mohamed, A. K. A., Bell, J. V. & Sikorowski, P. P. (1978). Field cage tests with *Nomuraea rileyi* against corn earworm larvae on sweet corn. *Journal of Economic Entomology*, **71**, 102-104.

Perry, D. F., Latteur, G. & Wilding, N. (1982). The environmental persistence of propagules of the Entomophthorales. *Proceedings of the 3rd International Colloquium of Invertebrate Pathology*, pp. 164-169.

Philipp, W.-D. & Hellstern, A. (1986). Biologische mehltanbekampfing mit *Ampelomyces quisqualis* bei reduzierler Luftfeuchtigkeit. *Zeitschrift für Pflanzenkrankheiten und Pflanzenschutz*, **93**, 348-391.

Prado, E. (1979). Bekampning av oronvivellarver (*Otiorhynchus sulcatus*) med hjalp av de insektspatogena svamparna *Beauvueria bassiana, Metarhizium anisopliae* och *Metarhizium flavoviride. Vaxtskyddsnotiser*, **44**, 160-167.

Prior, C., Jollands, P. & Le Patourel, G. (1988). Infectivity of oil and water formulations of *Beauveria bassiana* (Deuteromycotina: Hyphomycetes) to the cocoa weevil pest, *Pantorhytes plutus* (Coleoptera; Curculionidae). *Journal of Invertebrate Pathology*, **52**, 66-72.

Ramakers, P. M. J. & Samson, R. A. (1984). *Aschersonia aleyrodis*, a fungal pathogen of whitefly. II. Application as a biological insecticide in glasshouses. *Zeitschrift für Angewandte Entomologie*, **97**, 1-8.

Roberts, D. W. & Humber, R. A. (1981). Entomogenous fungi. In *Biology of Conidial Fungi*, vol. 2, ed. G. T. Cole & B. Kendrick, pp. 201-236. Academic Press: London.

Roberts, D. W. & Wraight, S. P. (1986). Current status on the use of insect pathogens as biocontrol agents in agriculture: fungi. In *Fundamental and Applied Aspects of Invertebrate Pathology*, ed. R. A. Samson, J. M. Vlak & D. Peters, pp. 510-513. Foundation of the 4th International Colloquium of Invertebrate Pathology: Wageningen, The Netherlands.

Rombach, M. C., Aguda, R. M., Shepard, B. M. & Roberts, D. W. (1986). Entomopathogenic fungi (Deuteromycotina) in the control of the black bug of rice, *Scotinophara coarctata* (Hemiptera, Pentatomidae). *Journal of Invertebrate Pathology*, **48**, 174-179.

Rombach, M. C. & Gillespie, A. T. (1988). Entomogenous Hyphomycetes for insect and mite control on greenhouse crops. *Biocontrol News & Information*, **9**, 7-18.

Rombach, M. C., Humber, R. A. & Evans, H. C. (1987). *Metarhizium album*, a fungal pathogen of leaf and plant hoppers of rice. *Transactions of the British Mycological Society*, **88**, 451-459.

Samsinakova, A. & Samsinak, K. (1970). Milben (Acari) als Verbrieter des pilzes *Beauveria bassiana* (Bals.) Vuill. *Zeitschrift für Parasitenk*, **34**, 351-355.

Samsinakova, A. & Kalalova, S. (1983). The influence of single spore isolate and repeated subculturing on the pathogenicity of conidia of the entomophagous fungus *Beauveria bassiana. Journal of Invertebrate Pathology*, **42**, 156-161.

Schabel, H.G, (1976). Green muscardine disease of *Hylobius pales* (Herbst) (Coleoptera: Curculionidae). *Zeitschrift für Angewandte Entomologie*, **81**, 413-421.

Shands, W. A., Hall, I. M. & Simpson, G. W. (1962). Entomophthoraceous fungi attacking the potato aphid in Northeastern Maine in 1960. *Journal of Economic Entomology*, **55**, 174-179.

Sikura, A. I. & Sikura, L. V. (1983). Use of biopreparations. *Zashchita Rastenii*, **5**, 38-39.

Soares, G. G., Marchal, M. & Ferron, P. (1983). Susceptibility of *Otiorhynchus sulcatus* (Coleoptera, Curculionidae) larvae to *Metarhizium anisopliae* and *Metarhizium flavoride* (Deuteromycotina, Hyphomycetes) at two different temperatures. *Environmental Entomology*, 12, 1886-1890.

Soper, R. S., Holbrook, F. R., Majchrowicz, I. & Gordon, C. C. (1975). Production of *Entomophthora* resting spores for biological control of aphids. *Technical Bulletin of the Maine Life Sciences and Agriculture Experimental Station*, 76, 15.

Soper, R. S., Smith, L. F. R. & Delyzer, A. J. (1976). Epizootiology of *Massospora levispora* in an isolated population of *Okanagana rimosa*. *Annals of the Entomological Society of America*, 69, 275-283.

Sopp, P., Palmer, A. & Gillespie, A. T. (1989). Application of *Verticillium lecanii* for the control of *Aphis gossypii* by a low-volume electrostatic rotary atomiser and a high volume hydraulic sprayer. *Entomophaga*, in press.

Sprenkel, R. K. & Brooks, W. M. (1975). Artificial dissemination and epizootic initiation of *Nomuraea rileyi*, an entomogenous fungus of Lepidopterous pests of soybeans. *Journal of Economic Entomology*, 68, 847-850.

St. Leger, R. J., Cooper, R. M. & Charnley, A. K. (1986). Cuticle degrading enzymes of entomopathogenic fungi: cuticle degradation *in vitro* by enzymes from entomopathogens. *Journal of Invertebrate Pathology*, 47, 167-177.

St. Leger, R. J., Cooper, R. M. & Charnley, A. K. (1987). Production of cuticle degrading enzymes by the entomopathogen *Metarhizium anisopliae* during infection of cuticles from *Calliphora vomitoria* and *Manduca sexta*. *Journal of General Microbiology*, 133, 1371-1382.

Storey, G. K. & Gardner, W. A. (1987). Vertical movement of commercially formulated *Beauveria bassiana* conidia through four Georgia soil types. *Environmental Entomology*, 16, 178-181.

Thomas, K. C., Khachatourians, G. G. & Ingledew, W. M. (1987). Production and properties of *Beauveria bassiana* conidia cultivated in submerged cultures. *Canadian Journal of Microbiology*, 33, 12-20.

Tillemans, F. & Coremans-Pelseneer, J. (1987). *Beauveria brongniartii* (fungus, Moniliales) as control agent against *Otiorhyncus sulcatus* (Coleoptera, Curculionidae). *Mededelingen Faculteit Landbouwwetenschappen RijksUniversiteit Gent*, 52, 379-384.

Tulloch, M. (1976). The genus Metarhizium. *Transactions of the British Mycological Society*, 66, 407-411.

Van Winkelhoff, A. J. & McCoy, C. W. (1984). Conidiation of *Hirsutella thompsonii* var. *synnematosa* in submerged culture. *Journal of Invertebrate Pathology*, 43, 59-68.

Veen, K. H. (1968). Recherches sur la maladie, due à *Metarhizium anisopliae*, chez le criquet pVlerin. *Mededelingen Landbouwhogeschool Wageningen*, 68, 1-117.

Walstad, J. D., Anderson, R. F. & Stambaugh, W. J. (1970). Effects of environmental conditions on two species of muscardine fungi (*Beauveria bassiana* and *Metarhizium anisopliae*). *Journal of Invertebrate Pathology*, 16, 221-226.

Wilding, N. (1973). The survival of *Entomophthora* spp. in mummified aphids at different temperatures and humidities. *Journal of Invertebrate Pathology*, 21, 309-311.

Wilding, N. (1982). Entomophthorales: Field use and effectiveness. In *Proceedings of the 3rd International Colloquium of Invertebrate Pathology*, 170-175.

Wilding, N. & Lauckner, F. B. (1974). *Entomophthora* infecting wheat bulb fly at Rothamsted, Hertfordshire, 1967-71. *Annals of Applied Biology*, **76**, 161-170.

Zacharuk, R. Y. (1970a). Fine structure of *Metarhizium anisopliae* infecting three species of larval Elateridae (Coleoptera) I. Dormant and germinating conidia. *Journal of Invertebrate Pathology*, **15**, 63-80.

Zacharuk, R. Y. (1970b). Fine structure of *Metarhizium anisopliae* infecting three species of larval Elateridae (Coleoptera) II. Conidial germ tubes and appressoria. *Journal of Invertebrate Pathology*, **15**, 81-91.

Zacharuk, R. Y. (1970c). Fine structure of *Metarhizium anisopliae* infecting three species of larval Elateridae (Coleoptera) III. Penetration of the host integument. *Journal of Invertebrate Pathology*, **15**, 372-396.

Zacharuk, R. Y. (1981). Fungal diseases of terrestrial insects. In *Pathogenesis of Invertebrate Microbial Diseases*, ed. E. W. Davidson, pp. 367-402. Allanheld, Osmun & Co.: Totowa, New Jersey.

Zimmermann, G, (1981). Gewachshausversuche zur Bekampfung des Gefurchten Dickmaulrulers, *Otiorhyncus sulcatus* F., mit dem Pliz *Metarhizium anisopliae* (Metsch.) Sorok. *Nachrichtenblatt des Deutschen Pflanzenschutzdienste*, **33**, 103-108.

Zimmerman, G. & Bode, E. (1983). Dispersal of the entomopathogenic fungus *Metarhizium anisopliae* (Fungi Imperfecti, Moniliales) by soil arthropods. *Pedobiologia*, **25**, 65-71.

Zimmerman, G. & Simons, W. R. (1986). Experiences with the biological control of the black vine weevil, *Otiorhynchus sulcatus* (F.). In *Fundamental and Applied Aspects of Invertebrate Pathology*, ed. R. A. Samson, J. M. Vlak & D. Peters, pp. 529-533. Foundation of the 4th International Colloquium of Invertebrate Pathology: Wageningen, The Netherlands.

Chapter 5

Mechanisms of fungal pathogenesis in insects

A. K. Charnley

*School of Biological Sciences, University of Bath, Claverton Down,
Bath BA2 7AY, UK*

Introduction

The spectre of insecticide resistance along with other problems associated with the development and use of synthetic chemical insecticides has
led to a renewed interest in the use of pathogens for control. Prominent
among the organisms involved are the fungi. The current status of entomopathogenic fungi as pest control agents is reviewed by Gillespie &
Moorhouse (Chapter 4).

Insect pathogenic fungi have a great potential for pest control which is
not being fully exploited. Slow kill is one of the constraints on their utilization. However, programmes for the selection of quick acting strains with
enhanced virulence are hampered by an ignorance of the determinants of
pathogenicity. Without this knowledge strain selection and improvement
are restricted to an empirical rather than to a rational approach.

Since the author last reviewed the subject (Charnley, 1984), there has
been a substantial improvement in our understanding of the mechanisms
of fungal pathogenicity in insects. The object of this contribution is to review the most recent developments and evaluate their significance.

Pre-infection development

Spore adhesion

Attachment of a fungal spore to the cuticle surface of a compatible host
represents the initial event in the establishment of mycosis. The propagules of most entomopathogenic fungi are passively dispersed and host
location is a random event. Exceptions are the motile zoospores of aquatic fungi like *Coelomomyces psorophorae* and *Aphanomyces astaci* which
employ chemotaxis (Ceranius & Söderhäll, 1984). The former fungus is
also singular in that it actively secretes an adhesive during encystment on
host cuticle (Travland, 1979). For most entomopathogenic fungi attachment is a passive process.

The slimy conidia of *Verticillium lecanii*, *Hirsutella thompsonii* and
some species of Entomophthorales use an amorphous mucus for attachment (Boucias & Latgé, 1986; Latgé *et al.*, 1986a). Dry spores such as

those of *Beauveria bassiana*, *Metarhizium anisopliae* and *Nomuraea rileyi* do not have a mucoid layer but instead are covered by layers of interwoven bundles of extremely hydrophobic rodlets. The composition of the rodlets is not yet known but it appears to be variable between species (Boucias & Latgé, 1986; Boucias, Pendland & Latgé, 1988).

The forces responsible for the interaction between fungal spores and the cuticle are not well understood. Fargues (1984) suggested that electrostatic forces are responsible for the initial contact and that these non-specific forces are complemented in time with more specific linkages. Such a biphasic system seems to operate concurrently in *Candida*-human cell interactions (Klotz & Penn, 1987). Lectins have been detected on the spore walls of *Beauveria bassiana*, *Metarhizium anisopliae* and *Nomuraea rileyi* and the mucus of *Conidiobolus obscurus* ballistospores (Boucias & Latgé, 1986). These carbohydrate binding glycoproteins could potentially provide the 'specific' linkages hypothesised by Fargues (1984). The specific binding of a lectin on the fungus to a sugar residue on the host cuticle (or *vice versa*, Kerwin & Washino, 1986) would, as a corollary constitute an initial recognition event. Such a phenomenon is established for fungus-nematode and fungus-fungus interactions (Nordbring-Hertz & Chet, 1986). However, competitive inhibition studies failed to demonstrate that lectins associated with the conidia of *B. bassiana*, *M. anisopliae* and *N. rileyi* are important in attachment. Inhibition of charge effects with polylysine did not neutralize binding to cuticle and adhesion of dry spores may be due mainly to hydrophobic forces exerted by the rodlets (Boucias *et al.*, 1988).

Adhesion to host cuticle appears to be a prerequisite for successful invasion. For example, conidia of hypovirulent strains of *Metarhizium anisopliae* failed to attach to the larval syphon of *Culex pipiens* (Al-Aidroos & Roberts, 1978). Host specificity may also correlate with spore attachment as with spores of *Coelomomyces psorophorae* on susceptible strains of mosquitoes (Zeobold *et al.*, 1979). However, reflecting the importance of non-specific hydrophobic forces, conidia of *Nomuraea rileyi* adhere indiscriminately to host and non-host cuticle while both aggressive and non-aggressive strains of *Conidiobolus obscurus* adhere to aphid cuticle (Boucias & Latgé, 1986).

Spore germination

Deuteromycete entomopathogens with relatively wide host ranges such as *Beauveria bassiana* and *Metarhizium anisopliae* produce conidia which germinate in response to non-specific sources of carbon and/or nitrogen such as amino acids, proteins, carbohydrates and a variety of lipids (Smith & Grula, 1981; St. Leger, Charnley & Cooper, 1986a). Although *M. anisopliae* (strain ME1) has a nutrient-independent pre-swelling growth phase of germination (Dillon & Charnley, 1986c), this strain along with those of most other Deuteromycete entomopathogens requires an exogenous nutrient source for conidial swelling and germ tube formation (Dillon & Charnley, 1985; Kerwin & Washino, 1986).

Despite the non-fastidious germination requirements of pathogenic Deuteromycetes like *Beauveria bassiana* and *Metarhizium anisopliae*, nutrient availability on the surface of insect cuticle may prove limiting (Grula *et al.*, 1978). Thus, while water soluble nutrients (mainly amino acids) on the surface of *Heliothis zea* larvae are sufficient for germination of *B. bassiana* (Woods & Grula, 1984) the same fungus will not germinate on sclerotized cuticle of *Dendroctonus ponderosae* without a nutrient supplement (Hunt *et al.*, 1984).

Entomopathogens with restricted host ranges may have more specific requirements for germination. *Nomuraea rileyi*, which primarily infects lepidopterans, responds to diacylglycerols and polar lipids (Boucias & Pendland, 1984). *Erynia variabilis* may be restricted to small dipterans, in part, by a requirement for oleic acid to induce conidial germination (Kerwin, 1982). Subtle differences in the germination behaviour of two groups of aggressive strains of *Conidiobolus obscurus* could have epidemiological significance (Latgé *et al.*, 1987). One group, which are stimulated to germinate by cuticular hydrocarbons and polar lipids, may be more prevalent in low aphid populations when individuals show little contamination with honeydew. Whereas a second group, which are stimulated by the water soluble components of honeydew, may take over at higher population densities.

In many instances ability to utilize epicuticular lipids may be fundamental to pathogenesis (Charnley, 1984). It is significant that pathogenic isolates of *Metarhizium anisopliae* and *Nomuraea rileyi* can grow *in vitro* on a variety of hydrocarbons including multi-branched alkanes which are refractory to most other microbes (St. Leger, Cooper & Charnley, 1988a).

Successful germination not only requires the ability to utilise available nutrients but also tolerance of potentially toxic substances. Correlations between the experimental removal of epicuticular lipids and enhanced susceptibility to fungal pathogens suggest the presence of antifungal lipids. However, alternative interpretations are possible and such experiments simply confirm the importance of the epicuticle as barrier to infection (Charnley, 1984: St. Leger, 1989). Antifungal fatty acids have been isolated from insect cuticle (Smith & Grula, 1982). However, evidence is lacking that they occur *in vivo* in inhibiting concentrations (Latgé, 1972). Apart from the work of Kerwin (1982, 1984) most studies with entomopathogenic fungi have focused on short chain unsaturated fatty acids. Though, as St. Leger (1989) has pointed out, greater antifungal activity occurs among longer chain saturated and unsaturated fatty acids (Kabara, 1978) which are present in all insect orders (Thompson, 1973).

Further progress in this area will be difficult. Given the differential solubility of fatty acids in polar and non polar solvents (St. Leger, 1989), there will be considerable problems in extracting all toxic compounds in the correct proportions. Assessment of the *in vivo* significance of these compounds will need to take into account cuticle pH (since a pH less than 7 is a prerequisite for the activity of toxic acids (Gershon & Shanks, 1978)), the ameliorating effects of other cuticular components, particularly nutrients (Fargues, 1981; Smith & Grula, 1982), and the possibility that the availability of fatty acids to the pathogen may be altered *in vivo* by their mode of binding to other cuticular components (St. Leger, 1989).

Antagonism from saprotrophic flora is a further hazard to a pathogenic fungus on insect cuticle. Antibiosis is suggested from studies which show enhanced germination on surface sterilised cuticle (Schabel, 1978; Uziel & Latgé cited in Latgé *et al.*, 1987). The best evidence comes from a recent study on the gut bacterial flora of the desert locust, *Schistocerca gregaria* (Dillon & Charnley, 1986a & b, 1988). Insect alimentary canals are generally inhospitable to fungi. As a consequence, invasion rarely occurs from them despite the fact that both foregut and hindgut are lined with cuticle which is largely unsclerotized (Bignell, 1984) and thus more susceptible than sclerite cuticle to the action of pathogen enzymes (see next section). It is generally thought that anaerobiosis, digestive enzymes, adverse pH, speed of food throughput and protection from the peritrophic membrane are primarily responsible for the gut barrier to fungi (for review see Dillon & Charnley, 1989). However, none of these factors can account for the failure of conidia of *Metarhizium anisopliae* to germinate in the guts of conventional (with a normal gut flora) desert locusts or *in vitro* in gut fluid (with or without a nutrient supplement) (Dillon & Charnley, 1986a). Since the guts of germ-free locusts do not have such inhibitory properties, Dillon & Charnley (1986a) suggested that an antifungal toxin was produced by the bacterial flora in conventional insects. Subsequently they found a number of antifungal phenols in faecal extracts from conventional but not from germ-free locusts (Dillon & Charnley, 1988). These phenols (hydroquinone, 3,4 dihydroxy- and 3,5 dihydroxy-benzoic acids) had similar properties to the antifungal toxins in the gut fluid and a cocktail of authentic forms of the phenols was inhibitory to conidia at a concentration estimated to occur in the faeces. When *Enterobacter agglomerans*, a prominent member of the hindgut flora of the desert locust (Hunt & Charnley, 1982), was introduced into germ-free insects the toxic phenols were found in the faeces in proportion to the size of the gut bacterial population (Dillon & Charnley, unpublished observations). *M. anisopliae* will penetrate the hindgut of starved germ-free locusts and induce symptoms of mycosis (Dillon & Charnley, 1986b). Invasion does not occur through the guts of fed or starved conventional locusts or fed germ-free locusts. Thus the antifungal toxins provide a significant contribution to the locust's defence against pathogenic fungi during periods of starvation when physical defences, such as gut movement and food through-put, prominent in fed insects, are less apparent.

The hindgut cuticle is also the location of the indigenous bacterial flora (Hunt & Charnley, 1981), which is significantly larger in starved than fed insects (Hunt & Charnley, unpublished observations). Thus, in starved conventional locusts, a locally high concentration of toxin and the exclusion of the fungus by the bacteria from suitable sites of attachment may combine to prevent gut penetration (Dillon & Charnley, 1986b).

The influence of cuticular components on the behaviour of entomopathogenic fungi has always been tested with extracts from non-infected intact insects. However, components secreted by the insect through the pore canals on the cuticle in response to the recognition of the presence of the fungal pathogen could also influence germination of spores (Latgé et al., 1987).

Significant amounts of the cuticle-degrading enzyme activity (notably endoprotease) are present on the outer walls of conidia of *Metarhizium anisopliae* (St. Leger, Charnley & Cooper, unpublished observations). Hydrolysis of the cuticle prior to germination may provide essential nutrients for initiation of swelling. Observations from a number of histochemical and ultrastructural studies are consistent with the initiation of cuticle degradation beneath attached fungal spores prior to the onset of germination (see Charnley, 1984).

Germ tube development and appressorium formation

Germ tubes of *Metarhizium anisopliae* adhere to the epicuticular surface by an amorphous mucus which is secreted by the hyphal tip (St. Leger et al., 1988b). A similar substance produced by *Nomuraea rileyi* has been identified as a mixture of β-1,3 glucans (Boucias, personal communication). For many entomopathogens it is not known what causes the cessation of germ tube elongation and the differentiation of penetrant structures (which may or may not include the production of a morphologically distinct appressorium, see Charnley, 1984). Appressorial formation by *M. anisopliae* (strain ME1) is induced preferentially over hair sockets of early instar larvae of *Manduca sexta* (St. Leger et al., 1988b). In contrast, appressorial formation by *M. anisopliae* (ME1) occurs much closer to the conidium on the cuticle of *Calliphora vomitoria*, *Schistocerca gregaria* and late instar *Manduca sexta* which lack the extensive microfolding of the epicuticle seen on early instar *Manduca* larvae (St. Leger et al., 1988b). These effects can be duplicated using polystyrene replicas of cuticles of early and late instar *Manduca* larvae indicating that the inhibitory effect of microfolds is due to surface topography rather than chemical differences (Goettel & St. Leger, unpublished observations). For some insect-fungus combinations surface growth appears most extensive on hard cuticles (see Charnley, 1984). This behaviour, whatever its cause, may aid location of thinner or softer cuticle or enhance invasiveness *via* synergism with other hyphae (Charnley, 1984). Thus it is interesting to note that Wraight *et al*. (1989) have found evidence for directional growth of *Erynia radicans* towards intersegmental cuticle of *Empoasca fabae*. Pekrul & Grula (1979) suggested that failure of germ tubes of certain *Beauveria bassiana* mutants to orientate towards the cuticle of *Heliothis zea* might reflect the importance of specific chemical cues. A specific recognition event such as lectin-ligand binding may be critical to the production of infection structures. On the other hand recent work on *M. anisopliae* suggests that differentiation of appressoria is strictly governed by the concentration of lipids and low molecular weight nitrogenous compounds (St. Leger *et al*., 1989). Thus, current evidence for *M. anisopliae* is consistent with an interaction between a thigmotropic response and surface nutrients in the control of appressorial formation. It remains to be seen whether other factors such as fungistatic compounds on the cuticle or specific chemical interactions also play a part (see Charnley, 1984).

Penetration

Host invasion can occur through wounds, thus by-passing the integumental barrier, and through sense organs and trachea. However, the principle route is through arthrodial membranes (at joints and between segments), segmental cuticle and mouth parts. As pointed out in the last section, often the alimentary canal is an inhospitable environment for entomopathogenic fungi. The many light and electron microscope studies on cuticular penetration by entomopathogenic fungi have led to the conclusion that both physical force and enzymic degradation are involved. The relative contributions of each presumably depending on many variables of host and pathogen (Charnley, 1984).

Epicuticle

The epicuticle is the first barrier encountered (Fig 5.1). The disappearance of the wax layer beneath appressoria of *Metarhizium anisopliae* on wireworm, Elaterid, cuticle indicates enzyme activity (Zacharuk, 1970) as does the presence of circular holes around germ tubes of *Beauveria bassiana* at the point of entry into larvae of *Heliothis zea* (Pekrul & Grula, 1979). Physical penetration is prominent, however, in host invasion by some entomophthoralean pathogens where characteristic triradiate and

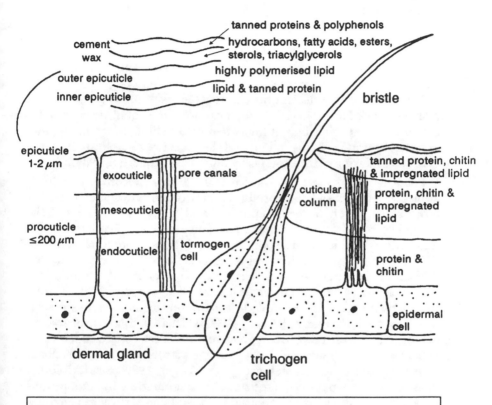

Fig. 5.1. The structure of insect cuticle.

tetraradiate fissures appear in the epicuticle (Brobyn & Wilding, 1983; Butt, 1987).

It is not yet known what enzymes are involved in the penetration of the epicuticle. The tanned proteins of the cement layer may resist the endo-protease (Pr 1) which is produced profusely during appressorial formation (see later), though even limited hydrolysis might facilitate forced entry. Lipases/esterases have been located histochemically in and around penetrant structures (Gabriel, 1968; Michel, 1981; St. Leger et al., 1987a); however, potential substrates for these enzymes, such as gly-cerides, form only a small fraction of the wax layer (Lockey, 1988; Fig 5.1).

The outer epicuticle is resistant to degradation (Hepburn, 1985) but in most insects it is probably fragile and thus susceptible to mechanical force (see St. Leger, 1989). In contrast, the inner epicuticle may yield to a combination of endoprotease and lipoprotein lipases which are produced by several entomopathogenic fungi (St. Leger, Charnley & Cooper, unpublished observations).

Procuticle

The procuticle constitutes the bulk of the cuticle and must provide a significant barrier to the invading fungus. This layer comprises chitin fibrils embedded in a protein matrix, together with lipids and quinones (Neville, 1984; Fig. 5.1). The mechanical properties of different cuticles depend on the proportions of the two main constituents, the nature and extent of hydration of the proteins and the degree of sclerotization or tanning (cross linking of the proteins by quinones) (Hillerton, 1984).

Passage of the fungus across the procuticle may be more or less direct or involve a degree of lateral proliferation between cuticular lamellae before or during vertical penetration (Charnley, 1984). Thickness of the procuticle correlates with disease resistance in that young larvae (with thin cuticles) are more susceptible than old larvae (with thick cuticles). The degree of cuticle sclerotization appears also to have a strong influence on penetrability. Although there are reports that sclerotized cuticle can be traversed, in the main it seems that insects with heavily sclerotized body segments are invaded via arthrodial membranes or spiracles (see Charnley (1984) and St. Leger (1989)). There is good reason for this behaviour. The resistance to compressive force of sclerotized cuticle (exocuticle see Fig. 5.1) suggests that it presents a stronger barrier than pliant nonsclerotized endo and mesocuticle (St. Leger, 1989). Sclerotized cuticle is also comparatively resistant to degradation by endogenous enzymes in moulting fluid (Hepburn, 1985) and fungal pathogen enzymes (St. Leger, Cooper & Charnley, 1986c, see later).

Initial fungal growth within the outer layers of the procuticle often occurs laterally. These subepicuticular expansions can cause fractures which favour penetration (Brey et al., 1986). Vertical penetrant hyphae may appear thin and constricted within the outer layers of the procuticle (exocuticle) presumably due to resistance to mechanical penetration, whereas the growing tip swells in the more pliant (more easily degraded) inner layers (endocuticle) (Robinson, 1966). Indentation or displacement of lamellae by lateral penetrant hyphae in both exo and endocuticles is another clear sign of a mechanical component to the penetration process (see Charnley 1984).

Enzyme production

Enzymic degradation of procuticle is suggested by changes in the staining reactions of cuticle around penetrant hyphae (see Charnley, 1984). Recently, Brey *et al*. (1986) reported wide zones of complete histolysis in cuticle beneath penetrant hyphae of *Conidiobolus obscurus* in *Acyrthosiphon pisum*. However, fine structural studies on many other insect-fungus interactions suggest that significant enzymolysis is not the norm. Absence of mechanical damage or displacement of lamellae suggests the action of enzymes, but without obvious zones of histolysis it must be assumed that fungal enzymes are usually limited to the vicinity of fungal structures in the initial stages of infection (Charnley, 1984). In recent ultrastructural studies, however, Hassan & Charnley (1989) and Goettel *et al*. (1989) noted clearing of the lamellar pattern but not complete histolysis around hyphae of *Metarhizium anisopliae* in the cuticle of *Manduca sexta* (Fig. 5.2 and Fig. 5.3). This is associated with dispersal of a pathogen protease (detected by an immunogold technique) through the cuticle and may be due to selective degradation of the cuticular components (see later).

Considerable advances have been made in the last few years, principally by St. Leger and his colleagues (St. Leger *et al*., 1986a-e, 1987a-c, 1988a-e) in our knowledge of cuticle-degrading enzymes produced by entomopathogenic fungi, and the first steps have been taken in determining the role of these enzymes in the invasion process. Pathogenic isolates of *Metarhizium anisopliae*, *Beauveria bassiana* and *Verticillium lecanii* grown in buffered or unbuffered liquid cultures containing 1% ground cuticle as sole carbon source produced a range of extracellular cuticle-degrading enzymes active for the major components of insect cuticles, namely protein, chitin and lipid (St. Leger *et al*., 1986a). Marked variations in enzyme levels between different isolates of *M. anisopliae* as well as between different species were found. These variations may in part be accounted for by differences in growth rate and the sequential phasing of enzyme production. Endoproteases were exceptional in being produced in large amounts by all the isolates.

Extracellular enzymes appeared sequentially in cultures of all three fungi. Esterase activity, endoprotease, aminopeptidase and carboxypeptidase appeared first (within 24 h) and increased rapidly after 48 h. The proteolytic enzymes were followed by *N*-acetylglucosaminidase (NAGase) which increased independently of chitinase. Only low levels of chitinase activity were found up to $3 \cdot 5$ days after inoculation, but then this activity increased dramatically. Lipase was not detected until day 5.

The order of appearance of the enzymes is paralleled by the sequence of cuticle constituents solubilized into the culture medium, where a rapid increase in amino sugars followed early release of amino acids. The late

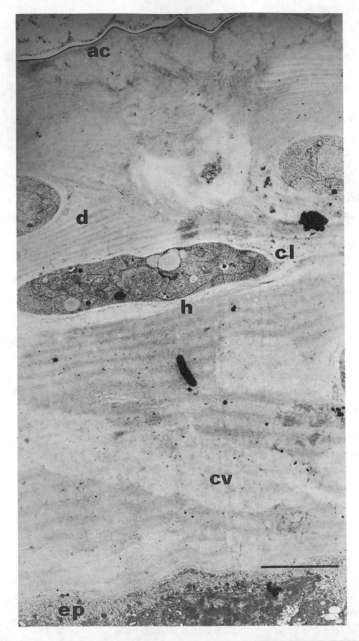

Fig. 5.2. Fine structure of the abdominal cuticle of 4th instar *Manduca sexta* larvae 48 h after inoculation with *Metarhizium anisopliae*. Note the deformation (d) of the cuticular lamellae around the hyphae (h) and the absence of the lamellar pattern in large areas of the cuticle in association with (cl) or in advance of penetrant hyphae (cv): ac = amorphous cuticle, ep = epidermis; bar marker = 3·1 μm. From Hassan & Charnley, 1989.

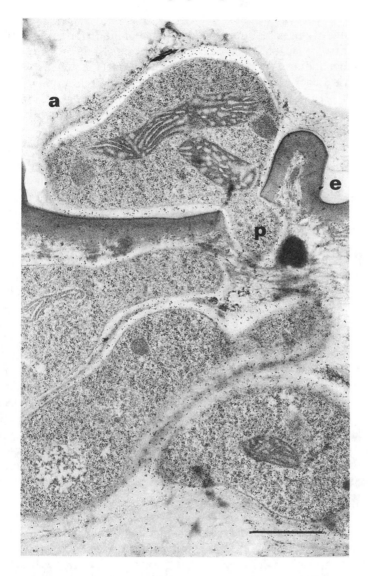

Fig. 5.3. Ultrastructural localization of the endoprotease Pr 1 using colloidal gold conjugated antibody. Shows *Metarhizium anisopliae* invading cuticle of 4th instar larvae of *Manduca sexta*. Note that enzyme (black dots) is present on the surface of the appressorium (a), in the vicinity of the penetrant hypha and extending into the surrounding cuticle: e = epicuticle, p = penetration peg; bar marker = 1·2 μm. From Goettels *et al.* 1989.

appearance of chitinase is presumably a result of induction as chitin eventually became available after degradation of encasing cuticle proteins. Chitinase is induced by chitin break-down products in *Metarhizium anisopliae* (St. Leger, Cooper & Charnley, 1986b) and in *Beauveria bassiana* (Smith & Grula, 1983). Fluorescence microscopy using calcofluor and FITC wheatgerm agglutinin (selective for β-1,3-glucans and chitin respectively) confirmed that proteolytic enzymes could rapidly remove proteins from unsclerotized cuticle which resulted in eventual (less than 4 days) exposure of chitin to enzymolysis. The late detection of lipase appears to be due to the fact that the enzyme is largely cell bound in young cultures (St. Leger, Charnley & Cooper, unpublished data).

Testing purified enzymes against locust cuticle *in vitro* showed that pretreatment or combined treatment with endoprotease (component Pr 1) (Table 5.1) was necessary for high chitinase activity (St. Leger *et al.*, 1986c). When exuviae (non-digested remains of old cuticle shed at ecdysis; exocuticle only) were used as the substrate instead of cuticle from larval sclerites (exo and endocuticle) little hydrolysis occurred. Similarly, even after several days growth of *Metarhizium anisopliae* and other entomopathogens on exuviae as the sole carbon source *in vitro* there was comparatively little disintegration, compared with the loss of integrity of sclerite cuticle (St. Leger *et al.*, 1986a). This does not prevent penetration of sclerotized locust cuticle by *M. anisopliae* (St. Leger, Charnley & Cooper, unpublished data); presumably mechanical pressure plays a greater role in this situation. This would account for the greater lateral growth in cuticles with substantial sclerotized areas than in more pliant cuticles. Sufficient digestion of protein in exocuticle, however, might still occur to loosen cuticular lamellae allowing an easier mechanical passage. Also, the extent of growth on exuviae *in vitro* confirms the potential of enzymes for obtaining nutrients during penetration of sclerotized sclerities.

Endoprotease activity in culture filtrates of *Metarhizium anisopliae* (strain ME1) may be resolved into 3 components (using hide protein azure as substrate), two with alkaline pH optima (Pr 1 and Pr 2) and a third with an acid pH optimum (Pr 3) (St. Leger, Charnley & Cooper, 1987b) (Table 5.1). Pr 1 is considerably more effective (25 ×) than Pr 2 at degrading cuticle and in contrast to Pr 2 it is also a good general protease with activity against a wide variety of proteins, including elastin, bovine serum albumen, collagen and denatured collagen. Both enzymes degrade casein. The kinetic properties determined against peptide anilide esters suggest that Pr 1 is chymoelastase-like and Pr 2 is trypsin-like in substrate specificity. Phenylmethyl sulphonyl fluoride (PMSF), a serine protease inhibitor, inhibited both enzymes. Pr 2 was also inhibited by the specific

Table 5.1. Properties of cuticle-degrading enzymes produced by *Metarhizium anisopliae* (strain ME1).

Enzyme	pH optimum	pI	Molecular weight	Substrate specificity	Inhibitors	Regulation
Protease (Pr 1)*	8	9·5	25,000	chymoelastase	PMSF, turkey egg white inhibitor, Boc-Gly-Leu-Phe-CH2Cl	C, CR
Protease (Pr 2)*	9	4·5	28,000	trypsin	PMSF, leupeptin, Tos-Lys-CH2Cl	C, CR
Aminopeptidase M	8	5·43	33,000	broad, alanine preferred	EDTA, amastatin, 1, 10-phenanthroline	C, CR
Dipeptidylaminopeptidase	8	4·6	74,000	post proline	PMSF, DFP	C, CR
Carboxypeptidase	7·5	–	–	–	–	–
Chitinase	5·2	6·3	34,000	Chitotetraose degraded to NAG	Li$^+$, Na$^+$	CR, I
N-acetylglucosamine	5·0	5·3	110,000-130,000	Chitobiose and chitotriose degraded to NAG	–	C, CR

Key: C = Constitutive, CR = catabolite repressible, I = Inducible, – = not determined, NAG = N-acetylglucosamine. * from St. Leger *et al.* (1987b), the rest from St. Leger, Cole, Charnley & Cooper (unpublished).

trypsin inhibitors leupeptin and Tos-Lys-CH$_2$Cl. Pr 1 was greatly inhibited by turkey egg white inhibitor (TEI) and Boc-Gly-Leu-Phe-CH$_2$Cl.

Two classes of aminopeptidase were isolated from culture of *Metarhizium anisopliae* grown on cuticle (Table 5.1). The enzymes were classified as an aminopeptidase (M) of broad specificity and a post proline dipeptidyl aminopeptidase (St. Leger, Cole, Charnley & Cooper, unpublished data). Neither peptidase alone hydrolysed intact locust cuticle. However, when combined with Pr 1 they effected enhanced release of amino acids. It seems reasonable to suppose that the participation of the carboxypeptidase (Table 5.1) would accelerate the breakdown of Pr 1 derived peptides, but this remains to be determined. The action of protein exohydrolases will provide nutrition for the fungus during penetration of the relatively massive barrier presented by insect cuticle.

St. Leger *et al*. (1987a) extracted protease and aminopeptidase from wings of the blowfly *Calliphora vomitoria* and abdominal cuticle of 5th instar larvae of *Manduca sexta* approximately 16h after inoculation with *Metarhizium anisopliae*. Endoprotease activity was separated into two components which closely resembled Pr 1 and Pr 2 in pI, substrate specificity and inhibitor spectrum. The third protease (Pr 3) produced in culture was not detected *in vivo* (≤ 40h post inoculation). Purified and non-purified extracts of infected blowfly wings tested by Ouchterlony gel diffusion against specific antiserum to Pr 1 gave a single precipitin line identical to that given by the pure enzyme, confirming the presence of Pr 1 during infection.

St. Leger *et al*. (1987a) also used translucent blowfly wings to localize proteolytic enzymes histochemically during penetration. Substrates and inhibitors specific for Pr 1 and Pr 2 established the production of these enzymes on appressoria, which developed 10-24h after inoculation. Aminopeptidase differed from endoprotease in that it was not present on immature appressoria, and the activity extended into the mucilage surrounding mature appressoria and appressorial plates.

Thus, Pr 1 appears to be a pathogenicity determinant by virtue of its ability to degrade cuticle extensively (St. Leger *et al*., 1986c, 1987b) and its production at high levels by the pathogen *in situ* during infection (St. Leger *et al*., 1987a). Further support for the importance of Pr 1 in penetration comes from the work of St. Leger *et al*. (1988b). They found that the treatment of tobacco hornworm larvae with TEI during infection significantly delayed mortality. The inhibitor also reduced melanization of cuticle (a response to infection, see next section) and invasion of the haemolymph as well as maintaining the host's growth rate. The inhibitor or antibodies to Pr 1 delayed penetration of the cuticle but did not affect spore viability or prevent growth and formation of appressoria on the cuticle surface. This suggests that inhibition of Pr 1 reduced infection by limiting fungal penetration of the insect cuticle. *In vitro* studies using the inhibitor showed that the accumulation of protein degradation products from cuticle, including ammonium, was dependent on active Pr 1. This confirms its major role in solubilizing cuticle proteins and making them available for nutrition.

Pr 1-like and Pr 2-like enzymes are produced in vitro by *Beauveria bassiana*, *Verticillium lecanii* (2 isolates), *Nomuraea rileyi*, and *Aschersonia aleyrodis* besides being a common feature of all isolates of *Metarhizium anisopliae* examined by St. Leger *et al.* (1987c). The consistent occurrence of these two classes of extracellular protease in five genera of entomopathogenic fungi implies an indispensable function for these enzymes. The role of Pr 2 enzymes is uncertain. Although present in large amounts in culture fluids and on cuticle of insects infected with *M. anisopliae*, Pr 2 enzymes have little cuticle-degrading ability. It is possible that they are involved with cellular control mechanisms. Pr 1 enzymes, as exemplified by that from *M. anisopliae* (ME1) have the potential to function for non-specific protein degradation during pathogenesis or saprotrophy.

Bidochka & Khachatourians (1987) found an elastolytic enzyme with similar properties to those of Pr 1 in culture filtrates of *Beauveria bassiana* (strain GK 2016). However, chymoelastases and trypsins are not the only proteases produced in culture by entomopathogenic fungi. Kucera (1980) isolated two different enzymes from a strain of *Metarhizium anisopliae*. PI (MW 71,000, pH optimum of 9 with haemoglobin as substrate) was inhibited by sulphydryl reagents. In contrast, PII (MW 35,000, pH optimum 6·5) was inhibited by PMSF. Collagenolytic enzymes have been purified from *Entomophthora coronata* (Hurion, Fromentin & Keil, 1979) and *Lagenidium giganteum* (Dean & Domnas, 1983). The enzyme from *E. coronata* had a broad specificity, wheras that from *L. giganteum* was specific for collagen.

Too little is known of the role of extracellular proteases in pathogenesis to assess whether differences in their properties influence virulence and host specificity. Not all the enzymes cited above may be involved with cuticular penetration. Although a specific collagenase is produced by *Lagenidium giganteum* (Dean & Domnas, 1983), collagen is found throughout insects particularly on basement membranes and in the neuronal sheath, but it is a minor component of cuticle. Those fungal enzymes that are cuticle-degrading may have additional functions *in vivo*, such as antagonism of host defences or activity as cytotoxic agents.

Electrostatic binding of the positively charged Pr 1 to negatively charged groups on locust cuticle is a prerequisite to activity (St. Leger, Charnley & Cooper, 1986d). Thus, it is interesting to note that Cox &

Willis (1985) found that in *Hyalophora cecropia* a group of highly acidic proteins was prominent in flexible cuticle (preferred site for penetration by entomopathogenic fungi, see above) but absent from rigid cuticle. Most isolates of *Metarhizium anisopliae, Beauveria bassiana, Verticillium lecanii* and *Nomuraea rileyi* also produce a minor (in most cases) acidic Pr 1 (St. Leger, Cooper & Charnley, 1987c). Locust cuticle is susceptible to these acidic Pr 1 enzymes suggesting that acidic and basic proteases may bind to different regions of locust cuticle, which is similar to that of *H. cecropia* in having a nonuniform distribution of charge (Hojrup, Andersen & Roepstorff, 1986).

Chitin constitutes 17 to 50% of the dry weight of cuticle; more pliant cuticles have a higher chitin content than stiff cuticles (Hillerton, 1984). In the main, chitin fibrils are laid down parallel to the cuticular surface and so present a potential barrier to penetration by entomopathogenic fungi. Preferential penetration of *Metarhizium anisopliae* down the vertical cuticular columns in *Manduca sexta* cuticle (where the chitin fibrils are perpendicular to the plane of the cuticle) supports this view (Hassan & Charnley, 1989) (Fig. 5.4A, B).

St. Leger *et al.* (1987a) failed to find evidence for the production of endochitinase during the first critical 40h after inoculation of *Calliphora vomitoria* wings or abdominal cuticle of *Manduca sexta* larvae with *Metarhizium anisopliae*. *N*-acetylglucosaminidase activity was extracted from infected cuticle 16h after inoculation, but this enzyme has only trace activity against polymeric chitin (St. Leger, Charnley & Cooper, unpublished data). The slow appearance of chitinase *in vitro* (St. Leger *et al.*, 1986a) and *in vivo* (St. Leger *et al.*, 1987a) appears to be due to the fact that chitinase is an inducible enzyme (St. Leger *et al.*, 1986b) and cuticular chitin is masked by protein (St. Leger *et al.*, 1986c). In contrast synthesis of Pr 1 and Pr 2 proteases in *M. anisopliae* (ME1) occurs rapidly by carbon and nitrogen derepression alone (St. Leger *et al.*, 1988c). The importance of chitin as a mechanical barrier to penetration and as a stabilizer of the cuticular protein matrix, in the absence of fungal chitinase, is evident from a recent study using the insecticide Dimilin, a specific inhibitor of chitin synthesis in insects (Fig. 5.5A, B). Hassan & Charnley (1983, 1989) showed that dual applications of Dimilin and *M. anisopliae* had a synergistic effect against larvae of *Manduca sexta*. Ultrastructural observations demonstrated that fungal penetration through Dimilin-affected cuticle was dramatically enhanced (Hassan & Charnley, 1989). Post-ecdysial Dimilin-affected cuticle (without chitin) was almost completely destroyed (Figs. 5.6 & 5.7) compared with pre-ecdysial cuticle (laid down prior to insecticide treatment) where hydrolysis was apparently selective (presumably protein only) and restricted to the vicinity of the fungal hyphae (Fig. 5.6). Consistent with these ultrastructural observa-

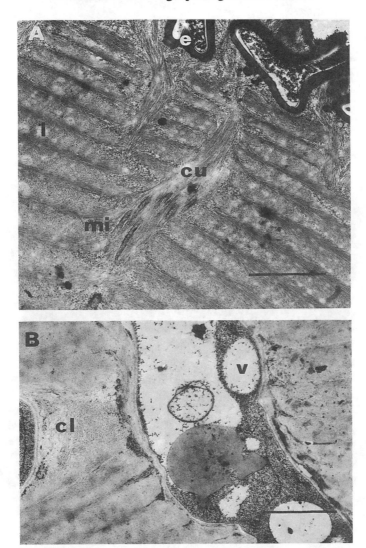

Fig. 5.4. A: Fine structure of the abdominal cuticle of an uninfected newly ecdysed 4th instar larva of *Manduca sexta*. Note the vertical cuticular column (cu) and long epidermal microvilli which give rise to the cuticular column (mi): l = lamella; bar marker = 1·5 μm.

B: Fine structure of the abdominal cuticle of 4th instar *Manduca sexta* larvae 48 h after inoculation with *Metarhizium anisopliae*, showing a penetrant hypha growing down a cuticular column. Note absence of lamellae in association with hyphae (cl): v = vacuole; bar marker = 1·8 μm. From Hassan & Charnley, 1989.

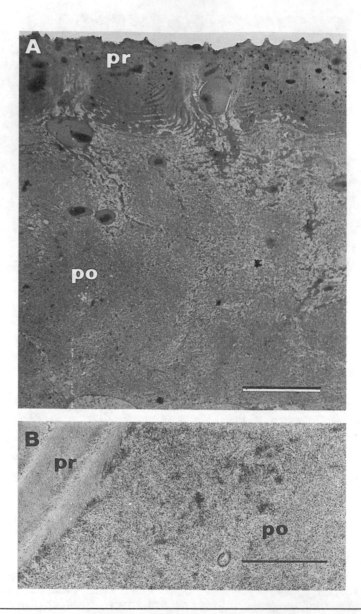

Fig. 5.5. A: Fine structure of the cuticle of 48 h old 4th instar *Manduca sexta* larvae. Insects fed on a Dimilin-treated diet. Note the disrupted non-lamellate post-ecdysial cuticle (po): pr = pre-ecdysial cuticle; bar marker = 8 μm. B: As for Fig. 5.5A, showing the interface between post-ecdysial (Dimilin-affected) and pre-ecdysial cuticles. Bar marker = 1·2 μm. A from Hassan & Charnley, 1989, B from Hassan & Charnley, unpublished.

tions, pharate fifth instar *M. sexta* cuticle, produced during treatment with Dimilin and thus completely disrupted by the insecticide (Hassan & Charnley, 1987), was considerably more susceptible to Pr 1 than control cuticle (St. Leger, Charnley & Cooper, unpublished observations).

Lipids form the third, minor, component of the procuticle. Doubts must be raised about the involvement of lipases in the penetration process by the inability to extract 'true' lipases (with activity against olive oil) from cuticle of *Manduca sexta* infected with *Metarhizium anisopliae* (St. Leger *et al.*, 1987a). However, this could reflect binding to either host cuticle or fungal cell walls. Non-specific esterases (assayed using chromogenic substrates or Tween and calcium salts) have been located on hyphae of *M. anisopliae* in the cuticle of the rhinoceros beetle, *Oryctes rhinoceros*, (Ratault & Vey, 1977). However, the role of esterases in cuticle degradation is far from clear, particularly as some of the substrates employed to assay esterase are also broken down by Pr 1 (St. Leger *et al.*, 1987a).

In contrast to the profuse *in vitro* production of extracellular enzymes on cuticle by entomopathogenic fungi, light histochemical and electron microscope studies indicate that *in vivo* enzyme activity is restricted until infection is well advanced and the cuticle breached. Recent confirmation of this has come from the ultrastructural localization of Pr 1 using an immunogold technique (Goettel *et al.*, 1989) (Fig. 5.3). A number of mechanisms may operate to inactivate or retain fungal enzymes to hyphae (Charnley, 1984; St. Leger, 1989). The influence of cuticle charge on enzyme binding as a prerequisite for activity has already been alluded to; however, it is possible that binding of charged enzymes to non-substrates may limit their mobility and activity (St. Leger, 1989). Pr 1 is too large to penetrate intact cuticular layers, though epicuticular filaments, pore canals and dermal ducts could aid dispersion (St. Leger, Cooper & Charnley, 1986e). Host-produced protease inhibitors may restrict enzyme activity of pathogens. Such compounds have been isolated from the cuticle of *Bombyx mori* larvae (Eguchi, Yamashita & Yoshida, 1988) and the crayfish *Aphanomyces astaci* (Häll & Söderhäll, 1982, 1983) (see next section). Host derived phenols provide a measure of resistance to fungi either by reducing the susceptibility of cuticle protein to degradation or by being antifungal (see next section).

Whatever the reason for localized cuticle degradation it could prove beneficial for the fungus by preventing premature dehydration of the host, restricting products of cuticle degradation (thus facilitating uptake for nutrition) and reducing the likelihood of invasion by competing microorganisms (St. Leger, 1989).

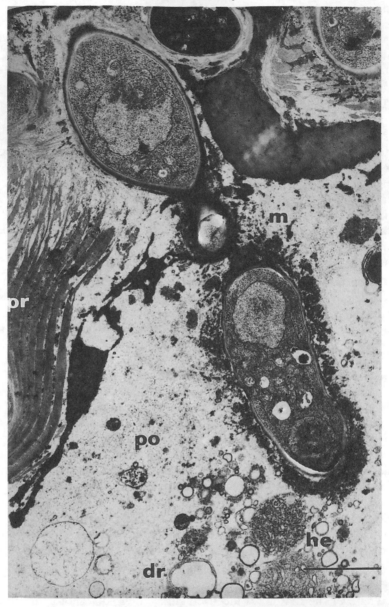

Fig. 5.6. Fine structure of the cuticle of 4th instar *Manduca sexta* larvae 48h after inoculation with *Metarhizium anisopliae*. Insect was fed on Dimilin-treated diet. Haemocytes (he) are probably discharging activated phenoloxidase and melanin (m) is being deposited on and around hyphal bodies: dr = cell debris, po = post-ecdysial cuticle, pr = pre-ecdysial cuticle; bar marker = 2·5 μm. From Hassan & Charnley, 1989.

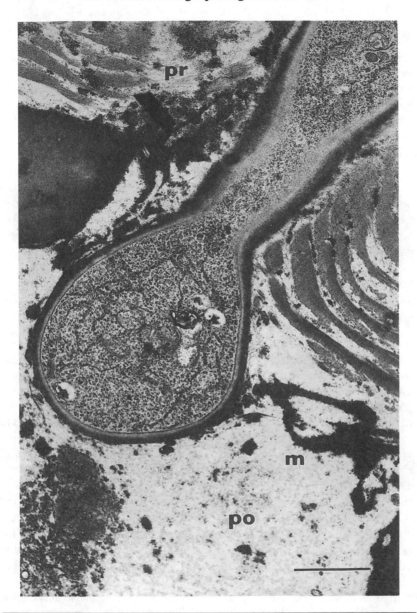

Fig. 5.7. As for Fig. 5.6, note the growth of the hypha down through the cuticular column. Bar marker = 1·3 μm. From Hassan & Charnley, unpublished.

Cause of Death

Once a penetrant hypha enters the haemocoel a number of options are available. Some fungi continue filamentous development, as with *Entomophthora coronata* in the termite *Reticulitermes flavipes* (Yendol & Paschke, 1965) and *Aspergillus flavus* in *Hyalophora cecropia* (Sussman, 1952). However, most entomopathogenic fungi convert to a yeast-like phase. In many instances an ability to make this transformation may be critical for pathogenesis (see next section).

Fungi also vary in the extent of their colonization of the insect prior to death. Proliferation of blastospores and/or hyphal fragments (hyphal bodies) in the haemolymph may be a prelude to early tissue invasion (see Charnley, 1984). Alternatively, the fungus may overcome its host after limited growth; a scenario which is usually attributed to the action of fungal toxins (Roberts, 1980).

The causes of death from mycosis are likely to vary with the behaviour of the pathogen. Extensive growth in the haemolymph and penetration of host tissues will disrupt host physiology and cause stress reactions possibly including autointoxication (see Charnley, 1984).

Toxin Production

There is considerable circumstantial evidence from Deuteromycete pathogens for the involvement of fungal toxins in host death. The action of cytotoxins is suggested by cellular disruption in advance of penetrant hyphae (Zacharuk, 1973). Behavioural symptoms such as partial or general paralysis, sluggishness and decreased irritability in mycosed insects are consistent with the action of neuromuscular toxins (Charnley, 1984). Several low molecular weight toxins have been identified from cultures of entomopathogenic fungi, particularly the Deuteromycetes (Roberts, 1980). However, most of these toxins have not been isolated from parasitized insects and their relevance to mycosis remains to be established (Charnley, 1984). An exception are the destruxins (DTX), a family of cyclic peptides produced by *Metarhizium anisopliae*. Injected doses of DTX cause incoordination, paralysis and death in Lepidopteran larvae and adult Diptera (Roberts, 1980; Samuels, Charnley & Reynolds, 1988a). Similar symptoms occur in larvae of *Bombyx mori* (Roberts, 1966) infected with *M. anisopliae* and lethal quantities of DTX have been extracted from diseased insects (Suzuki, Kawakami & Tamura, 1971). Comparable results have been achieved with *M. anisopliae* (ME1) in the tobacco hornworm, *Manduca sexta* (Samuels *et al.*, 1988a). However, two isolates of *M. anisopliae* (RS549 and RS1094) behaved very differently; they grew profusely in the haemolymph, did not induce paralysis in their hosts and produced considerably less DTX in culture than ME1. In addition, no DTX was found in the haemolymph of insects infected with RS549 and

RS1094. These two isolates were still pathogenic for *Manduca sexta* larvae despite the fact that they do not appear to produce paralysing doses of DTX. However, both isolates were significantly less virulent than ME1 and took longer to kill *M. sexta* larvae. High DTX production may not be essential for pathogenesis but may speed up the disease process.

Understanding the role of DTX in pathogenesis is complicated by the restricted susceptibility to DTX among the Insecta in comparison to the wide host range of *Metarhizium anisopliae*. A high 48h LD_{50} (injected dose that killed 50% of test insects) and rapid removal from the haemolymph of an injected dose of DTX in *Manduca sexta* may be contrasted with the low 48 h LD_{50}, and slow removal of DTX in *Bombyx mori* larvae (Samuels *et al.*, 1988a). Thus, at least for Lepidoptera, differences in ability to detoxify DTX could account for the distribution of DTX suscep-tibility. However, it is not clear why only Lepidoptera and adult Diptera respond by paralysis and death to injected DTX (Abalis, 1981; Samuels *et al.*, 1988a). The mode of action of DTX in causing paralysis in *Manduca*-larvae has been studied by Samuels, Reynolds & Charnley (1988b). Their results are consistent with DTX causing depolarization of the muscle membrane by activating calcium channels. It is interesting that low *in vivo* toxicity of DTX to Orthoptera such as *Schistocerca gregaria* correlates with *in vitro* insensitivity of muscles to DTX (Samuels, Reynolds & Charnley, unpublished observations), but other factors such as detoxification may also be important.

The dramatic effects of DTX on Lepidopteran muscles overshadow the fact that DTX exhibits cytotoxicity against other tissues from insects of a number of orders. Sloman & Reynolds (unpublished observations) found that 1 mg ml^{-1} of DTX inhibited ecdysone synthesis in culture by prothoracic glands of *Manduca sexta* while Cook & Charnley (unpublished observations) noted a substantial decrease in excretory activity of adult blowflies injected with sublethal doses of DTX. Lesions in cultured haemocytes of *Gromphadorhina laevigata* (Dictyopteran) exposed to 1 μg ml^{-1} DTX were characterised at the ultrastructural level by considerable vacuolization of the cytoplasm arising from dilation of the endoplasmic reticulum and mitochondria (Quiot, Vey & Vago, 1985). Intriguingly, low doses of DTX (1 μg ml^{-1}) inhibit DNA, RNA, protein synthesis and insect virus replication in cultured insect haemocytes (Quiot *et al.*, 1985).

It is not known whether these different effects of DTX may be attributed to a common mode of action. Although there are structural similarities between DTX and known neutral ionophores, e.g. valinomycin, DTX does not exhibit ionophoric activity in Pressman cells (Abalis, 1981; Samuels *et al.*, 1988b), human red blood cell ghosts (Samuels *et al.*, 1988b) or liposomes (Abalis, 1981). Abalis (1981) has suggested that DTX

108 A. K. Charnley

may act as a weak uncoupler of oxidative phosphorylation since, like dini-trophenol (a known uncoupler), it stimulates both F1 and Mg–ATPases from mitochondria. Such a mode of action could also account for the depolarizing action of DTX on Lepidopteran muscle as active transport appears to be involved in the maintenance of Lepidopteran muscle rest-ing potential (Huddart & Wood, 1966). However, preliminary experiments have failed to reveal any uncoupling activity of DTX in mi-tochondrial preparations from blowfly flight muscle (Charnley, unpublished).

At present it is difficult to relate the various cytotoxic effects (other than action on Lepidopteran muscle) to the symptoms and progress of disease caused by *Metarhizium anisopliae*. But clearly, failure of DTX to cause paralysis in a particular insect may not preclude a role for DTX in fungal pathogenesis. Indeed, injections of DTX do not evoke paralysis in Scarabaeid beetles but they potentiate the action of nonspecific strains of *Metarhizium* in *Cetonia aurata* and *Oryctes rhinoceros* respectively (Far-gues, Robert & Vey, 1985). This might be achieved, for example, by interfering with haemocyte function (see next section). It is possible that DTX has a similar role even in those insects like *Manduca sexta* which are susceptible to its myotoxic action. Effects of DTX on haemocytes and possibly other non-muscle targets might occur at levels too low for detec-tion by current methodology, and below the threshold of responsiveness of the muscle membrane. Thus even in RS549 and RS1094, DTX produc-tion may influence the cause of mycosis (Samuels *et al.*, 1988a).

Metarhizium anisopliae occurs in two varieties. *M. anisopliae* var *major* which has only been isolated from Scarabeid beetles and *M. anisopliae* var *anisopliae* which is more widespread. Whereas DTX may contribute to mycosis caused by var *anisopliae* the same may not be true for var *major*. Kaijiong & Roberts (1986) found that five isolates of var *major* produced only small amounts of DTX *in vitro* (about 1% of that produced by var *anisopliae*), while two isolates produced none at all *in vitro* and none was detected *in vivo* in mycosed wax moth larvae, *Galleria mellonella*.

Host defence responses

Melanization

The first overt response within the cuticle to fungal infection is often the appearance of a dark pigment, identified as melanin or melanin-protein complexes (see Charnley, 1984; St. Leger, 1989). St. Leger, Charnley & Cooper (1988d) showed that *Metarhizium anisopliae* produced a pheno-loxidase and a black pigment in media containing levels of phenols inhibitory to growth, thus holding out the intriguing possibility that cuticle melanin produced during invasion is not host-derived. However, St. Leger *et al.* (1988d) could only extract insect tyrosinase from melanized cuticle of *Manduca sexta* infected with *M. anisopliae*.

Melanic reactions appear to form an integral part of an insect's nonspecific response to wounding (Lai-Fook, 1966). However, melanization elicited by pathogenic fungi may not solely be triggered by physical damage. Certainly in crayfish cuticle, prophenoloxidase is activated by β-1,3-glucans on the fungal cell wall surface and glucan components of pathogen enzymes (Unestam & Söderhäll, 1977; Söderhäll & Unestam, 1979). Microbial components similarly trigger the enzyme cascade responsible for prophenoloxidase activation in insect and crustacean haemolymph (see later). The final step in the pathway in both animals is the limited proteolysis of the proenzyme by a trypsin-like serine protease. It is interesting to note that Pr 1 (chymoelastase) and not Pr 2 (trypsin) from *Metarhizium anisopliae* will activate haemolymph prophenoloxidase from *Manduca sexta in vitro*, opening up the possibility that the pathogen's enzymes may activate or interfere with the prophenoloxidase system of the host (St. Leger, 1989).

Given the known susceptibility of microorganisms to phenols it has generally been assumed that melanization reactions have antifungal effects (Charnley, 1984). The specificity of induction of fungal cell wall components implies an adaptive response (St. Leger, 1989). A few studies have demonstrated an inverse correlation between the degree of melanization and the success of the pathogen (see Charnley, 1984). However, often the melanic response fails to prevent infection (eg Hassan & Charnley, 1989; Butt *et al.*, 1988). Early induction of melanin production with sufficient magnitude appears to be critical to its effectiveness. Thus although melanin deposition starts in *Empoasca fabae* cuticle only 4 to 6h after inoculation, subsequent infection pegs of *Erynia radicans* grew faster than melanin could be produced (Butt *et al.*, 1988). Significant melanization in *Manduca sexta* cuticle infected with *Metarhizium anisopliae* did not occur until after the fungus had invaded the insect (48h) (Hassan & Charnley, 1989).

Both Butt *et al.* (1988) and St. Leger (1989) have suggested that melanization is primarily an effective defence against weak or slow growing pathogens, but is ineffective against more virulent fungi. This certainly appears to be true for the humoral encapsulation response in larval mosquitoes and blackflies, where melanin–like substances are deposited at the surface of the fungus apparently without participation of the haemocytes (Gotz & Vey, 1974). Likewise the melanization reaction in *Bombyx mori* is more extensive in response to challenge with the weak pathogen *Paecilomyces fumoso-rosea* than towards *Beauveria bassiana* (Aoki & Yanase, 1970). An ability to disable phenoloxidase activity may prove to be a virulence determinant. Certainly, DTX can suppress prophenoloxidase activation in locust haemocytes (Huxham & Lackie, 1986; Huxham, Lackie & McCorkindale, 1986).

Cellular reactions

Once the cuticle and epidermis have been breached the invading fungus is faced with the defence systems of the haemolymph. In the desert locust, *Schistocerca gregaria*, the presence of *Metarhizium anisopliae* within the cuticle elicits an inflammatory response some 12 h before the fungus enters the haemocoel (Gunnarsson, 1987). A multilayered haemocyte capsule builds up on the basement membrane of the epidermal cells beneath the infection site. The haemocytes also invade the basement membrane and enter the epidermis. Finally melanization occurs, starting with the endocuticle close to the epidermis and developing backwards through the capsule. Despite this intense activity on the part of the host, the fungus is not prevented from invading the haemocoel.

The cellular response to mycopathogens within the haemocoel results in phagocytosis, nodulation or encapsulation (Götz & Boman, 1985). Phagocytosis of experimentally introduced low doses of small fungal spores or yeast cells gives way to nodule formation with increased dosage. A process which has been described most completely for the interaction between *Galleria mellonella* and *Bacillus cereus* (for review see Ratcliffe *et al.*, 1985). On random contact with clumps of bacteria, granular haemocytes discharge a flocculent material which surrounds the bacteria and the haemocytes. The nodule starts to compact and melanize within 30 minutes and becomes attractive to plasmatocytes which form layers of cells around it. Cellular encapsulation of larger structures may also involve a biphasic process where initial contact is made by cystocytes and granular cells. These cells degranulate, lyse and form sites for the attachment of plasmatocytes which accumulate in layers around the intruder; the inner cells of the growing capsule become flattened (and sometimes melanized). The capsule or nodule stops growing presumably because the stimulus, coming either from the foreign object or the haemocytes making the initial contact, attracting the plasamatocytes is attenuated. The proximal cause of the cessation of haemocyte recruitment may be the coating of the capsule/nodule with basement membrane-like material containing glycosaminoglycans (Lackie, Takle & Tetley, 1985).

The fate of fungal elements which are phagocytosed, encapsulated or immobilised in nodules is uncertain. That the experience is not necessarily lethal is clear from reports that fungal spores can germinate within nodules and hyphal bodies can grow out of haemocytic capsules (Vey, 1984). The broad spectrum antibacterial proteins, the cecropins and attacins are ineffective against eukaryotic cells (Boman & Hultmark, 1987) and there is no evidence to date that production of specific antimycotic agents is induced by fungal cells or their metabolites. However, haemagglutinins, as observed in plant/pathogen interactions (Nordbring-Hertz & Chet, 1986), could act as mycostatic agents. Melanin production associated with haemocyte reactions and in cuticle penetration may have antifungal effects (St. Leger, 1989). Growth and production of cuticle-degrading proteases by *Metarhizium anisopliae* were far greater on unpigmented insect cuticles than on melanized-cuticle complexes (prepared by reacting L-DOPA or catechol with *Manduca sexta* cuticle) suggesting that phenolic oxidation products are fungitoxic (St. Leger, Cooper & Charnley, 1988e). Although phenols can inhibit microbial enzymes, activity of Pr 1 and Pr 2 from *Metarhizium anisopliae* (ME1), in contrast to bovine chymotrypsin, was not inhibited by incubation in a melanizing reaction mixture of *Manduca sexta* tyrosinase and L-DOPA, possibly reflecting enzymic adaptation by this pathogen (St. Leger *et al.*, 1988e). Melanin may provide a physical barrier constraining the fungus within the capsule or nodule. This appears to be the case in *Empoasca fabae* where humoral encapsulation of *Erynia radicans* results in contortion of some hyphae (Butt *et al.*, 1988). Melanin may chemically insulate the pathogen by restricting diffusion of the enzymes and toxins from the fungus and availability of nutrients, water and oxygen to the fungus (St. Leger, 1989). Melanized cuticle may be relatively resistant to enzyme attack, though Pr 1 from *Metarhizium anisopliae* (ME1) still releases melanin from melanized cuticle by hydrolyzing associated proteins (St. Leger *et al.*, 1988e). Finally, in the event of the hyphal walls themselves becoming melanized the fungus may lose the plasticity necessary for growth (St. Leger, 1989).

Recognition of 'non-self'

Recognition of 'non-self' is critical to the initiation of the haemocytic defence reaction. Physical properties such as charge and hydrophobicity directly influence the response of locust and cockroaches to abiotic implants (Lackie, 1988). However, the selective response to 'non-self' in insects is indicative of a specific chemical recognition on the part of the haemocytes. Serum and haemocyte cell membrane bound lectins have been found in many insects (eg Pereira, Andrade & Ribevo, 1976; Komano, Mizona & Natori, 1980; Jurenka, Manfredi & Harper, 1982; Suzuki & Natori, 1983). Since they agglutinate bacteria and parasitic protozoa and metazoa as well as fungi it is thought that they play a role in immune defence reactions (Ratcliffe *et al.*, 1985). However, evidence that agglutinins can act as opsonins facilitating the activity of haemocytes is generally poor. Indeed, Ratcliffe & Rowley (1983) failed to find enhanced phagocytosis of sheep red blood cells or bacteria in haemocyte monolayers from

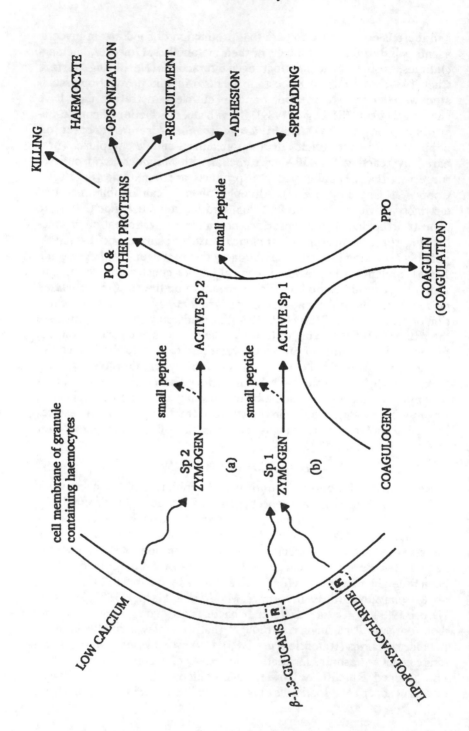

Clitumnus extradentatus and *Periplaneta americana* following pre-exposure to sera with haemagglutinin activity. Pendland, Heath & Boucias, (1988) have produced one of the few convincing pieces of evidence that an insect serum agglutinin may be opsonic. They showed that both serum and a purified galactose binding lectin from *Spodoptera exigua* opsonised hyphal bodies of *Paecilomyces farinosus*, resulting in considerably enhanced attachment of granulocytes in monolayers *in vitro*. Both haemocytes and hyphal bodies had exposed galactose residues. *Nomuraea rileyi* hyphal bodies, which lacked galactose residues, were not opsonised and were not cleared as effectively from the haemolymph of *S. exigua* larvae *in vivo* as were blastospores of *P. farinosus*. In the light of this it is interesting to note that Pendland & Boucias (1986) found that haemagglutinin activity in the haemolymph of *Anticarsia gemmatalis* was induced by intrahaemocoelic injection of blastospores of *N. rileyi*.

Although Ratcliffe & Rowley (1983) failed to find agglutinin opsonization of foreign particles, their experiments were consistent with other 'recognition factors' which were activated by constituents of microbial cell walls. Specific serum proteins such as vertebrate immunoglobulins have not yet been found in invertebrates. However, evidence is accumulating for Crustacea and insects that the prophenoloxidase (ppo) activating system may have a central role in 'non-self' recognition (for reviews see Ratcliffe *et al.*, 1985; Söderhäll & Smith, 1986). For some insects like *Sarcophaga bullata* (Saul & Sugumaran, 1988) the majority of ppo is in the plasma fraction. However, in most insect species studied the ppo system is located in two blood cell types: granular cells and cystocytes. These cells are the first to interact with an intruder and it is thought that they degranulate, coating the foreign surface with the 'sticky' proteins of the activated ppo system, where they promote the attachment of other haemocyte types. However, to date only a few species have been examined and it remains to be seen what interaction, if any, occurs between agglutinins and the ppo system or indeed whether the latter is universally involved in 'non-self' recognition in insects. In this context it is interesting to note that in the black cell mutant of *Drosophila melanogaster* the ppo-containing crystal cells are absent, yet the insect is still able to encapsulate foreign bodies effectively, producing amelanotic capsules (Rizki & Rizki, 1984).

Phenoloxidase represents the terminal component in a complex cascade of enzymes and other factors (Fig. 5.8) that constitute the activating

Fig. 5.8. (*Facing page*) Activation of the prophenoloxidase system of insects and crustacea and probable associated events. Sp 1 and Sp 2 = serine proteases, PPO = prophenoloxidase, PO = phenoloxidase, R = hypothetical receptors. Based on Ratcliffe *et al.* (1985) and Söderhäll & Smith (1986).

system. The cascade is best understood for crayfish *Astacus astacus* and the silkworm *Bombyx mori*, but recent work on other insects suggests that a similar system operates throughout the Arthropoda (for reviews see Ratcliffe *et al.*, 1985; Söderhäll & Smith, 1986; Söderhäll, 1986). Activation is triggered by β-1,3-glucans which are components of yeast and fungal cell walls, and lipopolysaccharides (LPS) and/or peptidoglycan from bacterial cell walls (Söderhäll & Unestam, 1979; Söderhäll & Häll, 1984). These microbial products initiate the activation of a serine protease (Sp 1), present as a zymogen (e.g. Yoshida & Ashida, 1986). In *B. mori* this calcium dependent enzyme converts inactive ppo into a smaller active form by cleavage of a small (5 kD) peptide from each sub-unit of the ppo (Ashida & Yoshida, 1988). An alternative activation pathway (most completely characterised for crayfish) involves a second serine protease (Sp 2). Like Sp 1, Sp 2 has a very restricted substrate specificity (only cleaving peptides with the structure R1-Gly-Arg-R2) but Sp 2 is activated by low calcium concentrations rather than by β-1,3-glucans or LPS (Söderhäll & Häll, 1984).

Role of protease inhibitors

The cascade involved in prophenoloxidase activation produces a number of physiologically very reactive molecules which participate in a variety of defence reactions, including opsonization (Fig. 5.8). It is thus in the best interests of the host to control the activation of the ppo cascade carefully. There have been many reports of endogenous protease inhibitors in various tissues (including haemolymph) of insects and arthropods (e.g. Sasaki, 1978; Eguchi, 1982; Häll & Söderhäll, 1982; Kucera 1982, 1984; Sugumaran, Saul & Ramesh, 1985; Saul & Sugumaran, 1986). The physiological role of these inhibitors is not clearly established. Changes in titre of protease inhibitor during metamorphosis of *Bombyx mori* prompted Eguchi & Kanabe (1982) to suggest that the inhibitors protected tissues from proteases leaking from the gut and other tissues during histolysis. Since isolated endogenous protease inhibitors prevent activation of ppo by serine proteases in *Manduca sexta* (Saul & Sugumuran, 1986) and *Musca domestica* (Namihara *et al.*, 1979), protease may confine the activation of ppo to the surface or immediate vicinity of an invading microorganism.

Cuticle-degrading proteases are implicated in host invasion by *Metarhizium anisopliae*. Extracellular fungal proteases are also toxic to insects by intrahaemocoelic injection (Kucera, 1980) and can activate prophenoloxidase *in vitro* (St. Leger, Charnley & Cooper, unpublished observations). Thus endogenous protease inhibitors may contribute to insect defence against pathogenic microorganisms. Consistent with this, Häll & Söderhäll (1983) found antigenically related protease inhibitors in haemocytes and cuticle of *Astacus astacus* which inhibited proteases from the pathogenic fungus *Aphanomyces astaci* but not the proteases from two closely related saprotrophic *Aphanomyces* species nor commercial preparations of trypsin or chymotrypsin (Häll & Söderhäll, 1982, 1983). Similarly, inhibitors from the haemolymph of *Galleria mellonella* and *Bombyx mori* were inhibitory to a *Metarhizium anisopliae* protease (Kucera, 1984) and *Aspergillus melleus* semialkaline protease (Eguchi, 1982) respectively. Eguchi, Ueda & Yamashita (1984) found many electrophoretic variants of haemolymph protease inhibitors among 126 silkworm, *B. mori*, strains. However, it is unlikely that the distribution of these variants correlates with strain resistance to entomopathogenic fungi as the protease inhibitors have broad specificity towards proteases from pathogenic and saprophytic fungi alike (St. Leger, 1989).

Evidence for an involvement of protease inhibitors in host defence comes from the work of Boucias & Pendland (1987). They showed that sixth instar *Anticarsia gemmatalis* larvae, which are resistant to infection by *Nomuraea rileyi*, contain high levels of serum protease inhibitor. Susceptibility of young *A. gemmatalis* larvae and larvae of all ages of *Trichoplusia ni* correlates with low titre or absence of protease inhibitor respectively.

Most entomopathogenic fungi are dimorphic, existing as yeast-like hyphal bodies in the host haemocoel during the pathogenic phase and converting to a mycelial form post-mortem prior to sporulation on the cadaver. The ability to convert to the yeast phase, at least for *Nomuraea rileyi*, may be a prerequisite for pathogenicity, since some attenuated strains which fail to produce hyphal bodies *in vitro* are non-pathogenic (Morrow & Boucias, unpublished, quoted in Morrow & Boucias, 1988). Interestingly, mycelia unlike hyphal bodies proved non-pathogenic upon injection into host insects (Heath & Boucias, unpublished, quoted in Morrow & Boucias, 1988); a condition which Boucias & Latgé (1989) attributed in part to the stimulation by the mycelia of an immediate host cellular response which resulted in their encapsulation.

Avoidance of host defence by switching to an alternative form of growth suggests either an altered cell wall structure/composition or that the production of an antihaemocyte chemical is consequent upon the change in morphology.

Unfortunately, we know little about the cell wall composition of entomopathogenic fungi. However, it is conceivable that masking or removal of β-1,3-glucans from the cell wall surface during mycelial-yeast transition prevents activation of the ppo system and thus one component of the 'non-self' recognition system may be missing. Changes in cell wall composition are certainly features of dimorphic human pathogens such as *Para-coccidioides brasiliensis* (Cole & Nozawa, 1981).

Some entomophthoralean fungi, e.g. *Entomophaga aulicae* (= *E. egressa*), produce wall-less protoplasts within the host haemocoel which do not elicit a host response (Dunphy & Nolan, 1980, 1981, 1982). Whereas, injection of walled cells of *E. aulicae* stimulate an encapsulation response in host (Dunphy & Nolan, 1982) and non host insects (Martin & Nolan, 1986). Cell walls of *E. aulicae* hyphal bodies are composed of unbranched β-1,3-glucans and chitin. Although protoplasts retain the synthetases responsible for β-1,3-glucan synthesis, factors in the haemolymph block transport of the finished product to the plasmalemma (Latgé, Beavais & Vey, 1986b).

Absence of a significant cellular response to fungal infection is by no means universal. In contrast to the above, Butt & Humber (1989) found that gypsy moth, *Lymantria dispar*, haemocytes *in vivo* recognize and encapsulate protoplasts of *Entomophaga grylli* specific for grasshoppers and protoplasts of *E. maimaga* and *E. aulicae* species pathogenic for lepidopteran larvae. There are also many reports of the encapsulation by insect haemocytes of elements of Deuteromycete fungi such as *Metarhizium anisopliae* and *Beauveria bassiana* (for reviews see Charnley, 1984; Vey, 1984). Although many of the studies quoted involve the unnatural injection of conidia, their results concur with descriptions of defence responses in topically inoculated insects that widespread cellular reactions are initiated and encapsulation may occur. Ability of a fungus to overcome host defence in such circumstances may not relate to strength of the cellular defence reaction but rather to the capacity of the fungus to prevent encapsulation or overcome the granuloma (Vey, 1984).Speed of fungal growth relative to the insect's defence response may contribute to the outcome of the confrontation (Götz & Boman, 1985). Circumstantial evidence from histological and fine-structural studies on *M. anisopliae* suggests that fungal toxins may substantially influence the fate of the fungus (Vey, 1984). Consistent with these observations, low doses of destruxins produced by *M. anisopliae* have profound effects *in vitro* on the behaviour of haemocytes of the desert locust *Schistocerca gregaria*, including inhibition of ppo (Huxham, Lackie & McCorkindale, 1986).With this insect-fungus interaction *in vivo*, although a substantial inflammatory reaction occurs on the basement membrane immediately below the site of infection (see above), few blood cells adhere to penetrant hyphae within the haemocoel (Gunnarsson, 1987).

Acknowledgements. I would like to thank Dr R. J. St. Leger for his helpful comments on the first draft of this manuscript.

References

Abalis, I. M. (1981). Biochemical and pharmacological studies of the insecticidal cyclodepsipeptides destruxins and bassianolide produced by entomopathogenic fungi. Ph.D. Thesis, Cornell University.

Al-Aidroos, K. & Roberts, D. W. (1978). Mutants of *Metarhizium anisopliae* with increased virulence towards mosquito larvae. *Canadian Journal of Genetics and Cytology*, **20**, 211-219.

Aoki, J. & Yanase, K. (1970). Phenoloxidase activity in the integument of the silkworm *Bombyx mori* infected with *Beauveria bassiana* and *Spicaria fumoso-rosea*. *Journal of Invertebrate Pathology*, **16**, 459-464.

Ashida, M. & Yoshida, H. (1988). Limited proteolysis of prophenoloxidase during activation by microbial products in insect plasma and effect of phenoloxidase on electrophoretic mobilities of plasma proteins. *Insect Biochemistry*, **18**, 11-19.

Bidochka, M. J. & Khachatourians, G. H. (1987). Purification and properties of an extracellular protease produced by the entomopathogenic fungus *Beauveria bassiana*. *Applied and Environmental Microbiology*, **53**, 1679-1684.

Bignell, D. E. (1984). The arthropod gut as an environment for microorganisms. In *Invertebrate-Microbial Interactions*, ed. J. M. Anderson, A. D. M. Rayner & D. W. H. Walton, pp. 205-229. Cambridge University Press.

Boman, H. G. & Hultmark, D. (1987). Cell-free immunity in insects. *Annual Review of Microbiology*, **41**, 103-126.

Boucias, D. G. & Latgé, J. P. (1986). Adhesion of entomopathogenic fungi on their host cuticle. In *Fundamental and Applied Aspects of Invertebrate Pathology*, ed. R. A. Samson, J. M. Clark & R. D. Peters, pp. 432-433. Foundation of the 4th International Colloquium of Invertebrate Pathology: Wageningen.

Boucias, D. G. & Latgé, J. P. (1989). Fungal elicitors of invertebrate cell defense system. In *Fungal Antigens: Isolation, Purification and Detection*, ed. E. Drouchet, G. T. Cole, L. DeRepentigny, J. P. Latgé & B. Dupont, in press. Plenum Press: New York.

Boucias, D. G. & Pendland, J. C. (1984). Nutritional requirements for conidial germination of several host range pathotypes of the entomopathogenic fungus *Nomuraea rileyi*. *Journal of Invertebrate Pathology*, **43**, 288-293.

Boucias, D. G. & Pendland, J. C. (1987). Detection of protease inhibitors in the haemolymph of resistant *Anticarsia gemmatalis* which are inhibitory to the entomopathogenic fungus *Nomuraea rileyi*. *Experientia*, **43**, 336-339.

Boucias, D. G., Pendland, J. C. & Latgé, J. P. (1988). Nonspecific factors involved in attachment of entomopathogenic deuteromycetes to host insect cuticle. *Applied and Environmental Microbiology*, **54**, 1795-1805.

Brey, P. T., Latgé, J. P. & Prevost, M. C. (1986). Integumental penetration of the pea aphid *Acyrthosiphon pisum* by *Conidiobolus obscurus* (Entomophthoraceae). *Journal of Invertebrate Pathology*, **48**, 34-41.

Brobyn, P. J. & Wilding, N. (1983). Invasive and developmental processes of *Entomophthora muscae* infecting houseflies (*Musca domestica*). *Transactions of the British Mycological Society*, **81**, 1-8.

Butt, T. M. (1987). A fluorescence microscopy method for the rapid localization of fungal spores and penetration sites on insect cuticle. *Journal of Invertebrate Pathology*, **50**, 72-74.

118   A. K. Charnley

Butt, T. M. & Humber, R. A. (1989). Response of gypsy moth haemocytes to natural fungal protoplasts of three *Entomophaga* species (Zygomycetes: Entomophthorales). *Journal of Invertebrate Pathology*, 53, 121-123.

Butt, T. M., Wraight, S. P., Galani-Wraight, S., Humber, R. A., Roberts, D. W. & Soper, R. S. (1988). Humoral encapsulation of the fungus *Erynia radicans* (Entomophthorales) by the potato leafhopper *Empoasca fabae* (Homoptera: Cicadellidae). *Journal of Invertebrate Pathology*, 52, 49-57.

Cerenius, L. & Söderhäll, K. (1984). Chemotaxis in *Aphanomyces astaci*, an arthropod-parasitic fungus. *Journal of Invertebrate Pathology*, 43, 278-282.

Charnley, A. K. (1984). Physiological aspects of destructive pathogenesis in insects by fungi: a speculative review. In *Invertebrate-Microbial Interactions*, ed. J. M. Anderson, A. D. M. Rayner & D. W. A. Walton, pp. 229-270. Cambridge University Press.

Cole, E. T. & Nozawa, Y. (1981). Dimorphism. In *Biology of Conidial Fungi*, vol I, ed. G. T. Cole & B. Kendrich, pp. 97-133. Academic Press: New York.

Cox, D. L. & Willis, J. H. (1985). The cuticular proteins of *Hyalophora cecropia* from different anatomical regions and metamorphic stages. *Insect Biochemistry*, 15, 349-362.

Dean, D. D. & Domnas, A. J. (1983). Isolation and partial characterization of collagenolytic enzyme from the mosquito-parasitizing fungus, *Lagenidium giganteum*. *Archives of Microbiology*, 136, 212-218.

Dillon, R. J. & Charnley, A. K. (1985). A technique for accelerating and synchronising germination of conidia of the entomopathogenic fungus *Metarhizium anisopliae*. *Archives of Microbiology*, 142, 204-206.

Dillon, R. J. & Charnley, A. K. (1986a). Inhibition of *Metarhizium anisopliae* by the gut bacterial flora of the desert locust, *Schistocerca gregaria*: evidence for an antifungal toxin. *Journal of Invertebrate Pathology*, 47, 350-360.

Dillon, R. J. & Charnley, A. K. (1986b). Invasion of the pathogenic fungus *Metarhizium anisopliae* through the guts of germ-free desert locusts, *Schistocerca gregaria*. *Mycopathologia*, 96, 59-66.

Dillon, R. J. & Charnley, A. K. (1986c). Germination physiology of conidia of *Metarhizium anisopliae*. In *Fundamental and Applied Aspects of Invertebrate Pathology*, ed. R. A. Samson, J. M. Vlak & D. Peters, p. 255. Foundation of the 4th International Colloquium of Invertebrate Pathology: Wageningen.

Dillon, R. J. & Charnley, A. K. (1988). Inhibition of *Metarhizium anisopliae* by the gut bacterial flora of the desert locust: Characterisation of antifungal toxins. *Canadian Journal of Microbiology*, 34, 1075-1082.

Dillon, R. J. & Charnley, A. K. (1989). The fate of fungal spores in the insect gut. In *The Fungal Spore and Disease Initiation in Plants and Animals*, ed. G. T. Cole & H. C. Hoch, in press. Plenum Press: New York.

Dunphy, G. B. & Nolan, R. A. (1980). Response of Eastern Hemlock Looper hemocytes to selected stages of *Entomophthora egressa* and other foreign particles. *Journal of Invertebrate Pathology*, 36, 71-84.

Dunphy, G. B. & Nolan, R. A. (1981). A study of the surface proteins of *Entomophthora egressa* protoplasts and of larval spruce budworm haemocytes. *Journal of Invertebrate Pathology*, 38, 352-361.

Dunphy, G. B. & Nolan, R. A. (1982). Cellular immune responses of spruce budworm larvae to *Entomophthora egressa* protoplasts and other test particles. *Journal of Invertebrate Pathology*, **39**, 81-92.

Eguchi, M. (1982). Inhibition of the fungal protease by haemolymph protease inhibitors of the silkworm, *Bombyx mori*. (Lepidoptera: Bombycidae). *Applied Entomology and Zoology*, **17**, 589-590.

Eguchi, M & Kanbe, M. (1982). Changes in haemolymph protease inhibitors during metamorphosis of the silkworm *Bombyx mori* L. (Lepidoptera, Bombycidae). *Applied Entomology and Zoology*, **17**, 179-187.

Eguchi, M, Ueda, K. & Yamashita, M. (1984). Genetic variants of protease inhibitors against fungal protease and chymotrypsin from haemolymph of the silkworm, *Bombyx mori*. *Biochemical Genetics*, **22**, 1093-1102.

Eguchi, M., Yamashita, M. & Yoshida, S. (1988). Protein inhibitors from the integument and haemolymph of the silkworm; *Bombyx mori* (Lepidoptera: Bombycidae) against fungal proteases. In *Proceedings of the XVIII International Congress of Entomology*, abstracts p. 130.

Fargues, J. (1981). Specificité des hyphomycètes entomopathogens a résistance interspécifique des larves d'insectes. Thèse de Doctorat d'état, Université de Paris VI.

Fargues, J. (1984). Adhesion of the fungal spore to the insect cuticle in relation to pathogenicity. In *Infection Processes of Fungi*, ed D. W. Roberts & J. R. Aust, pp. 90-109. The Rockefeller Foundation: New York.

Fargues, J., Robert, P. H. & Vey, A. (1985). Effet des destruines A,B et dans la pathogenèse de *Metarhizium anisopliae* chez les larves de Coleopteres Scarabaeidae. *Entomophaga*, **30**, 353-364.

Gabriel, B. P. (1968). Histochemical study of the insect cuticle infected by the fungus *Entomophthora coronata*. *Journal of Invertebrate Pathology*, **11**, 82-89.

Gershon, H. & Shanks, L. (1978). Antifungal activity of fatty acids and derivatives: structure activity relationships. In *The Pharmacological Effects of Lipids*, ed. J. J. Kabara, pp. 51-62. American Society of Oil Chemists: USA.

Goettel, M. S., St. Leger, R. J., Rizzo, N. W., Staples, R. C. & Roberts, D. W. (1989). Ultrastructural localization of a cuticle degrading protease produced by the entomopathogenic fungus *Metarhizium anisopliae* during penetration of host cuticle. *Journal of General Microbiology*, in press.

Götz, P. & Boman, H. G. (1985). Insect Immunity. In *Comprehensive Insect Physiology, Biochemistry and Pharmacology*, ed. G. A. Kerkut & L. I. Gilbert, pp. 453-485. Pergamon Press: New York.

Götz, P. & Vey, A. (1974). Humoral encapsulation in Diptera (Insecta), defense reactions of *Chironomus* larvae against fungi. *Parasitology*, **68**, 193-205.

Grula, E. A., Burton, R. L., Smith, R., Mapes, T. L. Cheung, P. Y. K., Pekrul, S., Champlin, F. R., Grula, M., & Abegaz, B. (1978). Biochemical basis for the pathogenicity of *Beauveria bassiana*. In *Proceedings, First Joint USA/USSR Conference on the Production, Selection and Standardization of Entomopathogenic Fungi*, ed. C. M. Ignoffo, pp. 192-216. American Society for Microbiology: Washington, D.C.

Gunnarsson, S. (1987). Cellular immune reactions in the desert locust *Schistocerca gregaria* infected by the fungus *Metarhizium anisopliae*. Acta Universitatis Upsaliensis, PhD Thesis.

Häll, L. & Söderhäll, K. (1982). Purification and properties of a protease inhibitor from crayfish hemolymph. *Journal of Invertebrate Pathology*, **39**, 29-37.

Häll, L. & Söderhäll, K. (1983). Isolation and properties of a protease inhibitor in crayfish (*Astacus astacus*) cuticle. *Comparative Biochemistry and Physiology*, **76B**, 699-702.

Hassan, A. E. M. & Charnley, A. K. (1983). Combined effects of diflubenzuron and the entomopathogenic fungus *Metarhizium anisopliae* on the tobacco hornworm *Manduca sexta*. In *Proceedings, 10th International Congress of Plant Protection*, vol. 3, p. 790.

Hassan, A. E. M. & Charnley, A. K. (1987). The effect of Dimilin on the ultrastructure of the integument of *Manduca sexta*. *Journal of Insect Physiology*, **33**, 669-676.

Hassan, A. E. M. & Charnley, A. K. (1989). Ultrastructural study of the penetration by *Metarhizium anisopliae* through Dimilin-affected cuticle of *Manduca sexta*. *Journal of Invertebrate Pathology*, in press.

Hepburn, H.R. (1985). Structure of the integument. In *Comprehensive Insect Physiology, Biochemistry and Pharmacology*, vol. 3, ed. G. A. Kerkut & L. I. Gilbert, pp. 1-58. Pergammon Press: Oxford.

Hillerton, J. E. (1984). Cuticle: mechanical properties. In *Biology of the Integument: I, Invertebrates*, ed. J. Bereite-Hahn, A. G. Matoksy & K. S. Richards, pp. 626-637. Springer-Verlag: Berlin.

Hojrup, P., Andersen, S. O. & Roepstorff, P. (1986). Isolation, characterisation and N-terminal sequence studies of cuticular proteins from the migratory locust, *Locusta migratoria*. *European Journal of Biochemistry*, **154**, 153-159.

Huddart, H. & Wood, D. W. (1966). The effect of DNP on the resting potential and ionic content of some skeletal muscle fibres. *Comparative Biochemistry and Physiology*, **18**, 681-688.

Hunt, D. W. A., Borden, J. M., Rame, J. E. & Whitney, M. S. (1984). Nutrient mediated germination of *Beauveria bassiana* cuticle on the integument of the bark beetle, *Dendroctonus ponderosae* (Coleoptera: Scolytidae). *Journal of Invertebrate Pathology*, **44**, 304-314.

Hunt, J. & Charnley, A. K. (1981). Abundance and distribution of the gut flora of the desert locust *Schistocerca gregaria*. *Journal of Invertebrate Pathology*, **38**, 378-385.

Hurion, N., Fromentin, H. & Keil, B. (1979). Specificity of the collagenolytic enzyme from the fungus Entomophthora coronata: comparison with the bacterial collagenase from *Achromobacter iophagus*. *Archives of Biochemistry and Biophysics*, **192**, 438-445.

Huxham, I. M. & Lackie, A. M. (1986). A simple visual method for assessing the activation and inhibition of phenoloxidase production by haemocytes *in vitro*. *Journal of Immunological Methods*, **94**, 271-277.

Huxham, I. M., Lackie, A. M. & McCorkindale, N. J. (1986). An *in vitro* assay to investigate activation and suppression by a pathogenic fungus of prophenoloxidase by insect haemocytes. In *Fundamental and Applied Aspects of Invertebrate Pathology*, ed. R. A. Samson, J. M. V. Lak & D. Peters, p. 463. Foundation of the 4th International Colloquium of Invertebrate Pathology: Wageningen.

Jurenka, R., Manfredi, K. & Harper, K. D. (1982). Haemagglutin activity in Acrididae (Grasshopper) Haemolymph. *Journal of Insect Physiology*, **28**, 177-181.

Kabara, J. J. (1978). Fatty acids and derivatives as antimicrobial agents - a review. In *The Pharmacological Effects of Lipids*, ed. J. J. Kabara, pp. 1-14. American Society of Oil Chemists: USA.

Kaijiong, L. & Roberts, D. W. (1986). The production of destruxins by the entomogenous fungus *Metarhizium anisopliae* var *major*. *Journal of Invertebrate Pathology*, 47, 120-122.

Kerwin, J. L. (1982). Chemical control of the germination of asexual spores of *Entomophthora culicis*, a fungus parasitic on Dipterans. *Journal of General Microbiology*, 128, 2179-2186.

Kerwin, J. L. (1984). Fatty acid regulation of the germination of *Erynia variabilis* conidia on adults and puparia of the lesser housefly, *Fannia canicularis*. *Canadian Journal of Microbiology*, 30, 158-161.

Kerwin, J. L. & Washino, R. K. (1986). Cuticular regulation of host regulation and spore germination by entomopathogenic fungi. In *Fundamental and Applied Aspects of Invertebrate Pathology*, ed. R. A. Samson, J. M. Vlak, & D. Peters, pp 423-425. The Foundation of the 4th International Colloquium of Invertebrate Pathology: Wageningen.

Klotz, S. A. & Penn, R. L. (1987). Multiple mechanisms may contribute to the adherence of *Candida* yeasts to living cells. *Current Microbiology*, 16, 119-122.

Komano, H., Mizona, D. & Natori, S. (1980). Purification of lectin induced in the haemolymph of *Sarcophaga peregrina* larvae on injury. *Journal of Biological Chemistry*, 225, 2919-2924.

Kucera, M. (1980). Proteases from the fungus *Metarhizium anisopliae* toxic for *Galleria mellonella* larvae. *Journal of Invertebrate Pathology*, 35, 304-310.

Kucera, M. (1982). Inhibition of the toxic proteases from *Metarhizium anisopliae* by extracts of *Galleria mellonella* larvae. *Journal of Invertebrate Pathology*, 40, 299-300.

Kucera, M. (1984). Partial purification and properties of *Galleria mellonella* larvae proteolytic inhibitors acting on *Metarhizium anisopliae*. *Journal of Invertebrate Pathology*, 43, 190-196.

Lackie, A. M. (1988). Immune mechanisms in insects. *Parasitology Today*, 4, 98-105.

Lackie, A. M., Takle, G. B. & Tetley, L. (1985). Haemocytic encapsulation in the locust *Schistocerca gregaria* and the cockroach *Periplaneta americana*. *Cell and Tissue Research*, 290, 343-351.

Lai-Fook, J. (1966). The repair of wounds in the integument of insects. *Journal of Insect Physiology*, 12, 195-226.

Latgé, J. P. (1972). Contribution a l'étude du *Cordyceps militaris* (Fr.) Link. Systematique, biologie, physiologie. Doctorat de specialité, Université de Toulouse.

Latgé, J. P., Beauvais, A. & Vey, A. (1986). Wall synthesis in the entomophthorales and its role in the immune reaction of infected insects. *Developmental and Comparative Immunology*, 10, 639.

Latgé, J. P., Cole, G. T., Horisberger, M. & Provost, M. C. (1986). Ultrastructure and chemical composition of the ballistospore wall of *Conidiobolus obscurus*. *Experimental Mycology*, 10, 99-113.

Latgé, J. P., Sampedo, L,, Brey, P. & Diaquin, M. (1987). Aggressiveness of *Conidiobolus obscurus* against the pea aphid: influence of cuticular extracts on ballistospore germination of aggressive and non-aggressive strains. *Journal of General Microbiology*, 133, 1989-1997.

Lockey, K. H. (1988). Lipids of the insect cuticle: origin, composition and function. *Comparative Biochemistry and Physiology*, **89**, 595-645.

Martin, F. R. & Nolan, R. A. (1986). American cockroach (*Periplaneta americana*) hemocyte response to *Entomophaxgax aulicae* protoplasts. *Canadian Journal of Zoology*, **64**, 1369-1372.

Michel, B. (1981). Recherches experimentales sur la pénétration des champignons pathogens chez les insectes. Thèse 3e cycle, Université Montepellier.

Morrow, B. J. & Boucias, D. G. (1988). Comparative analysis of the *in vitro* growth of the hyphal body and mycelial stage of the entomopathogenic fungus *Nomuraea rileyi*. *Journal of Invertebrate Pathology*, **57**, 197-206.

Namihira, G., Ejima, T., Inaba, T. & Funatsu, M. (1979). A protein factor inhibiting activation of prophenoloxidase with natural activator and its purification. *Agricultural Biology and Chemistry*, **43**, 471-476.

Neville, A. C. (1984). Cuticle: organisation. In *Biology of the Integument, I. Invertebrates*, ed. B. J. Hahn, A. G. Maroksy & K. S. Richards, pp. 611-625. Springer-Verlag: Berlin.

Nordbring-Hertz, B. & Chet, I. (1986). Fungal lectins and agglutinins. In *Microbial Lectins and Agglutinins*, ed. D. Mirelman, pp. 393-408. Wiley: New York.

Pekrul, S. & Grula, E. A. (1979). Mode of infection of the corn earworm (*Heliothis zea*) by *Beauveria bassiana* as revealed by scanning electron microscopy. *Journal of Invertebrate Pathology*, **34**, 238-247.

Pendland, J. C. & Boucias, D. G. (1986). Characteristics of a galactose-binding haemagglutinin (lectin) from haemolymph of *Spodoptera exigua* larvae. *Developmental and Comparative Immunology*, **10**, 477-487.

Pendland, J. C., Heath, M. A. & Boucias, D. G. (1988). Function of a galactose-binding lectin from *Spodoptera exigua* larval haemolymph: opsonization of blastospores from entomogenous hyphomycetes. *Journal of Insect Physiology*, **34**, 533-540.

Pereira, M. E. A., Andrade, A. F. B. & Ribevo, S. H. (1976). Lectins of distinct specificity in *Rhodnius prolixus* interact selectively with *Trypanosoma cruzi*. *Science*, **211**, 597-600.

Quiot, J. M., Vey, A. & Vago, C. (1985). Effects of mycotoxins on invertebrate cells *in vivo*. In *Advances in Cell Culture*, ed. K. Maramarosch, pp. 199-212. Academic Press: New York.

Ratcliffe, N. A. & Rowley, A. F. (1983). Recognition factors in insect haemolymph. *Developmental and Comparative Immunology*, **7**, 653-656.

Ratcliffe, N. A., Rowley, A. F., Fitzgerald, S. W., Rhodes, C. P. (1985). Invertebrate immunity - basic concepts and recent advances. *International Review of Cytology*, **97**, 184-350.

Ratault, C. & Vey, A. (1977). Production d'esterases et de d' N-acetyl-β-D-glucosaminidase dans le tégument du Coleoptère *Oryctes rhinoceros* par le champignon entomopathogène *Metarhizium anisopliae*. *Entomophaga*, **22**, 289-294.

Rizki, T. M. & Rizki, R. M. (1984). The cellular defense system of *Drosophila melanogaster*. In *Insect Ultrastructure*, vol. 2. ed. R. C. King & H. Akai, pp. 579-604. Plenum Press: New York.

Roberts, D. W. (1966). Toxins from the entomogenous fungus *Metarhizium anisopliae*. II Symptoms and detection in moribund hosts. *Journal of Invertebrate Pathology*, **8**, 222-227.

Roberts, D. W. (1980). Toxins of entomopathogenic fungi. In *Microbial Control of Insects, Mites and Plant Diseases*, ed. H. D. Burges, pp. 441-463. Academic Press: New York.

Robinson, R. K. (1966). Studies on penetration of insect integument by fungi. *Pest Articles & News Summaries*, 12, 131-142.

Samuels, R. I., Charnley, A. K. & Reynolds, S. E. (1988a). The role of destruxins in the pathogenicity of 3 strains of *Metarhizium anisopliae* for the tobacco hornworm *Manduca sexta*. *Mycopathologia*, 104, 51-58.

Samuels, R. I., Reynolds, S. E. & Charnley, A. K. (1988b). Calcium channel activation of insect muscle by destruxins, insecticidal compounds produced by the entomopathogenic fungus *Metarhizium anisopliae*. *Comparative Biochemistry and Physiology*, 90C, 403-412.

Sasaki, T. (1978). Chymotrypsin inhibitors from haemolymph of the silkworm *Bombyx mori*. *Journal of Biochemistry*, 84, 267-276.

Saul, S. J. & Sugumaran, M. (1986). Protease inhibitor controls phenoloxidase activation in *Manduca sexta*. *Federation of European Biochemical Societies Letters*, 208, 113-117.

Saul, S. J. & Sugumaran, M. (1988). Prophenoloxidase activation in the haemolymph of *Sarcophaga bullata* larvae. *Archives of Insect Biochemistry and Physiology*, 7, 91-103.

Schabel, H. G. (1978). Percutaneous infection of *Hylobius pales* by *Metarhizium anisopliae*. *Journal of Invertebrate Pathology*, 31, 180-187.

Smith, R. J. & Grula, E. A. (1981). Nutritional requirements for conidial germination and hyphal growth of *Beauveria bassiana*. *Journal of Invertebrate Pathology*, 37, 222-230.

Smith, R. J. & Grula, E. A. (1982). Toxic components on the larval surface of the corn earworm (*Heliothis zea*) and their effects on germination and growth of *Beauveria bassiana*. *Journal of Invertebrate Pathology*, 39, 15-22.

Smith, R. J. & Grula, E. A. (1983). Chitinase is an inducible enzyme in *Beauveria bassiana*. *Journal of Invertebrate Pathology*, 42, 319-326.

Söderhäll, K. (1986). The cellular immune system in crustaceans. In *Fundamental and Applied Aspects of Invertebrate Pathology*, ed. R. A. Samson, J. M. Vlak & D. Peters, pp. 417-420. The Foundation of the 4th International Colloquium of Invertebrate Pathology: Wageningen.

Söderhäll, K. & Häll, L. (1984). Lipopolysaccharide-induced activation of prophenoloxidases activating system in crayfish haemocyte lysate. *Biochimica et Biophysica Acta*, 797, 99-104.

Söderhäll, K. & Unestam, T. (1979). Activation of serum prophenoloxidase in arthropod immunity. The specificity of cell wall glucan activation and activation by purified fungal glycoproteins of crayfish phenoloxidase. *Canadian Journal of Microbiology*, 25, 406-414.

Söderhäll, K. & Smith, V. J. (1986). The prophenoloxidase activating system: the biochemistry of its activation and role in arthropod cellular immunity, with special reference to crustaceans. In *Immunity in Invertebrates*, ed. M. Brehelin, pp. 208-223. Springer-Verlag: New York.

St. Leger, R. J. (1989). The integument as a barrier to microbial infections. In *The Physiology of Insect Epidermis*, ed. A. Retnakaran, & K. Binnington, in press. Inkata Press: Australia.

St. Leger, R. J., Charnley, A. K. & Cooper, R. M. (1986a). Cuticle-degrading enzymes of entomopathogenic fungi: synthesis in culture on cuticle. *Journal of Invertebrate Pathology*, **48**, 85-95.

St. Leger, R. J., Cooper, R. M. & Charnley, A. K. (1986b). Cuticle-degrading enzymes of entomopathogenic fungi: regulation of production of chitinolytic enzymes. *Journal of General Microbiology*, **132**, 1509-1517.

St. Leger, R. J., Cooper, R. M. & Charnley, A. K. (1986c). Cuticle-degrading enzymes of entomopathogenic fungi: cuticle degradation *in vitro* by enzymes from entomopathogens. *Journal of Invertebrate Pathology*, **47**, 167-177.

St. Leger, R. J., Charnley, A. K., Cooper, R. M. (1986d). Cuticle-degrading enzymes of entomopathogenic fungi: mechanisms of interaction between pathogen enzymes and insect cuticle. *Journal of Invertebrate Pathology*, **47**, 295-302.

St. Leger, R. J., Cooper, R. M. & Charnley, A. K. (1986e). Restriction of fungal depolymerases by insect host cuticle. In *Natural Antimicrobial Systems*, ed. G. W. Gould, M. E. Rhodes Roberts, A. K. Charnley, R. M. Cooper & R. G. Board, p. 316. Bath University Press: Bath, England.

St. Leger, R. J., Cooper, R. M. & Charnley, A. K. (1987a). Production of cuticle-degrading enzymes by the entomopathogen *Metarhizium anisopliae* during infection of cuticles from *Calliphora vomitoria* and *Manduca sexta*. *Journal of General Microbiology*, **133**, 1371-1382.

St. Leger, R. J., Charnley, A. K., & Cooper, R. M. (1987b). Characterization of cuticle-degrading proteases produced by the entomopathogen *Metarhizium anisopliae*. *Archives of Biochemistry and Biophysics*, **253**, 221-232.

St. Leger, R. J., Cooper, R. M. & Charnley, A. K. (1987c). distribution of chymoelastases and trypsin-like enzymes in five species of entomopathogenic Deuteromycetes. *Archives of Biochemistry and Biophysics*, **258**, 123-131.

St. Leger, R. J., Cooper, R. M. & Charnley, A. K. (1988a). Utilization of alkanes by entomopathogenic fungi. *Journal of Invertebrate Pathology*, **52**, 356-360.

St. Leger, R. J., Durrands, P. K., Charnley, A. K. & Cooper, R. M. (1988b). The role of extracellular chymo-elastase in the virulence of *Metarhizium anisopliae* for *Manduca sexta*. *Journal of Invertebrate Pathology*, **52**, 285-294.

St. Leger, R. J., Durrands, P. K., Charnley, A. K. & Cooper, R. M. (1988c). Regulation of production of proteolytic enzymes by the entomopathogen *Metarhizium anisopliae*. *Archives of Microbiology*, **150**, 413-416.

St. Leger, R. J., Charnley, A. K. & Cooper, R. M. (1988d). Production of pigments and phenoloxidase by the entomopathogen *Metarhizium anisopliae*. *Journal of Invertebrate Pathology*, **52**, 215-221.

St. Leger, R. J., Cooper, R. M. & Charnley, A. K. (1988e). The effect of melanization of *Manduca sexta* cuticle on growth and infection of *Metarhizium anisopliae*. *Journal of Invertebrate Pathology*, **52**, 459-471.

St. Leger, R. J., Butt, T. M., Roberts, D. W. & Staples, R. C. (1989). *In vitro* studies and the production of infection structures and a cuticle-degrading protease by the entomopathogenic fungus *Metarhizium anisopliae*. *Experimental Mycology*, in press.

Sugumaran, M., Saul, S. J. & Ramesh, N. (1985). Endogenous protease inhibitors prevent undesired activation of prophenoloxidase in insect hemolymph. *Biochemistry and Biophysics Research Communications*, **132**, 1124-1129.

Sussman, A. S. (1952). Studies of an insect mycosis. III. Histopathology of an asper-
gillosis of *Platysamia cecropia*. *Annals of the Entomological Society of America*,
45, 233-245.

Suzuki, A., Kawakami, K. & Tamura, S. (1971). Detection of destruxins in silkworm
larvae infected with *Metarhizium anisopliae*. *Agricultural and Biological Chemistry*, **35**, 1641-1643.

Suzuki, T. & Natori, S. (1983). Identification of a protein having hemagglutinating
activity in the haemolymph of the silkworm, *Bombyx mori*. *Journal of Biochemistry*, **93**, 583-590.

Thompson, S. N. (1973). A review and comparative characterization of the fatty acid
compositions of seven insect orders. *Comparative Biochemistry and Physiology*,
45, 467-482.

Travland, B. (1979). Structures of the motile cells of *Coelomomyces psorophorae*
and function of the zygote in encystment on host. *Canadian Journal of Botany*,
57, 1021-1035.

Unestam, T. & Söderhäll, K. (1977). Soluble fragments from fungal cell walls elicit
defence reactions in crayfish. *Nature*, **267**, 45-46.

Vey, A. (1984). Cellular antifungal reaction in Invertebrates. In *Infection Processes
of Fungi*, ed. D. W. Roberts & J. R. Aust, pp. 168-175. The Rockefeller Foundation: New York.

Woods, S. P. & Grula, E. A. (1984). Utilizable surface nutrients on *Heliothis zea*
available for growth of *Beauveria bassiana*. *Journal of Invertebrate Pathology*, **43**,
259-269.

Wraight, S. P., Butt, T. M., Galaini-Wraight, S., Allee, L. L. & Roberts, D. W. (1989).
Germination and infection processes of the entomophthoralean fungus *Erynia
radicans* on the potato leafhopper *Empoasca fabae*. *Journal of Invertebrate Pathology*, in press.

Yendol, W. G. & Paschke, J. D. (1965). Pathology of an *Entomophthora* infection
in the Eastern subterranean termite *Reticulitermes flavipes*. *Journal of Invertebrate Pathology*, **7**, 414-422.

Yoshida, A. & Ashida, M. (1986). Microbial activation of two serine enzymes and
prophenoloxidase in the plasma fraction of haemolymph of the silkworm *Bombyx
mori*. *Insect Biochemistry*, **16**, 539-545.

Zacharuk, R. Y. (1970). Fine structure of the fungus *Metarhizium anisopliae* infecting three species of larval Elateridae. II. Conidial germ tubes and appressoria.
Journal of Invertebrate Pathology, **15**, 81-91.

Zacharuk, R. Y. (1973). Electron-microscope studies of the histopathology of fungal infections by *Metarhizium anisopliae*. *Miscellaneous Publications of the Entomological Society of America*, **9**, 112-119.

Zebold, S. L., Whistler, H. C., Shemanchuk, J. A., Travland, L. B. (1979). Host specificity and penetration in the mosquito pathogen *Coelomyces psorophorae*. *Canadian Journal of Botany*, **57**, 2766-2770.

Chapter 6

Improvement of fungi to enhance mycoherbicide potential

George E. Templeton & Dana K. Heiny

Department of Plant Pathology, University of Arkansas, Fayetteville, AR 72701, USA

Introduction

Mycoherbicide research and development over the past two decades has led to commercial use of several indigenous plant pathogenic fungi for weed control, and several others are in development that will be available for commercial use within the next 2 to 5 years. It has been predicted that thirty weeds might be controlled by mycoherbicides by the year 2000 (R. S. Soper, personal communication).

An array of constraints, biological, economic, or regulatory, have been overcome to achieve success with mycoherbicides. Many other barriers to commercialization have been identified and appear to be surmountable by research on disease and fungus biology and by application of recent advances in biotechnology. In this paper we discuss the status of mycoherbicide development, the barriers to their development and some anticipated benefits of biotechnology for improvements of fungi to enhance mycoherbicide potential.

Mycoherbicide development and commercialization

Current status of mycoherbicides

The mycoherbicide Collego® has been used commercially since 1982 in Arkansas to control northern jointvetch (*Aeschynomene virginica*), a leguminous weed in rice (*Oryza sativa*) fields (Templeton, 1987). It is marketed as a dry formulation consisting of 15% viable, dry conidia of *Colletotrichum gloeosporioides* f. sp. *aeschynomene* and 85% inert ingredients. Each formulation batch is assayed or packaged to contain $7 \cdot 54 \times 10^{11}$ viable conidia per bag, the amount required for treatment of 4·05 hectares at the rate of $93 \cdot 5$ l ha^{-1} (Bowers, 1986). This amounts to $1 \cdot 8 \times 10^7$ viable spores m^{-2}. The pathogen infects, but does not adequately control, a closely related species, Indian jointvetch (*Aeschynomene indica*), a serious localized weed in the gulf coast rice production area of Louisiana and Texas.

Grower acceptance of Collego has been good, and most growers that use it do so repeatedly. A true comparison of its ability to compete with a chemical herbicide is not available, however, since the only chemical herbicide for this weed, 2,4,5-T, was removed from the market as Collego was being introduced.

Some difficulty has been encountered with integration of Collego into the pest management program of short-season rice varieties that may require application of fungicides for control of rice blast disease. The period between Collego application and fungicide application is so brief that activity of the Collego fungus is curtailed (TeBeest & Templeton, 1985; Smith, 1986). A major restraint to greater use of Collego has been its small market potential by comparison to that of broad-spectrum, chemical herbicides and other products of the pharmaceutical company or the chemical company that has owned Collego. Market penetration is very low but is expected to increase under the management of Ecogen Corporation, the new owner who has fewer products vying for sales resources.

The mycoherbicide Devine® has been used commercially since 1981 in Florida citrus (Rutaceae) groves to control milkweed vine, *Morrenia odorata* (Ridings, 1986). It is marketed as a wet formulation of chlamydospores of *Phytophthora palmivora* with a shelf life of six weeks in refrigerated storage. It is applied at the rate of 8×10^4 chlamydospores m^{-2} to the surface of moist soil under citrus trees. A safety zone of $1 \cdot 6$ km must be maintained between site of application and plantings of several susceptible vegetable crops and ornamental plants. It is feasible to distribute and market this relatively labile product because of the limited marketing area. Had it been a product that was to have been sold over a large area, improvement in formulation for longer shelf-life at ambient temperatures would have been necessary (Kenney, 1986).

No damage to non-target crops or citrus have been noted in the seven years since first commercial use of this mycoherbicide despite mild infection of citrus in controlled environments with high inoculum rates (200 × field rates). In treated areas, there has been no detectable change in virulence of the pathogen or related pathogens (*P. nicotiana* var. *parasitica*) with which it might intercross (R. Charudattan, personal communication). This affirms the original assertion that the fungus, although capable of interspecific hybridization in culture, does not do so in nature with sufficient frequency to create sustainable strains with undesirable attributes.

Grower acceptance of Devine has been good. There has been less sold than expected because of residual weed control by the fungus. Milkweed seedlings emerging under treated trees for up to 5 years after treatment have been killed by the fungus in some orchards. Like Collego, Devine has a small market potential compared to other products owned by the company so marketing efforts are not given high priority.

The persimmon wilt fungus, *Acremonium diospyri*, is not commercially available but is routinely used as a mycoherbicide to control persimmon trees (*Diospyros virginiana*) in rangeland in south-central Oklahoma. It has been used since 1960 to control trees up to 10 cm in diameter (Wilson-1965). All trees in a grove may be killed within three years following hand inoculation of 80% or more of the trees with a conidial suspension of the fungus (Griffith, 1970). The fungus is provided free to local ranchers by the Noble Foundation near Ardmore, Oklahoma. Since treatments are most effective when made as leaves begin to emerge, heavy suspensions of conidia provided in plastic squirt bottles are applied in the spring to wounds made in the cambium of tree trunks with a hand axe.

Loss of virulence in the fungus during culture has occurred frequently and requires reisolation of virulent strains each season from diseased trees. The geographic separation and localized area of use plus the requirement for wound inoculation has negated concern about the potential for spread of this pathogen to rare, scattered, cultivated persimmons in the state or to forests where persimmon is an economic species (Griffith, 1970). Furthermore, persimmon sites in Arkansas and Oklahoma inoculated in 1963 still have abundant persimmon groves and no evidence of localized establishment of the disease.

Velgo® is a potential mycoherbicide for velvetleaf (*Abutilon theophrasti*) control in corn (*Zea mays*) and soybeans (*Glycine max*) in the U.S. corn belt and southern Ontario, Canada. It is a strain of *Colletotrichum coccodes* (Wymore, Watson & Gotlieb, 1987). Commercialization is anticipated in 3 to 5 years as a combination treatment with reduced rates of several chemical herbicides that will not control velvetleaf alone. The fungus alone at the rate of 1×10^9 spores m^{-2} kills 46% of the plants inoculated at the two- to three-leaf stage. Lack of complete kill is due to premature shedding of infected leaves and regrowth of new leaves. The host range of this isolate of *C. coccodes* is restricted compared to other isolates, and although it infects several other plants, it does not cause significant damage to them. Production of sclerotia that may permit seasonal carry-over in some regions is not considered a factor in velvetleaf control in subsequent seasons but will need to be addressed if susceptible crops are to be rotated in the intended use area.

Luboa 2 is a selected strain of *Colletotrichum gloeosporioides* f. sp. *cuscutae* that is used as a mycoherbicide in the People's Republic of China to control dodder (*Cuscuta chinensis* and *Cuscuta australis*) parasitic on broadcast-planted soybeans. It has been used since its discovery in 1963 for practical control of this parasitic weed. Spore concentrations of

2×10^7 spores ml^{-1} are applied with a hand sprayer until run-off. Best results are obtained when spraying is done at 16.00 to 17.00 h on days when humidity is high, usually in late July to early August. Luboa 2 is an improved strain necessitated by loss of virulence in Luboa 1. Virulence is positively correlated with spore size. Greater use is expected when new formulations with longer shelf life are available (Yang Han Li, personal communication).

Biomal® is a potential mycoherbicide for control of round-leaved mallow (*Malva pusilla*) in wheat (*Triticum aestivum*) and lentils (*Lens culinaris*) in the provinces of Manitoba and Saskatchewan, Canada, and the northern tier of wheat-producing states in the United States. It is a selected strain of *Colletotrichum gloeosporioides* f. sp. *malvae*. It is applied in spore suspensions containing 2×10^9 spores l^{-1} at the rate of 3×10^2 l ha^{-1} (Mortensen, 1988). This amounts to 6×10^7 viable spores m^{-2}. The pathogen infects, but does not adequately control, velvetleaf, a serious malvaceous weed widely distributed in corn and soybeans.

Successful control of round-leaved mallow has been achieved in regional field tests in the wheat and lentil growing region, a relatively arid area, where the disease normally occurs at endemic levels. No technical difficulties have been encountered with spore production and formulation or with environmental fate and toxicology tests that would preclude its registration and commercial use for round-leaved mallow control (Mortensen, 1988). Much greater market potential could be realized if strains are selected or developed that would control velvetleaf throughout the corn and soybean growing region of the midwestern United States.

Casst® is a potential mycoherbicide for control of sicklepod (*Cassia obtusifolia*) and coffee senna (*Cassia occidentalis*) in soybeans and peanuts (*Arachis hypogaea*) in the southern region of the United States (Walker & Riley, 1982). It is a strain of *Alternaria cassiae* and is applied at the rate of $1 \cdot 1$ kg ha^{-1} in $76 \cdot 7$ l of water with an oil-based adjuvant (J. S. Bannon, personal communication). The spore concentration in the spray tank is $7 \cdot 5 \times 10^4$ spores ml^{-1}, amounting to application of $1 \cdot 4 \times 10^6$ spores m^{-2}. Commercial release is planned for Casst in 1990 for use alone or in combination with several chemical herbicides that do not, when used alone, control either weed. Spores are currently produced on solid substrates with some modification of the procedure of Walker & Connick (1983).

A mycoherbicide for control of American blackcherry (*Prunus serotina*) in forests in the Netherlands is being developed with a strain of the fungal pathogen *Chondrostereum purpureum* (Scheepens, 1980). It is a native fungus being developed to control an introduced weed and, like persimmon wilt, requires wound inoculation to initiate the disease process. Weed trees and sprouts are cut mechanically, and the cut surfaces of the stumps are painted or sprayed with mycelial fragments in agar suspensions. Mycelial concentrations equivalent to 20 to 200 μg dry weight per stump are effective, and applications made in spring or autumn are similarly effective in removing blackcherry saplings from forests or resprouting stumps in reforestation plantings.

Fructification and sporulation of the fungus on naturally infected and treated trees and stumps may provide periodic inoculum threats to certain cultivated *Prunus* species within a 5 km radius during pruning seasons (de Jong, 1988). Risk to non-target native trees and fruit trees is regarded as acceptable to the Dutch Plant Quarantine Authority because of the spatial separation of forest plantations and fruit growing regions. Low market potential is a deterrent to commercialization, but the ease with which the organism can be produced and marketed in a localized area raises expectations that it will be available commercially in the future.

A potential mycoherbicide for control of Bathurst burr (spiny cocklebur or spiny clotbur = *Xanthium spinosum*) is being developed in New South Wales, Australia, with a strain of *Colletotrichum orbiculare* that is indigenous in the area where the weed is an economic problem (Auld, McRae & Say, 1988). Commercialization is expected in 3 to 5 years. Small-scale field tests with the pathogen (not formulated or dried) indicate that it kills large weed plants in relatively dry conditions when heavily inoculated. The pathogen will not kill common cocklebur (*Xanthium pennsylvanicum*) when spray inoculated, but if injected into young stems it causes death. The fungus is known as a pathogen of species in the family Cucurbitaceae (Sutton, 1980). However, since it is poorly adapted to spread and cucurbit production is not practical where the weed is a problem, the lack of specificity is unlikely to be a major setback to the development of the fungus for widespread use.

Another host-pathogen interaction that has been examined in detail for mycoherbicide potential is *Cercospora rodmanii* on waterhyacinth (*Eichhornia crassipes*)(Charudattan *et al.*, 1985). Mycelium of this leaf spot-inducing pathogen was grown for 3 weeks on potato-dextrose broth containing 5% yeast extract, blended, and applied at the rate of 1·1 g wet weight m^{-2}. Following inoculation of the plants grown in buckets of nutrient-supplemented water, disease stress caused a significant reduction in the net rate of leaf production. However, success of this biocontrol strategy was limited by a low rate of plant kill and the ability of the host to compensate for disease loss by rapid leaf production. Charudattan *et al.* (1985) concluded that *C. rodmanii* could be an effective control for waterhyacinth when used with multiple applications of inoculum early in the growing season or under conditions favouring low to moderate host growth rates, including combining with sublethal rates of chemical herbicides or insect biocontrol agents. Abbott Laboratories of Chicago, Illinois, formulated *C. rodmanii* as a wettable powder and obtained a U. S. Environmental Protection Agency Experimental Use Permit for evaluation of the mycoherbicide (Charudattan, 1986). The fungus has not yet been registered for commercial use.

Biological, economic and regulatory constraints to mycoherbicide development

An array of constraints has been identified during research and development of mycoherbicides. Many are innate biological deficiencies of particular pathogens; others are economic or regulatory restraints based either upon some biologic shortcoming or some perception of risk.

Paucity of commercial-scale technology for bulk production and conservation of inoculum is a pervasive constraint to development of potential mycoherbicides. Many virulent, host-specific fungi have not been induced to sporulate in submerged fermentation, the most common commercial-scale technology for production of biologicals. New, solid state technology has been required for bulk production of the *Alternaria* in the mycoherbicide Casst for sicklepod control (Walker & Boyette, 1985). Success of such technology will broaden immensely the opportunities for mycoherbicide development and control of other pests with hyphomycetous fungi.

Comparable needs exist for innovative biotechnology that will permit improved storability of certain fungi. Many, like the pathogen in Devine and *Chondrostereum* for blackcherry control in the Netherlands, can be produced in submerged fermentation but as yet have not been dried or formulated to retain viability for the 18- to 24-month period required in most market channels. Success of perishable preparations of fungi is restricted to those that control site-specific weeds within limited geographic regions.

Inadequate virulence and loss of virulence in culture are often constraints that affect development of specific fungi as mycoherbicides. Loss of virulence has been a particular problem with the *Acremonium* species used to control persimmon in Oklahoma rangeland. Overcoming this constraint would enable development of a product for use on persimmon in pastures and rangelands over a much larger geographic region than is now possible. Loss of virulence has also been encountered with strains of the *Colletotrichum* species used as the mycoherbicide Luboa for control of dodder in China (Yang Han Li, personal communication).

Inadequate virulence is also a common constraint. Collego has inadequate virulence to control Indian jointvetch. Biomal, Velgo and strains of

Colletotrichum malvarum incite disease of velvetleaf, but none appear sufficiently virulent alone to control this widespread, important weed.

In some instances development of a weed pathogen may be restricted because of too broad a host range, presenting risks to crop plants or other economically important plants if they are grown where weed control is needed. Collego, Devine, Velgo, the spiny cocklebur pathogen and the blackcherry pathogen all have economically important hosts but have been deemed safe because of geographic separation between use areas and susceptible species. An extensive study of spore dissemination with the blackcherry pathogen in the Netherlands has defined a safety zone of 5000 metres around the treated area beyond which the inoculum level of the pathogen does not exceed levels that can occur from natural infections in the area (de Jong, 1988). This monumental epidemiological study serves as a model for determining safety zones for other wind disseminated pathogens that may be useful for biological weed control.

Development of mycoherbicides for crop mimics such as red rice (*Oryza sativa*), wild oats (*Avena fatua*), and broomcorn (*Sorghum vulgare* var. *technicum*) has not been successful because of excessively broad host ranges of several candidate pathogens. Red rice is highly susceptible to *Pyricularia oryzae*, a variable and serious pathogen of cultivated rice. This fungus is not a reasonable candidate for a mycoherbicide because of its pathogenicity to common rice cultivars. Isolates from grass (*P. grisea*) are candidates for control of crabgrass (*Digitaria sanguinalis*) in lawns or agricultural areas separated from rice-growing regions. A product containing *P. grisea* is not forthcoming because crabgrass isolates can potentially damage rice or cross with rice isolates of the pathogen (Mackill & Bonman, 1986; Yaegashi & Hebert, 1976). Similar constraints apply to several pathogens of broomcorn and pathogens of wild oats.

Requirements for particular environmental conditions for infection and disease development is another pervasive constraint to development of pathogens as mycoherbicides. Use of indigenous pathogens generally assures that proper environmental conditions exist in the region for infection and disease development; however, occurrence of ideal conditions may not coincide with the period when weed control is necessary. The indigenous pathogen, *Colletotrichum malvarum*, for example, will control prickly sida (*Sida spinosa*) during cool seasons, early spring or early fall, but control of this weed is needed in relatively warm conditions after emergence of the warm-season crops — cotton (*Gossypium hirsutum*) and soybeans.

Too little moisture for disease development is considered a major constraint to development of many mycoherbicides. Free moisture in the form of dew or from irrigation is needed for spore germination and must coincide with the period when weed control is necessary. It is not possible to generalize about availability of moisture for spore germination from regional climatic data since the microclimate at the infection court is the key environment. The mycoherbicide Biomal in Canada and the pathogen for spiny cocklebur control in Australia are intended for use in relatively arid climates but at times when microclimates are suitable for infection and disease development.

High specificity for one or two weed species is a general biological constraint that constitutes an economic barrier to commercial development of many potential mycoherbicides. Many existing mycoherbicides would be more successful commercially if they controlled one or more additional weeds. Collego would be more economically successful for example if it controlled, in addition to northern jointvetch, the related species, Indian jointvetch. In this instance the barrier is inadequate virulence or inability to overcome disease resistance since Indian jointvetch is infected by the *Colletotrichum* strain in Collego. Each weed is a problem in different rice-producing areas in the southern United States. The anthracnose pathogen being developed in New South Wales, Australia, for control of spiny cocklebur is another example. It would have substantially more economic potential if it controlled the related common cockleburs in the U.S. cotton and soybean crops. In this instance the constraint is apparently related to establishment of the pathogen in the host since wound inoculation leads to disease development and death of cocklebur.

Geographic biotypes of weeds can occur and constitute an economic constraint to mycoherbicide development. For example, biotypes of prickly sida and johnsongrass (*Sorghum halepense*) are resistant to potential mycoherbicides. In some instances market sizes are adequate when taken as a whole but consist of numerous, widely separated areas. The potential of Luboa for dodder control in the United States is limited because dodder is such a widely separated problem: in alfalfa (*Medicago sativa*) seed production and in tomatoes (*Lycopersicon esculentum*) in California, in carrot (*Daucus carota* var. *sativa*) production in Maryland, in cranberry (*Vaccinium macrocarpon*) production in Wisconsin and in the northeastern United States. However, the high value of these crops may be enough incentive for some enterprising company to further develop this mycoherbicide.

Limited market size is offset in many cases by the utility of a mycoherbicide in any crop where the weed host occurs and by low development and registration costs. Diminishing market size is an increasing economic problem for chemical herbicides as well. Long-term use of broad-spectrum chemicals has led naturally to succession of resistant, often site-specific weed problems.

Regulatory constraints to mycoherbicide development have been centered mainly on exotic pathogens and genetically engineered pathogens. United States authorities consider these to be equivalent risks. Restrictions on importation of exotic pathogens for application to introduced weeds requires excessive evaluation in containment facilities, which are not generally available in the United States. These restrictions do not take into account the innate deficiency for dissemination of most fungi with good potential as mycoherbicides nor the pervasive environmental constraints on disease development in general. It is expected that constraints on introduction of genetically engineered organisms will diminish as microbes genetically engineered for other purposes are found to be safe in field trials now in progress.

Anticipated benefits of biotechnology to mycoherbicide development

Inoculum production, preservation and formulation

The application of large-scale fermentation to production of biocontrol agents is relatively new. Although media and growth conditions may be generalized for groups of similar organisms, ultimately each fungus has peculiar requirements for optimum yields. Even different isolates within the same pathogenic species can have widely different habits and requirements for media and environment. Knowledge about microbial metabolism, growth and sporulation is a prerequisite to a scientific selection and adaptation of media for fermentation (Miller & Churchill, 1986). Two good sources for media requirements are Demain & Solomon (1986) and Zabriskie *et al*. (1982). The pharmaceutical industry has fermentation expertise that could be utilized for mass production of agents for biocontrol of weeds (Churchill, 1982).

Low cost and high productivity of top quality inoculum are the goals of fermentation. Liquid culture techniques are generally considered the most efficient and economical. Unfortunately, many filamentous fungi do not sporulate well in liquid culture. The Collego fungus may have been an exception among fungi. Development of a procedure for producing high numbers of spores in liquid culture for the Collego fungus took some effort and refinement (Daniel *et al*., 1973), but it was comparatively easy compared to attempts with other fungi, including other isolates of *Colletotrichum* (Templeton, 1976; Boyette, Templeton & Smith, 1979; Churchill, 1982).

Many of the fungi that produce pycnidia and some Hyphomycetes such as *Alternaria* species are more adaptable to methods of solid fermentation than to the less labour-intensive liquid fermentation methods. In liquid culture these fungi produce mycelia but few, if any, conidia. Surface culturing is used to produce fungal pathogens of insects commercially, such as *Beauveria bassiana*, *Metarhizium anisopliae* (Ignoffo, 1981), and *Acrostalagmus aphidum* [presumably = *Verticillium lecanii*] (Kristiansen & Bu'lock, 1980). A method described for producing spores of *Alternaria macrospora*, a pathogen of spurred anoda (*Anoda cristata*), involves growth of the fungus in trays of vermiculite under lights (Walker, 1981). This procedure requires extra handling and space for incubation and harvesting relative to liquid fermentation. Similar problems are encountered in production of spores of *Alternaria crassa*, a pathogen of jimsonweed (*Datura stramonium*), on trays of mycelial homogenate (Boyette, 1986). Other solid substrate fermentation methods include increasing the surface area with pleated paper. Rotating tube or drum systems and disc fermenters take advantage of the tendency of filamentous fungi to attach to and colonize immersed solid surfaces (Anderson & Blain, 1980). Rapid, synchronized sporulation occurs on the tube or disc surfaces upon removal of the medium. This technique may be ideal for producing large quantities of spores. A large surface area is available, and spore harvesting can be done aseptically (Anderson & Blain, 1980). Spores from pycnidia may be harvested by scraping into water from agar cultures, homogenizing pycnidia grown on removable cellophane membrane over agar media or by growing pycnidia on water-softened granules that can be ground into solution.

Future alternatives to solid fermentation include developing or conditioning new strains of fungi to sporulate when submerged, designing media or media additives that stimulate sporulation, and discovery of particular physical treatments that enhance sporulation, such as temperature variation, irradiation, agitation and aeration.

Many of the goals of preserving pathogens in mycoherbicide formulations are the same as those for preserving stock cultures in the laboratory: to maintain cell viability; to prevent contamination; to prevent changes in genetic, biochemical or pathogenic properties; and to ensure ease of transport and handling (Hill, 1981; Dietz & Churchill, 1985). It is therefore reasonable that many of the principles of culture preservation can be applied to spore preparation for commercial use in killing weeds.

Dehydration is the most effective method of reducing the metabolic state of organisms (Hill, 1981). Methods of drying spores include airdrying, freeze-drying, spray-drying, vacuum-tumbler drying and chemical desiccation (Churchill, 1982). Often additives such as skim milk or sodium glutamate improve survival of microorganisms during dehydration (Hill, 1981). The dry-spore product must be sealed from the air due to its hygroscopic properties (Hill, 1981). Cell suspensions of bacteria or fungi preserved on anhydrous silica gel granules can be stored up to 3·5 years at 4°C (Trollope, 1975). Storage at room temperature usually reduces the minimum survival period (Trollope, 1975). Of fungi examined for survival after drying which also have potential as biocontrol agents of weeds, *Fusarium* species have the shortest length of survival (Trollope, 1975; Antheunisse, de Bruin-Tol & van der Pol-van Soest, 1981).

Preservation and formulation are often coincident processes. Mycoherbicides can be formulated and marketed as wettable powders, dusts, granular products or flowable liquids just as microbial insecticides and chemical pesticides are formulated and marketed (Bowers, 1986; Ignoffo, 1981; Ridings, 1986; Walker & Connick, 1983). Depending on the nature of the spores, drying may be done with the 'naked' spore or an agent may be added to facilitate separation of individual spores, which improves mixing or handling properties. An example of the former is vacuum harvesting of *Alternaria crassa* conidia, which are dried under reduced pressure over $CaSO_4$ for 48 h at 4°C prior to storage at 4°C in bottles (Boyette, 1986). An example of the latter is mixing *Colletotrichum gloeosporioides* f. sp. *malvae* spores with kaolin (hydrated aluminum silicate) to produce a thick slurry prior to air-drying (Mortensen, 1988). Resuspension and application of this mixture is no more difficult than use of a wettable powder herbicide formulation.

Walker & Connick (1983) developed a pelletized formulation using sodium alginate for dry preparation of fungal inoculum. Production of spores on the gel pellet surface is repetitive after drying and rehydration. For weeds that grow in a rosette, scattering granules on the foliage favours repeated infections without reapplication. The method is adaptable to a number of genera, and the pelleting formula can be modified for each organism. Weidemann & Templeton (1988) incorporated nutritional substrates into the granules with mycelium and microconidia of *Fusarium solani* f. sp. *cucurbitae* for use in pre-emergence application to control Texas gourd (*Cucurbita texana*). Control with granular formulations, in which granules were broadcast on the soil surface, was generally better than control with aqueous conidial applications incorporated into the soil before planting. A granular formulation that floats also has potential for use in aquatic environments (Boyette, Walker & Sciumbato, 1983).

Another innovation with possible application to the formulation of fungi for mycoherbicides is the development of membranes for encapsulating individual spores. Methods are available for immobilizing an enzyme within a semipermeable membrane, such as hollow cellulose acetate fibre (Barfoed, 1981). Similar methods have been used for entrapping microbial cells (Chibata, Tosa & Sato, 1986). Resin gels and collagen membranes are also suitable for encapsulating living spores (Chibata *et al.*, 1986). Moisture drawn inside the membrane would be retained long enough to promote germination of the enclosed spore. Of course, initial studies must rule out production by the spore of germination self-inhibitors that would be restricted from diffusing outward by the same artificial membrane, preventing germination. Such a membrane around a hyaline spore might also contain a pigment such as fungal melanin, a substance known to protect spores against ultraviolet light (Durrell, 1964) and gamma-radiation (Zhdanova, 1974).

As discussed previously, the mycoherbicide Devine is sold as a liquid formulation. If it were to be marketed in a larger area a shelf life longer than the usual six weeks would be required (Kenney, 1986). Spores in liquid formulations might be protected using antibiotics and preservatives common in the food and cosmetics industry. Reversible germination inhibitors might be employed to maintain fungistasis in the suspension until dilution or neutralization of the inhibitor. A concentrated liquid formulation of this type would be diluted sufficiently upon mixing with water for spray application as to be undetectable in the environment. Chemicals that might be employed include sodium benzoate, benzalkonium chloride, sorbic acid, and propionate. Preservation might also be accomplished by lowering the pH or using sugar concentrates. A germination inhibitor such as gloeosporone, an endogenous self-inhibitor produced by conidia of *Colletotrichum gloeosporioides* f. sp. *jussiaea* (Lax, Templeton & Meyer, 1985; Adam, Zibuck & Seebach, 1987), might be of use in this system. Crude extracts from leaves of some plants are also known to inhibit spore germination (Allen, 1976), and leaves could be a source from which to obtain inhibitors to use in liquid formulations. Of course, preservatives and inhibitors would have to be selected to complement particular pathogens.

Technology for formulating and applying fungal spores in the absence of dew has been developed and is continually being improved. Quimby *et al.* (1988) reported preparation of a viscous invert (water-in-oil) emulsion that retarded evaporation. Droplet deposits greater than 1 mm in diameter were produced by an air-assist atomizing nozzle. The deposits enhanced kill of sicklepod seedlings by *Alternaria cassiae* (88%) relative to seedlings without invert (0%) in the absence of dew (Quimby *et al.*, 1988). Problems that must be overcome with this system include foliar toxicity to some plant species and the requirement for special spraying apparatus to handle the viscous material. Enclosing spores in insoluble, water permeable polymers, such as polyacrylamide, which could retain water like a sponge, might be another solution to the common problem of inadequate dew deposition and duration. Cell immobilization with polyacrylamide has been accomplished with several fungi (Chibata *et al.*, 1986).

In some cases it may be possible to enhance disease following mycoherbicide application by the use of additives in the inoculum suspension.

Surfactants, nutrients, stimulants and other types of chemicals should be considered when the pathogen alone is not completely effective. Winder & Van Dyke (1987) studied the effects of twelve adjuvants on conidial germination and symptom development with *Bipolaris sorghicola* on johnsongrass (*Sorghum halepense*). They observed a significant stimulation of germination with 1% Soydex and an increase in germ tube number with 1% Soydex or 0·02% Tween 80. Highest disease ratings resulted using Soydex in the inoculum suspension. The use of herbicides and growth regulators is commonly associated with increased disease. Suggested reasons for increased susceptibility include changes in host composition and structure, leakage of metabolites to the host surface, and changes in natural host defence mechanisms (Griffiths, 1981). Herbicides can also stimulate spore germination and growth of germ tubes (Griffiths, 1981). Tank mix applications of *Colletotrichum coccodes* and thidiazuron, a cotton defoliant, increased velvetleaf mortality significantly compared to *C. coccodes* or thidiazuron individually (Wymore *et al.*, 1987). Since thidiazuron apparently has little effect on corn or soybean, velvetleaf might be controlled in these crops with a mixture of thidiazuron and *C. coccodes*.

Volatile flavour compounds stimulate germination in members of several fungal genera (French, 1985). Compounds such as nonanal, 2,4-hexadienal and 2,4-nonadienal are effective in stimulating germination in *Penicillium*, *Fusarium* and *Alternaria* species. *Colletotrichum musae* and *Phytophthora palmivora* also respond to several identified flavour compounds (French, 1985). The use of stimulators such as these at the time of mycoherbicide application could cause more rapid and complete germination, circumventing the problem of brief dew periods.

Ultraviolet light screens might also be added to mycoherbicide formulations if they have demonstrated usefulness. Ultraviolet absorbers include ethyl dihydroxypropyl *p*-aminobenzoate (ethyl dihydroxypropyl PABA), glycerol PABA, amyl dimethyl PABA and octyl dimethyl PABA (Flick, 1984). With ultraviolet light screens and dew enhancers in the formulation, mycoherbicide application might be done when it is convenient during the day, rather than in the evening. Formulations with non-specific additives may have to be used with greater care since additives that enhance disease caused by a weed pathogen might also enhance infection by a pathogen of the crop.

Combining more than one fungus in mycoherbicide formulations could also increase market potential. Boyette *et al.* (1979) combined the Collego fungus with *Colletotrichum gloeosporioides* f. sp. *jussiaea* to control successfully two unrelated weed species with one application in a rice field. Combining the Biomal fungus, which killed 75% of velvetleaf plants under controlled conditions (Mortensen, 1988), with *C. coccodes* (Velgo) might result in effective velvetleaf control. Pathogens could be applied together to kill one or several weed species in a field. Several isolates of one pathogen or several species of pathogens, each having slightly different environmental requirements, could be mixed in the formulation to ensure that at least one would encounter the optimum environmental window. Biological control of weeds with multiple pathogens comprises a research opportunity that has been neglected.

Understanding the biology, physiology and biochemistry of host-pathogen interactions

The biology, physiology and biochemistry of host-pathogen interaction require greater understanding if advances are to be made in mycoherbicide research. Historically, basic research has been done in plant pathology with the general goal of finding ways of protecting crops from disease. Developing means of enhancing disease in weeds takes a different perspective of the problem. The biology of a pathogen on its host in response to the environment is directly related to the physiology and biochemistry of the interaction under those conditions. Knowing the mechanisms behind the success or failure of attempts to cause disease in particular weeds would enhance our ability to exploit those systems for practical biocontrol.

Separately from the host, information about the physiology of fungi could assist efforts to improve production and formulation of inoculum. For example, a number of rust fungi are known to be inhibited by methyl-cis-3,4-dimethoxycinnamate (Macko et al., 1972). Inactivation of enzymes responsible for hydrolyzing the germination pore plug is the apparent mode of action of this inhibitor (Allen, 1976). Germination inhibitors are easily washed away with water, allowing immediate digestion of the plug and emergence of the germ tube. More knowledge about germination inhibitors of fungi with mycoherbicide potential could help in developing liquid formulations. Light-induced sporulation of fungi has been associated with endogenous compounds called P-310s due to their maximum absorption at 310 nm (Trione, 1981). Addition of these compounds to growing cultures can replace the stimulus of light by inducing sporulation of certain fungi in the dark (Trione, 1981). P-310s might be useful in liquid fermentation of filamentous fungi. Additionally, the requirements for growth of a number of obligate parasites in axenic culture are still unknown. Development of mycoherbicides with several devastating pathogens that happen to be biotrophic would be more feasible if they could be cultured.

Fundamental knowledge about penetration processes in weed-pathogen systems might suggest substances that could be added to inoculum suspensions to stimulate or augment enzymes involved in penetration of the plant. *Alternaria crassa* normally has only minor effects on hemp sesbania (*Sesbania exaltata*), but it was induced to kill this weed species when spores were suspended in water soluble filtrates from hemp sesbania or jimsonweed (Boyette, 1987). Fungal conidia in pectin solutions gave similar results. The suggestion was that pectolytic hydrolysis was involved.

Cutin is a major structural component of the cuticle that forms the first penetration barrier of a plant. Cutinase, the enzyme that degrades cutin, is secreted by a number of fungi (Kollatakudy, 1985). Cutinase has been identified from at least three *Colletotrichum* species, notably *C. gloeosporioides* from papaya (*Carica papaya*). Cutin hydrolysate and monomers of cutin induce cutinase. Low levels of cutinase are probably preformed in spores (Kollatakudy, 1985). A reasonable first attempt at stimulating *Colletotrichum orbiculare*, the spiny cocklebur pathogen, to kill common cocklebur after spray inoculation, would be to utilize cutin monomers as cutinase inducers or to add cutinase to the inoculum suspension. The lack of a commercial source of cutin and cutinase is a deterrent to this line of research. Cutin in itself and the catalytic specificity of cutinase do not contain elements of host-pathogen specificity, but minor components that might be present in cutin could influence host specificity (Kollatakudy, 1985). It may be that *C. orbiculare* produces cutinase on spiny cocklebur but for some reason is unable to secrete cutinase on common cocklebur. However, given the complexity of host-nonhost relationships to fungi (Heath, 1987), it is likely that *C. orbiculare* will have to overcome more than one barrier in the common cocklebur plant before it can establish itself as a pathogen.

Too little is known about the biochemistry in the disease process of the Collego fungus on Indian jointvetch to be able to predict precisely what factors could be manipulated to achieve control. Candidates for study in this and other weed-pathogen systems include enzymes, such as pectolytic enzymes, cellulose-degrading enzymes, hemicellulose-degrading enzymes, phospholipases, and proteases; toxin activity (Walker & Templeton, 1978); and phytoalexins, elicitors and suppressors (Misaghi, 1982). Ultrastructural comparisons of resistant and susceptible reactions in weed hosts would give clues to the nature of the interaction, the timing of significant disease events, and the involvement of cell-wall-degrading enzymes (O'Connell, Bailey & Richmond, 1985). More information is needed concerning pathogen recognition of a host by sensing of plant surface topography. The mechanisms underlying recognition of specific surface signals, as elucidated by Hoch *et al.* (1987) with *Uromyces appendiculatus* on *Phaseolus vulgaris*, may suggest methods of artificially broadening the host range of particular mycoherbicides. Host-pathogen interactions are complex, and a number of approaches should be considered for future mycoherbicide development.

Genetic manipulation of host-pathogen systems

It is possible to produce protoplasts from the cells of a number of fungal genera (Peberdy, 1979; Hocart & Peberdy, Chapter 11). The procedure involves enzymatic digestion of cell walls in the presence of an osmotic stabilizer. Conditions are optimized to provide the greatest yield, health and uniformity of protoplasts. Somatic hybridization resulting from protoplast fusion can be manipulated both within and between species. In cases where natural events of hyphal fusions or cell fusions of compatible mating types are difficult to achieve, protoplasts provide a means of producing diploids or multinucleate heterokaryons (Peberdy, 1983). Protoplast fusion crosses between closely related species of *Aspergillus* or closely related species of *Penicillium* resulted in hybrid heterokaryotic and diploid (alloploid) colonies upon regeneration. Upon haploidization a range of nonparental segregants were recovered (Peberdy, 1983). Leslie (1983) induced mutation by UV irradiation in *Gibberella zeae* and identified morphological, auxotrophic, and temperature sensitive variants. Protoplast fusion was performed with four of the auxotrophic mutants, resulting in prototrophy by heterokaryosis as long as selective pressure was maintained. It is clear that protoplast fusions could facilitate the combining of desirable traits of potential mycoherbicide fungi. Enhanced virulence, toxin production, sporulation, tolerance to environmental stresses and shifts in specificity could be the benefits of these techniques. The production and regeneration of protoplasts from mycelium of *Alternaria crassa*, potential biocontrol agent of jimsonweed, has been demonstrated (Brooker, 1988).

Charudattan (1985) has aptly identified the areas of mycoherbicide development that could benefit from genetic alteration of the pathogen. Charudattan (1985) believes our immediate needs for improving fungal strains for weed control can be met by random mutation and selection. Pesticide resistance is one example. The efficacy of Collego on northern jointvetch is reduced if benomyl is applied within 3 weeks after Collego treatment for control of rice blast caused by *Pyricularia oryzae* or for control of soybean anthracnose diseases (Smith, 1986). TeBeest (1984) induced mutation to benomyl resistance in spores of the Collego fungus by treatment with ethyl methanesulphonate. The mutants may eventually be used in a new and improved Collego with pesticide resistance.

Mutation to change survival capability of pathogens may be desirable in some instances. For example, fungi might be altered from sclerotium-formers to non-sclerotium-formers (Miller, Ford & Sands, 1987) so that rotation crops in a field in which a mycoherbicide was applied would not be at risk from a previous season's application. However, an understanding of the nature of the mutation would be necessary so that measures could be taken to ensure that reverse mutation to a sclerotium-former was not likely.

Mutation and selection might be adequate for recovering isolates with preferred capabilities in particular environments. For example, an ability to germinate and penetrate the host more rapidly when moisture is available would alleviate much concern over the duration of dew periods following inoculation. Midseason control of prickly sida by *Colletotrichum malvarum* could be achieved if mutants could be selected that express high virulence at high temperatures.

Virulence of pathogens might be increased by mutation and selection. Candidates for improved virulence include the Collego fungus on Indian jointvetch and *Colletotrichum coccodes* on velvetleaf. The current problem is that too little is known about the virulence factors of these fungi and consequently it is not known whether genetic material exists in the nuclei or cytoplasm that could contribute to dramatic changes in virulence. Of course, any time a mutation is selected for any trait the complete host range must be retested to detect other possible changes in phenotype of the pathogen.

Recombinant DNA technology provides a more elegant means of altering the DNA of a pathogen. Transformation of filamentous fungi requires the use of protoplasts (Turgeon & Yoder, 1985). Several vectors for gene transfer are available and have been used successfully in fungi. A number of fungi have been transformed, including several plant pathogens (Rodriguez & Yoder, 1987). Although incorporation of genetic material by these techniques is still somewhat random, greater control over the site and extent of genetic changes can be exercised compared to random mutation. It is possible to select isolates with desirable characteristics and evaluate the DNA chemically and physically. The accurate and subtle manipulations favour the prospects of taking virulence genes from one fungus and inserting them into another while still retaining specificity of the second. However, as far as we know, no genes for virulence have yet been cloned in plant pathogenic fungi.

Turgeon & Yoder (1985) postulate that mycoherbicides may be improved using genetic engineering techniques to increase virulence by manipulation of genes controlling toxin production. Transfer of genes controlling production of host-nonspecific toxins, along with genes for resistance to the toxin to protect recipient cells from self-destruction, could result in enhanced virulence in a pathogen that is specific for a target weed. So far, no genes controlling toxin production have been cloned successfully in plant pathogenic fungi, so the applications are years away.

These techniques may also be useful for increasing sporulation capacity and directing adaptation to adverse environments. Pesticide resistance has already been stably incorporated into pathogens by transformation techniques that did not alter pathogenicity. The β-tubulin genes cloned from *Colletotrichum graminicola* and *Neurospora crassa benomyl-resistant mutants were used by Panaccione, McKiernan & Hanay (1988) to transform C. graminicola*, a pathogen of corn and sorghum, to benomyl resistance. Extensive changes in an organism may not be worthwhile, since research and development costs of DNA cloning can be expensive. However, it should be more practical to insert genes from one organism into another organism that already produces the product, bypassing a rate-limiting step or improving growth characteristics of the producing organism (Dale, 1983).

Loss of virulence in culture is a pervasive problem in plant pathology. Undoubtedly, some reports of loss of virulence actually involve loss of viability or improper handling of pathogens and plant material. Some aspects may be controllable by manipulating physiological influences such as media, temperature, and illumination conditions under which cultures are maintained. Reversible loss of virulence has been attributed to changes in characters determined by cytoplasmic inheritance (Waller, 1977). Irreversible changes in isolates may be due to chromosomal mutations. The degree to which mycoviruses or plasmids are involved in loss of virulence is unknown. Alteration of nuclear ratios in vegetative heterokaryons may result in phenotypic changes in isolates, such as loss of virulence. Subculturing of hyphal tips could easily result in unconscious selection of variants due to unequal distribution of nuclei (Waller, 1977). The problem of loss of pathogenicity or virulence in culture can only be solved with certainty if the following are known:

- the particular characters that contribute to pathogenicity or virulence;
- which of those characters have been lost;
- how those characters are regulated (e.g. feedback inhibition, repression, induction);
- what genes are directly or indirectly involved in altering those characters;
- how the genes have been changed or regulated (e.g. repression, mutation, complementation).

Perhaps the new gene technology will help to resolve the problem of loss of virulence.

The paradox of mycoherbicides is that the specificity sought is often too narrow for the market aspirations. Luboa, Collego, Biomal and

Colletotrichum orbiculare would all be more valuable if they controlled additional weeds in the same genus or family as the target weed. Transformation technology may allow scientists to broaden selectively the number of hosts a pathogen can control, or at least enhance disease on hosts it naturally infects. Once again, success depends on an understanding of the mechanisms involved in resistance and susceptibility of the particular weed-pathogen system. A pathogenicity factor could be the absence of a molecule or process that might otherwise trigger active defence responses (Heath, 1987).

Another problem with potential mycoherbicides is a host range that is too broad. For example, *Pyricularia grisea* on crabgrass may affect rice, and *Gloeocercospora sorghi*, a pathogen of shattercane (*Sorghum bicolor*) and johnsongrass (*S. halepense*), can cause disease in corn and sorghum (*Sorghum bicolor*)(Mitchell, 1987). Species such as *Rhizoctonia*, *Sclerotium*, and *Pythium* have little host specificity, killing weed and crop plants without discrimination. In cases such as these it might be possible to genetically engineer crops with resistance to a mycoherbicide. The seed and mycoherbicide could be sold together as is currently proposed for some chemical herbicides. Currently, four permit applications for release into the environment of genetically engineered plants with resistance to chemical herbicides await review by the U.S. Animal and Plant Health Inspection Service (Anonymous, 1988). The plants are tobacco and tomatoes with tolerance to glyphosate or sulphonylurea. The approach of altering crops to be resistant to mycoherbicides does not have the implications of planned chemical overuse that has alarmed some environmentalists concerned about chemical herbicide resistance (Doyle, 1985). However, in the cases of *Rhizoctonia*, *Pythium* and *Sclerotium*, it does require a commitment of the land to the planting of crops that have incorporated resistance and a genetically manipulated inability of the fungi to overwinter. Conventional breeding might be used to incorporate this mycoherbicide resistance in crops, but tissue culture screening and recombinant DNA techniques would probably be faster. Durable resistance might be achieved by introducing several new defence mechanisms into the plant that are unmatched by fungal pathogenicity factors or by introducing one or two defence mechanisms that the fungus is incapable of evolving a means to overcome (Heath, 1987).

Concluding remarks

Considerable progress has been made during the past decade in use of fungi as mycoherbicides, and the pace of this progress appears to be accelerating. Thus far, only naturally occurring strains, indigenous to limited geographic sites, have been tried successfully. Many naturally occurring strains remain to be examined for potential, and the need for biological pesticides is increasing. There are many more site-specific weed problems arising as a result of the widespread use of broad-spectrum, chemical herbicides. Resistant species become weedy as competition from other weeds is removed. These resistant species are often closely related to the crop, and thus require a higher order of selectivity for control than existing products provide. An increasing role for biologicals can be expected from pesticide resistance problems and the need for better integrated pest management strategies that utilize biologicals. In addition, the need for biologicals is increasing because of the progressively more stringent regulatory requirements for registration and re-registration of chemical pesticides. Many products facing re-registration, especially those approaching patent expiration, do not have enough market potential to warrant the expense of re-registration. At the same time, the regulatory climate for biologicals is becoming clearer and more conducive to advances in biotechnology.

Perhaps a more important reason to develop biological pesticides is the persistent lack of confidence in synthetic chemicals by a discerning society. Pesticide contamination of food and ground water is undeniably an emotional issue that continues to mount despite the fact that society is faced daily with other more hazardous products in common use. Fear is not equated with risk. Our almost total dependence upon synthetic chemicals for pest control can be significantly reduced by increases in fundamental knowledge about pests and their natural enemies and by use of advances in biotechnology to enhance utility. It is thought by some that biologicals can and should become the primary means of pest control (Anonymous, 1987).

Modest progress in development of natural strains of fungi as mycoherbicides can be anticipated; there are many to explore and develop. However, all are likely to have innate biological limitations that affect their utility. Currently this is the case with those already commercialized or nearing commercialization. There have been a great number of naturally occurring strains researched for possible use as mycoherbicides, but only a small proportion have been developed to commercial products. Sometimes this is due to the absence of effective technology transfer or poor economic potential, but more often it is due to one or more biological deficiencies that may be surmountable by biotechnology. A fruitful area of focus in biotechnology research would be the commercial-scale production and conservation of spores of fungi that now can only be produced on solid substrates and have only been conserved with culture collection techniques. Considerable advances are expected from use of existing biotechnology in this area. Goals to improve virulence, host specificity, and environmental capability with genetic engineering techniques hold much promise. An understanding of the molecular basis of disease and intensive fundamental studies of the biology of pathogens at the organism and ecosystem levels will be needed to exploit rationally the existing opportunities in genetic engineering for pest control. The expectation of society to have non-chemical alternatives for pest control appears to be broadly based and deeply rooted. Thus it seems reasonable to expect greater public and private efforts in biotechnology to enhance fungi for control of weeds.

References

Adam, G., Zibuck, R. & Seebach, D. (1987). Total synthesis of (+)-gloeosporone: assignment of absolute configuration. *Journal of the American Chemical Society*, **109**, 6176-6177.

Allen, P. J. (1976). Spore germination and its regulation: control of spore germination and infection structure formation in the fungi. In *Encyclopedia of Plant Physiology*, vol 4, ed. R. Heitefuss & P. H. Williams, pp. 51-85. Springer-Verlag: New York.

Anderson, J. G. & Blain, J. A. (1980). Novel developments in microbial film reactors. In *Fungal Biotechnology*, ed. J. E. Smith, D. R. Berry & B. Kristiansen, pp. 125-152. Academic Press: New York.

Anonymous, (1987). *Research Briefings 1987*, Report of the Research Briefing Panel on Biological Control in Managed Ecosystems, National Academy of Sciences. National Academy Press: Washington, D.C.

Anonymous, (1988). APHIS receives permit applications for release of genetically engineered organisms. *Phytopathology News*, **22**, 72.

Antheunisse, J., de Bruin-Tol, J. W. & van der Pol-van Soest, M. E. (1981). Survival of microorganisms after drying and storage. *Antonie van Leeuwenhoek*, **47**, 539-545.

Auld, B. A., McRae, C. F. & Say, M. M. (1988). Possible control of *Xanthium spinosum* by a fungus. *Agriculture, Ecosystems and Environment*, **21**, 219-223.

Barfoed, H. C. (1981). Production of enzymes by fermentation. In *Essays in Applied Microbiology*, ed. J. R. Norris & M. H. Richmond, pp. 5/1-5/31. John Wiley & Sons: New York.

Bowers, R. C. (1986). Commercialization of Collego™ - an industrialist's view. *Weed Science*, **34** (suppl. 1), 24-25.

Boyette, C. D. (1986). Evaluation of *Alternaria crassa* for biological control of jimsonweed: host range and virulence. *Plant Science*, **45**, 223-228.

Boyette, C. D. (1987). Biocontrol of hemp sesbania (*Sesbania exaltata* (Raf.) Cory.) by an induced host range alteration of *Alternaria crassa*. *Weed Science Society of America Abstracts*, **27**, 48.

Boyette, C. D., Templeton, G. E. & Smith Jr, R. J. (1979). Control of winged waterprimrose (*Jussiaea decurrens*) and northern jointvetch (*Aeschynomene virginica*) with fungal pathogens. *Weed Science*, **27**, 497-501.

Boyette, C. D., Walker, H. L. & Sciumbato, G. L. (1983). Production of floating inocula of *Rhizoctonia solani* and *R. oryzae*. *Phytopathology*, **73**, 842 (abstract).

Brooker, N. L. (1988). Production and regeneration of *Alternaria crassa* protoplasts. *M.S. Thesis, University of Arkansas, Fayetteville*.

Charudattan, R. (1985). The use of natural and genetically altered strains of pathogens for weed control. In *Biological Control in Agricultural IPM Systems*, ed. M. A. Hoy & D. C. Herzog, pp. 347-372. Academic Press, Inc.: New York.

Charudattan, R. (1986). Integrated control of waterhyacinth (*Eichhornia crassipes*) with a pathogen, insects, and herbicides. *Weed Science*, 34 (suppl. 1), 26-30.

Charudattan, R., Linda, S. B., Kluepfel, M. & Osman, Y. A. (1985). Biocontrol efficacy of *Cercospora rodmanii* on waterhyacinth. *Phytopathology*, 75, 1263-1269.

Chibata, I., Tosa, T. & Sato, T. (1986). Methods of cell immobilization. In *Manual of Industrial Microbiology and Biotechnology*, ed. A. L. Demain & N. A. Solomon, pp. 217-229. American Society for Microbiology: Washington, D.C.

Churchill, B. W. (1982). Mass production of microorganisms for biological control. In *Biological Control of Weeds with Plant Pathogens*, ed. R. Charudattan & H. L. Walker, pp. 139-156. John Wiley & Sons: New York.

Dale, J. (1983). Application of the principles of microbial genetics to biotechnology. In *Principles of Biotechnology*, ed. A. Wiseman, pp. 56-93. Chapman & Hall: New York.

Daniel, J. T., Templeton, G. E., Smith Jr, R. J. & Fox, W. T. (1973). Biological control of northern jointvetch in rice with an endemic fungal disease. *Weed Science*, 21, 303-307.

de Jong, M. D. (1988). Risk to fruit trees and native trees due to control of black cherry (*Prunus serotina*) by silverleaf fungus (*Chondrostereum purpureum*). *Dissertation, Landbouwuniversiteit te Wageningin, Wageningen, Netherlands.*

Demain, A. L. & Solomon, N. A. (1986). *Manual of Industrial Microbiology and Biotechnology.* American Society for Microbiology: Washington, D.C.

Dietz, A. & Churchill, B. W. (1985). Culture preservation and stability. In *Comprehensive Biotechnology*, vol 2, *The Principles of Biotechnology: Engineering Considerations*, ed. C. L. Cooney & A. E. Humphrey, pp. 37-49. Pergamon Press: New York.

Doyle, J. (1985). Biotechnology's harvest of herbicides. *GeneWatch*, 2, 1-5.

Durrell, L. W. (1964). Composition and structure of walls of dark fungus spores. *Mycopathologia et Mycologia Applicata*, 23, 339-345.

Flick, E. W. (1984). *Cosmetic and Toiletry Formulations.* Noyes Publications: Park Ridge, New Jersey.

French, R. C. (1985). The bioregulatory action of flavor compounds on fungal spores and other propagules. *Annual Review of Phytopathology*, 23, 173-199.

Griffith, C. A. (1970). *Annual Report.* Samuel Roberts Noble Foundation, Inc.: Ardmore, Oklahoma.

Griffiths, E. (1981). Iatrogenic plant diseases. *Annual Review of Phytopathology*, 19, 69-82.

Heath, M. C. (1987). Host vs. nonhost resistance. In *Molecular Strategies for Crop Protection*, ed. C. J. Arntzen & C. Ryan, pp. 25-34. Alan R. Liss, Inc.: New York.

Hill, L. R. (1981). Preservation of microorganisms. In *Essays in Applied Microbiology*, ed. J. R. Norris & M. H. Richmond, pp. 2/1-2/31. John Wiley & Sons: New York.

Hoch, H. C., Staples, R. C., Whitehead, B., Comeau, J. & Wolf, E. D. (1987). Signaling for growth orientation and cell differentiation by surface topography in *Uromyces*. *Science*, 235, 1659-1662.

Ignoffo, C. M. (1981). Living microbial insecticides. In *Essays in Applied Microbiology*, ed. J. R. Norris & M. H. Richmond, pp. 7/1-7/31. John Wiley & Sons: New York.

Kenney, D. S. (1986). Devine®—The way it was developed—an industrialist's view. *Weed Science*, **34** (suppl. 1), 15-16.

Kollatakudy, P. E. (1985). Enzymatic penetration of the plant cuticle by fungal pathogens. *Annual Review of Phytopathology*, **23**, 223-250.

Kristiansen, B. & Bu'lock, J. D. (1980). Developments in industrial fungal biotechnology. In *Fungal Biotechnology*, ed. J. E. Smith, D. R. Berry & B. Kristiansen, pp. 203-223. Academic Press: New York.

Lax, A. R., Templeton, G. E. & Meyer, W. L. (1985). Isolation, purification, and biological activity of a self-inhibitor from conidia of *Colletotrichum gloeosporioides*. *Phytopathology*, **75**, 386-390.

Leslie, J. F. (1983). Some genetic techniques for *Gibberella zeae*. *Phytopathology*, **73**, 1005-1008.

Mackill, A. O. & Bonman, J. M. (1986). New hosts of *Pyricularia oryzae*. *Plant Disease*, **70**, 125-127.

Macko, V., Staples, R. C., Renwick, J. A. A. & Pierone, J. (1972). Germination self-inhibitors of rust uredospores. *Physiological Plant Pathology*, **2**, 347-355.

Miller, T. L. & Churchill, B. W (1986). Substrates for large-scale fermentations. In *Manual of Industrial Microbiology and Biotechnology*, ed. A. L. Demain & N. A. Solomon, pp. 122-136. American Society for Microbiology: Washington D.C.

Miller, R. V., Ford, E. J. & Sands, D. C. (1987). Induced auxotrophic and non-sclerotial isolates of *Sclerotinia sclerotiorum*. *Phytopathology*, **77**, 1720 (abstract).

Misaghi, I. J. (1982). *Physiology and Biochemistry of Plant-Pathogen Interactions*. Plenum Press: New York.

Mitchell, J. K. (1987). Host and environmental factors used to model johnsongrass control with the bioherbicides *Colletotrichum graminicola* and *Gloeocercospora sorghi*. *Phytopathology*, **77**, 1771 (abstract).

Mortensen, K. (1988). The potential of an endemic fungus, *Colletotrichum gloeosporioides*, for biological control of round-leaved mallow (*Malva pusilla*) and velvetleaf (*Abutilon theophrasti*). *Weed Science*, **36**, 473-478.

O'Connell, R. J., Bailey, J. A. & Richmond, D. V. (1985). Cytology and physiology of infection of *Phaseolus vulgaris* by *Colletotrichum lindemuthianum*. *Physiological Plant Pathology*, **27**, 75-98.

Panaccione, D. G., McKiernan, M. & Hanay, R. M. (1988). *Colletotrichum graminicola* transformed with homologous and heterologous benomyl-resistance genes retains expected pathogenicity to corn. *Molecular Plant-Microbe Interactions*, **1**, 113-120.

Peberdy, J. F. (1979). Fungal protoplasts: isolation, reversion, and fusion. *Annual Review of Microbiology*, **33**, 21-39.

Peberdy, J. F. (1983). Genetic recombination in fungi following protoplast fusion and transformation. In *Fungal Differentiation: A Contemporary Synthesis*, vol 4, ed. J. E. Smith, pp. 559-581. Marcel Dekker, Inc.: New York.

Quimby Jr., P. C., Fulgham, F. E., Boyette, C. D. & Connick Jr., W. J. (1988). An invert emulsion replaces dew in biocontrol of sicklepod - a preliminary study. In *Pesticide Formulations and Application Systems* Volume 8, ASTM STP 980, ed.

150   George E. Templeton & Dana K. Heiny

D. A. Hovde & G. B. Beestman, pp.264-270. American Society for Testing and Materials: Philadelphia.

Ridings, W. H. (1986). Biological control of stranglervine in citrus—a researcher's view. *Weed Science*, **34** (suppl. 1), 31-32.

Rodriguez, R. J. & Yoder, O. C. (1987). Selectable genes for transformation of the fungal plant pathogen *Glomerella cingulata* f. sp. *phaseoli* (*Colletotrichum lindemuthianum*). *Gene*, **54**, 73-81.

Scheepens, P. C. (1980). Bestrijding van de Amerikaanse volgelkers met pathogene schimmels; een perspectief. *Verslag*, vol **29**. CABO: Wageningen, The Netherlands.

Smith Jr., R. J. (1986). Biological control of northern jointvetch (*Aeschynomene virginica*) in rice (*Oryza sativa*) and soybeans (*Glycine max*)—a researcher's view. *Weed Science*, **34** (suppl. 1), 17-23.

Sutton, B. C. (1980). *The Coelomycetes: Fungi Imperfecti with Pycnidia, Acervuli and Stromata*. Commonwealth Mycological Institute: Kew, Surrey, UK.

TeBeest, D. O. (1984). Induction of tolerance to benomyl in *Colletotrichum gloeosporioides* f. sp. *aeschynomene* by ethyl methanesulfonate. *Phytopathology*, **74**, 864, (abstract).

TeBeest, D. O. & Templeton, G. E. (1985). Mycoherbicides: progress in the biological control of weeds. *Plant Disease*, **69**, 6-10.

Templeton, G. E. (1976). *Colletotrichum malvarum* spore concentration and agricultural process. U. S. Patent 3,999,973.

Templeton, G. E. (1987). Mycoherbicides - achievements, developments and prospects. In *Proceedings of the Eighth Australian Weeds Conference*, September 21-25 1987, pp. 489-497. Weed Society of New South Wales: Sydney, Australia.

Trione, E. J. (1981). Natural regulators of fungal development. In *Plant Disease Control: Resistance and Susceptibility*, ed. R. C. Staples & G. H. Toenniessen, pp. 85-102. John Wiley & Sons: New York.

Trollope, D. R. (1975). The preservation of bacteria and fungi on anhydrous silica gel: an assessment of survival over four years. *Journal of Applied Bacteriology*, **38**, 115-120.

Turgeon, G. & Yoder, O. C. (1985). Genetically engineered fungi for weed control. In *Biotechnology: Applications and Research*, ed. P. N. Cheremisinoff & R. P. Ouellette, pp. 221-230. Technomic Publishing Co., Inc.: Lancaster, Pennsylvania.

Walker, H. L. (1981). Granular formulation of *Alternaria macrospora* for control of spurred anoda (*Anoda cristata*). *Weed Science*, **29**, 342-345.

Walker, H. L. & Boyette, C. D. (1985). Biocontrol of sicklepod (*Cassia obtusifolia*) in soybeans (*Glycine max*) with *Alternaria cassiae*. *Weed Science*, **33**, 212-215.

Walker, H. L. & Connick Jr., W. J. (1983). Sodium alginate for production and formulation of mycoherbicides. *Weed Science*, **31**, 333-338.

Walker, H. L. & Riley, J. A. (1982). Evaluation of *Alternaria cassiae* for the biocontrol of sicklepod (*Cassia obtusifolia*). *Weed Science*, **30**, 651-654.

Walker, H. L. & Templeton, G. E. (1978). *In vitro* production of a phytotoxic metabolite by *Colletotrichum gloeosporioides* f. sp. *aeschynomene*. *Plant Science Letters*, **13**, 91-96.

Waller, J. M. (1977). The culture of fungal pathogens of plants. In *Crop Protection Agents—Their Biological Evaluation*, ed. N. R. McFarlane, pp. 519-529. Academic Press: New York.

Weidemann, G. J. & Templeton, G. E. (1988). Efficacy and soil persistence of *Fusarium solani* f. sp. *cucurbitae* for control of Texas gourd (*Cucurbita texana*). *Plant Disease*, **72**, 36-38.

Wilson, C. L. (1965). Consideration of the use of persimmon wilt as a silvicide for weed persimmons. *Plant Disease Reporter*, **49**, 789-791.

Winder, R. S. & Van Dyke, C. G. (1987). The effect of various adjuvants in biological control of johnsongrass (*Sorghum halepense* (L.) Pers.) with the fungus *Bipolaris sorghicola*. *Weed Science Society of America Abstracts*, **27**, 128.

Wymore, L. A., Watson, A. K., & Gotlieb, A. R. (1987). Interaction between *Colletotrichum coccodes* and thidiazuron for control of velvetleaf (*Abutilon theophrasti*). *Weed Science*, **35**, 377-383.

Yaegashi, H. & Hebert, T. T. (1976). Perithecial development and nuclear behavior in *Pyricularia*. *Phytopathology*, **66**, 122-126.

Zabriskie, D. W., Armiger, W. B., Phillips, D. H. & Albano, P. A. (1982). *Trader's Guide to Fermentation Media Formulation*, 2nd edition. Trader's Protein: Memphis, Tennessee.

Zhdanova, N. N. & Pokhodenko, V. D. (1974). The protective properties of fungal melanine pigment of some soil Dematiaceae. *Radiation Research*, **59**, 221 (abstract).

Chapter 7

Fungi as biological control agents for plant parasitic nematodes

B. R. Kerry

*AFRC Institute of Arable Crops Research, Rothamsted Experimental
Station, Harpenden, Hertfordshire AL5 2JQ, UK*

Introduction

Plant parasitic nematodes spend part of their life cycle in soil or on the
root surface where they are exposed to parasitism and predation by a wide
range of organisms. In general, research on the effects of natural enemies
on nematode populations has concentrated on microbial parasites and pathogens and, in particular, nematophagous fungi (Kerry, 1987; Stirling,
1988). Because of increasing concern over health and environmental
hazards due to the production and use of nematicides, biological control
represents an attractive alternative, but has not received substantially increased support compared to other methods of nematode management.
Stirling (1988) states in his review of the subject that 'biological control is
in the invidious position of needing a successful example of the contrived
use of an antagonist to demonstrate to administrators and funding agencies that further work is worthy of support, while lacking the resources to
make that advance'. In recent years some commercially produced biological control agents have been available to growers but, in general, the large
application rates required, and their attendant costs, or unpredictable efficacy have restricted their use. Research on biological control of
nematodes generally lacks in-depth quantitative studies and little is known
of modes of action or of the epidemiology of the parasitic agent. Such information is essential if rational use is to be made of these control agents.

There have been many recent reviews on the biological control of
nematodes (Kerry, 1987; Sayre & Starr, 1988; Stirling 1988) and this paper
is not intended to provide a comprehensive account of published work.
Rather it is a report of the progress and problems that have occurred in
about the last 15 years with emphasis on studies on the biological control
of cyst nematodes in arable crops, completed mainly at Rothamsted.

In discussions on the biological control of plant parasitic nematodes it
is useful to separate two types of control, natural and induced (Kerry,
1987). Natural control occurs in soils where indigenous nematophagous

fungi are sufficiently numerous that they prevent susceptible nematode species from multiplying. The control is not the result of a specific introduction but results from an increase in the density of fungal parasites in soil over a period of several years. Such control is usually associated with nematodes on perennial crops or those grown in monocultures in which the continuous presence of the nematode host supports the development of its fungal parasites. Biological control is induced when fungal parasites are applied once and successfully established in soil, or applied repeatedly as a microbial pesticide, to control a specific nematode pest. Fungi that are effective natural control agents may have few characteristics required of an agent that is added to soil to induce control (Kerry, 1987). The latter agent must be (a) easy and cheap to produce and its application require little deviation from standard farm practices, (b) capable of rapid colonization and/or persistence in soil, (c) capable of killing several major pest species so that potential markets are large enough to encourage commercial development, but should have no deleterious effects on non-target organisms, and (d) able to retain viability in storage without special facilities. In common with natural agents, fungi used to induce biological control must be virulent and provide predictable control below economic thresholds, be compatible with agrochemicals, and be safe. No organism studied so far has all of these characteristics but some show considerable potential for inducing biological control (Morgan-Jones & Rodriguez-Kabana, 1987) and have proved capable of providing long-term natural control of cyst and root-knot nematodes (Stirling, McKenry & Mankau, 1979; Kerry, Crump & Mullen, 1982).

Nematode target

The worldwide economic importance of cyst and root-knot nematodes, has meant that most attempts at biological control have concerned these pests. Few studies have been done on other groups, such as the ectoparasitic species that spend all their life in soil and presumably are continuously exposed to natural enemies and more easily controlled than endoparasites that spend part of their life protected within roots and have greater multiplication rates. The obligate parasite, *Hirsutella rhossiliensis* has been associated with declining populations of *Criconemella xenoplax*, an ectoparasitic nematode, on peach but it is a weak saprotrophic competitor that has not established in soil from introductions (Jaffee & Zehr, 1982, 1984). It is essential to consider the nematode target for control in the selection of an agent that may only kill one specific stage in the pest's life cycle. For example, nematode-trapping fungi have frequently been tested for their efficacy in the control of cyst and root-knot nematodes. However, they are only capable of killing second-stage juveniles or males as they migrate through soil; both are short-lived stages and it has proved difficult to manipulate the fungi so that the periods of nematode migration and trap formation coincide. Control has tended to be erratic and it would appear that other agents are likely to be more reliable (Morgan-Jones & Rodriguez-Kabana, 1987).

Cyst and root-knot nematodes have similar life cycles in that second-stage juveniles invade the roots and establish specialised feeding cells in susceptible hosts on which the nematode depends for its nutrition. Once the nematode has initiated such a response it becomes sedentary and moults three times before becoming adult. Females are saccate and, in the case of cyst nematodes, they swell to such an extent that they rupture the root cortex and are exposed in the soil; females of root-knot nematodes are embedded in gall tissue that results from nematode infection and usually only the egg mass is exposed (Southey, 1978). The number of generations in a growing season depends on the nematode species, host crop and soil temperature. The fecundity of females is also affected by environmental factors but, in general, populations of cyst nematodes can increase 50-fold and root-knot nematodes 1000-fold in favourable conditions in a growing season. Such factors are important in determining the efficacy of a control agent. For example, the parasitic fungus *Dactylella oviparasitica* may effectively control root-knot nematodes (*Meloidogyne* spp.) on peaches in California where females produce approximately 120 eggs per egg mass, but not in adjacent vineyards where the nematodes produce egg masses containing up to 900 eggs (Stirling *et al.*, 1979).

The close relationship between the nematode and its host, in which the second-stage juvenile dies if it fails to initiate a feeding cell in the root, has meant that cyst nematodes tend to have restricted host ranges among agricultural crops. Although some root-knot nematodes tend to have wider host ranges than cyst nematodes, there are marked differences between species. Hence, the rotation of susceptible and non-host crops is an effective method of control and remains the main management tool for many of these pests.

Second-stage juveniles invade close to the root tip and, if present in sufficient numbers, the root may cease to grow. The root systems of damaged plants are reduced in size and may fail to penetrate beyond the surface layers of the soil. These root systems are less efficient at taking up water and nutrients from soil and infected plants are stunted, often chlorotic, and may wilt in dry conditions. Most damage is done when the second-stage juveniles invade young seedlings and affect root morphology. Nematode feeding and the invasion in more mature plants is less damaging than in seedlings. Hence, yield losses are related to the preplanting nematode population density in the soil, which is the result of previous cropping history (Jones, 1978). At small population levels, plants are able to compensate for any damage caused to roots but further increases in the pre-planting population density may result in significant yield losses. The pre-cropping nematode population at which a plant is no longer able to compensate for damage to its root system is defined as the tolerance limit; crop tolerance to cyst nematodes is usually low, less than 5 eggs g^{-1} soil (Seinhorst, 1986). All management strategies for these nematodes are aimed at reducing this initial inoculum so that yield losses are minimised.

Fungal parasites

More than 150 fungal species have been isolated from cyst and root-knot nematodes (Kerry, 1986; Stirling, 1988). Fungi in the plant or soil may colonise the nematode female through its natural openings or by penetrating the cuticle (Kerry, 1988a). More cysts than females of *Heterodera glycines* were colonised by fungi and a wider range of species was isolated from cysts on roots and in soil than from developing females on roots (Gintis, Morgan-Jones & Rodriguez-Kabana, 1983). Few of these fungi have been tested for their parasitic status or their potential as biological control agents.

Tribe (1977a) separated the fungi that were isolated from cyst nematodes into those that parasitised females and those that attacked eggs. In general, those that were isolated from females were obligate parasites. However, fungi such as *Verticillium chlamydosporium*, *Paecilomyces lilacinus* and *Dactylella oviparasitica* that have been isolated from the eggs of root-knot and cyst nematodes have all been found to attack young females; these fungi are all facultative parasites capable of saprophytic growth in soil and in the rhizosphere. Clearly, distinguishing between fungal parasites on the basis of the stage of nematode attacked can be misleading, particularly as the main effect of *V. chlamydosporium* on nematode multiplication was considered to be dependent on the number of females parasitised and not on the number of eggs (Kerry, 1988b).

Three species of Mastigomycotina have been identified from cyst nematode females: *Catenaria auxiliaris*, *Nematophthora gynophila*, and an undescribed Lagenidiaceous fungus. Their taxonomic status and life cycles have been partially described (Tribe, 1977b; Kerry & Crump, 1980); all are believed to be obligate parasites that infect by motile zoospores. Parasitism by these fungi usually results in the total destruction of the female whose body content is eventually replaced by a mass of resting spores. Because the body wall of the female is also destroyed, these spores are exposed in soil and are readily dispersed by water movement and the activities of other soil organisms. Most observations have been made on *N. gynophila* because this fungus is known to be an important parasite involved in the natural control of the cereal cyst nematode in many soils in Northern Europe. In studies made in transparent observation chambers

(Crump & Kerry, 1977), *N. gynophila* destroyed females of the cereal cyst nematode in about seven days at 13°C and all parasitised females broke up and disappeared from the root surface. Infection of cereal cyst nematode in the field was much affected by rainfall in June, July and August when females were exposed to parasitism on roots (Kerry *et al.*, 1982). Moisture levels in soil have a marked effect on the movement of the zoospores, which are most active in pores that are filled with water. The fungus has been found in many European and some soils from USA infested with cyst nematodes but not in the Australian wheat belt, where conditions were probably too dry (Kerry, 1988a). *N. gynophila* attacks females of *Heterodera* but not *Globodera* species and has only been recorded from root-knot nematodes that produce small galls, so that females are exposed on the root surface.

Nematophthora gynophila has also been found in soils suppressive to beet and soybean cyst nematodes but the extent of the control exerted is not clear (Crump & Kerry, 1981; Kerry, 1987). The estimation of parasitism is laborious and time consuming because of the fragility of infected females and their rapid destruction, which makes them difficult to extract intact from soil. The fungus produces thick-walled resting spores that remain in soil for several years. These spores can be extracted from soil and were found to be more numerous in soils suppressive to cereal and beet-cyst nematodes than in non-suppressive soils (Crump & Kerry, 1981; Kerry & Andersson, 1983).

The lack of an *in vitro* method of production for any of the obligately parasitic Mastigomycotina pathogens attacking cyst nematodes has made them difficult to study and research on their potential as biological control agents is limited. Their role in the natural control of some cyst nematode pests, as outlined above, has been established but the difficulties in obtaining sufficient inoculum to induce biological control has meant that there has been only one report where these fungi have been added to soil (Stirling & Kerry, 1983). In contrast, facultative parasites are readily grown on a range of media and as a consequence are more easily manipulated and have been studied in more detail than the obligate parasites. *Verticillium chlamydosporium* is the major facultative parasite involved in the decline of cereal-cyst nematode populations under monocultures of susceptible cereals (Kerry *et al.*, 1982). The fungus is frequently found in soils containing *Nematophthora gynophila* and has been recorded throughout Europe, and in Australia and the USA (Kerry, 1988a). *V. chlamydosporium* has been isolated from females or eggs of *Heterodera-avenae* (Kerry & Crump, 1977), *H. schachtii* (Bursnall & Tribe, 1974), *H. glycines*, (Gintis, Morgan-Jones & Rodriguez-Kabana, 1983), *Globodera rostochiensis* (Roessner, 1987) and *Meloidogyne arenaria* (Godoy, Rodriguez-Kabana & Morgan-Jones, 1983a), and in laboratory tests has infected eggs of these nematodes and *G. pallida* (Crump, personal communication) and *M. incognita* (Kerry, unpublished). The fungus produces characteristic chlamydospores that can be extracted using the same method as that for *N. gynophila* resting spores (Crump & Kerry, 1981). It is more abundant in suppressive than in non-suppressive soils but is present in soil as hyphae and conidia as well as chlamydospores and can grow saprotrophically in soil and rhizosphere in the absence of nematode hosts (Kerry, Simon & Rovira, 1984). When applied to soil it significantly reduced populations of *M. arenaria* on squash (Godoy *et al.*, 1983a) and *H. avenae* on wheat (Kerry *et al.*, 1984).

Paecilomyces lilacinus is the only facultative parasite of females and eggs that has been tested in the field. It is generally accepted that if these fungi are introduced into soil they will require an energy source to help them overcome competition from the residual soil microflora (Papavizas & Lewis, 1981). Conidia and hyphae of the actively growing fungus have been added to soil on rice grain as an energy source at rates equivalent to $0 \cdot 4$ t ha^{-1}. Significant reductions in populations of *Globodera rostochiensis* (Davide & Zorilla, 1983), *Meloidogyne* spp. and *Tylenchulus semipenetrans* (Jatala, 1986) have sometimes been achieved. However, too often trials have been done with inadequate controls (see below). The application of biological control agents with the planting material allows their application rates to be reduced in field crops, where large areas have to be treated. Facultative parasites that colonise the rhizosphere, such as *V. chlamydosporium* (Kerry *et al*, 1984) and *P. lilacinus* (Hewlett *et al.*, 1988) offer the possibility of being applied as seed treatments. Application of *P. lilacinus* on potato tubers reduced infestations of *Meloidogyne incognita* (Canto-Saenz & Kaltenbach, 1984).

Paecilomyces lilacinus is being produced in the Philippines and sold under the trade name, Bioact®. However, this species has been associated with human eye and skin disorders (Minogue *et al.*, 1984). Although the isolate parasitizing nematodes may be different from that affecting humans, there is an urgent need for proper toxicological tests. Natural control does not necessarily imply safe control.

Natural biological control

Populations of at least six species of *Heterodera* and *Globodera* and one of *Meloidogyne* have been reported to decline under monocultures or perennial crops (Kerry, 1987). Fungal parasites of females and eggs have been considered responsible for the decline in nematode populations on susceptible crops. The importance of these parasites has been demonstrated directly, by the extraction of large numbers of infected nematodes from soil, and indirectly, by suppressing their activity through the use of selective biocides. Such methods were used to determine the nature and extent of the natural control of the cereal-cyst nematode (*Heterodera avenae*). This enabled Kerry, Crump & Mullen (1980) to conclude that fungal parasites were responsible for killing about 95% of the females and eggs of this species and were the major factor limiting multiplication in suppressive soils. Methods used to determine the nature and extent of nematode suppression in soil have been reviewed elsewhere (Kerry, 1986).

Although natural control may be very effective in a range of soil types, it is slow to establish and it may take up to five years before the nematode is adequately controlled. In microplot studies at Rothamsted, cereal-cyst nematode infestations in four soils have declined from > 400 eggs g^{-1} soil to < 5 eggs g^{-1} in five years of continuous susceptible cropping and populations have remained small under a further three cereal crops (Kerry, unpublished data). Control below the economic threshold has only been demonstrated for the cereal-cyst nematode in Northern Europe. Other nematode species such as *Heterodera schachtii* establish an equilibrium population with their parasites that is above this threshold. Hence, natural control must be supported by other control methods if non-damaging nematode populations are to be maintained.

In cereal-cyst nematode suppressive soils, female nematodes develop on roots but few of these survive to form cysts full of healthy eggs. Parasitism occurs on the root surface where most females are killed (Kerry, 1980). Further parasitism of eggs in cysts dispersed in soil by cultivations after harvest was considered insignificant. The role of the plant in determining the level of fungal parasitism has not been investigated but it is known that hosts differ markedly in the distribution and numbers of nematode females that develop on their roots. Both of these factors could significantly affect the efficacy of biological control agents.

Induced biological control

Compared with the numbers of compounds (approximately 10,000) screened each year for nematicide activity by several major chemical companies, efforts to identify useful biological control agents are insignificant and few organisms have been adequately tested (Kerry, 1987). There is an urgent need to make extensive surveys, particularly in the centres of origin of pest species, and to develop proper bioassay procedures.

Stirling (1988) identified three important components for testing the efficacy of fungi introduced into soil:

- The test should include adequate controls; if the fungus has to be added to soil on an energy source then the substrate without the fungus, and preferably the autoclaved, colonised substrate or the substrate colonised by an ineffective isolate of the fungus, should also be compared with the untreated control. Too often in the literature reports of efficacy have proved impossible to verify because

the effect of the fungus cannot be separated from that of the substrate. Although the final nematode population can be used as a
measure of the activity of nematophagous fungi applied to soil, it
does not necessarily indicate that the fungus has caused the reduction in nematode multiplication. The energy source can have
marked effects on the soil microflora and these effects will differ
with soil conditions making interpretation of experiments difficult
(Baker, Elad & Chet, 1984).

• The fungus should be re-isolated from nematodes at the end of the
experiment. As induced biological control of a nematode is still
rarely predictable, it is incumbent upon nematologists to verify that
any control in tests is related to the level of parasitism observed.

• It would be useful to re-isolate the fungus from soil to determine
its growth and survival during the test.

Few experiments described in the literature have satisfied all these
criteria and most have not satisfied any. This may not be surprising
because of the many difficulties involved.

At Rothamsted over 100 different isolates of *Verticillium
chlamydosporium* have been collected from cyst and root-knot nematode
infested soils and have proved to vary greatly in many aspects of their biology including their parasitic ability. Exposing nematode eggs to the
fungus on water agar and measuring the numbers colonised after three
weeks at 18°C gave reliable estimates of the relative parasitic abilities of
different isolates (Irving & Kerry, 1986). However, levels of parasitism on
agar are unlikely to be similar to those in soil and *V. chlamydosporium* is
known to infect female nematodes as well as eggs. When oilseed rape infected with *Heterodera schachtii* was grown in an organic potting compost
extensively colonised by three isolates of *V. chlamydosporium*, several effects were observed in the development of females and eggs of the
nematode in comparison with uncolonised compost (Table 7.1). The fungus significantly reduced the numbers and size of females, and the
numbers of eggs produced on the roots, and increased the levels of parasitism. Kerry (1988b) showed that the proportion of *H. schachtii* cysts
colonised by the fungus on oilseed rape roots was not related to the level
of biological control but that the timing of infection was critical. If 40%
of females were infected within two weeks of their emergence on the root
surface then the population failed to multiply. As *V. chlamydosporium* can
affect the development of female cyst nematodes at a number of stages
during maturation, it is often difficult to relate the final nematode population to the level of fungal parasitism observed without detailed studies
which would be inappropriate for a routine screening test. Also, the effect of biological agents, in contrast to chemicals, is likely to be density

Table 7.1. Effect of three isolates of *Verticillium chlamydosporium* on the development of females and eggs of the beet cyst nematode on oilseed rape grown in compost colonized by the fungus.

Isolate	No. of females produced	Infected females (arcsin transformed %)	Eggs per. female	Infected eggs (arcsin transformed %)	Size of female (μm) length	breadth
control	106	12	401	6	830	630
2	21	29	32	79	550	375
35	34	42	123	87	635	440
66	66	33	220	34	690	485
±SED	16	7	47	5	45	37
	**	**	***	***	***	***

The control lacked fungus. Data analyzed by analysis of variance. Entries are the means of 5 replicates; ** & *** denote significant difference between means at $P = 0·01$ and $P = 0·001$ respectively. SED = standard error of the difference between means.

dependent and so the level of control achieved may depend on the size of the initial nematode infestation. There is a need for detailed studies of the epidemiology of the nematode-fungus interaction so that proper bioassays can be developed.

Some characteristics of obligate and facultative parasitic fungi greatly affect their potential as biological control agents (Table 7.2). Obligate parasites tend to have narrow host ranges that limit their commercial potential. However, in order to survive periods when hosts are scarce they often have thick-walled resting spores which are resistant to desiccation and are simpler to store and handle than a fungus that must be applied as a mycelium or conidial suspension. Obligate parasites are difficult to grow in large quantities and none of those found parasitizing cyst nematodes have been grown *in vitro*. As these fungi remain dormant until a host comes in contact with them there is no proliferation or colonization in soil and therefore no need for an energy source. The amount of inoculum required for nematode control must be added to soil and thoroughly mixed to ensure that large numbers of nematodes are killed. Because of the difficulties in producing sufficient inoculum, few tests have been made with obligate parasites. However, some tests have been done with the obligate bacterial parasite, *Pasteuria penetrans* (Stirling, 1984). Results with this organism suggest that if the spores can be thoroughly incorporated in soil, levels of control equivalent to those of current nematicides can be achieved. Also, the control with obligate parasites tends to be more pre-

Table 7.2. Characteristics affecting the biological control potential of obligate and facultative parasites

Parasite	Biological control potential
Obligate e.g. *Nematophthora gynophila* *Catenaria auxiliaris*	1. Host specific; may attack only a limited number of pest species. 2. Thick-walled resting spores produced that may be resistant to desiccation, have a long shelf-life, and be easy to handle. 3. Difficult to culture *in vitro*. 4. No active growth in soil and so the amount of inoculum required for control must all be added and be thoroughly incorporated. 5. No energy source required with inoculum.
Facultative e.g. *Verticillium chlamydosporium* *Paecilomyces lilacinus* *Dactyella oviparasitica*	1. Usually have a wide host range. 2. Some may produce resting structures e.g. chlamydospores. 3. Easily cultured *in vitro* including liquid fermentation. 4. Grow actively in soil and inoculum can be increased saprotrophically in soil and the rhizosphere; however, growth is affected by soil conditions. 5. An energy source is required to establish the fungus in soil.

dictable than with facultative parasites that are active in soil and more likely to be affected by soil conditions (Kerry, 1988b).

Verticillium chlamydosporium grows readily in shaken liquid culture and up to 7×10^7 conidia ml^{-1} were produced in 22 days at 19°C using standard laboratory media (Kerry, Irving & Hornsey, 1986). In such cultures few chlamydospores were formed although they were readily produced on solid media. There has been little research on the nutrient requirements of nematophagous fungi (see Kerry, 1984) and details of minimal media suitable for large scale production have not been published. Similarly, little work has been done on the development of formulations, and most fungi have been added to soil as actively growing mycelia and/or conidia on a suitable organic carrier (Kerry, 1987). *V. chlamydosporium* hyphae and conidia incorporated in alginate granules containing kaolin supported little growth in soil compared to those containing wheat bran (Kerry, 1988b). However, the fungus remained viable in both types of air-dried granules for at least six months at 5°C and produced colonies when placed on water agar (Kerry, unpublished). Few studies have examined the survival and growth of nematophagous fungi in soil. However, until the factors that affect facultative parasites during

Table 7.3. Proliferation and survival of two isolates of *Verticillium chlamydosporium* introduced into soil as aqueous suspensions of chlamydospores.

Soil	Colony forming units × 500 per g soil						
	Sampling occasion (weeks)						
	0	1	2	4	6	8	Mean±SED
Isolate 2							
silty loam	14	10	20	21	18	22	17·5
organic	32	37	107	96	95	49	69·3 ±4·7 ***
silty loam	34	4	34	20	14	30	22·7
Mean	26·7	17·0	53·7	45·7	42·3	33·7	
±SED			6·7				
Isolate 13							
silty loam	53	108	118	90	79	98	82·7
organic	58	156	163	57	95	81	101·7 ±9·2 NS
silty loam	40	131	102	89	98	99	93·2
Mean	50·3	131·7	127·7	78·7	90·7	92·3	
±SED			13·0				

Data analyzed by analysis of variance; entries are the means of three replicates. NS and *** denote no significant difference and a significant difference at $P=0·001$ respectively. SED = standard error of the difference between means.

their saprotrophic stage in soil are understood it seems unlikely that application rates can be significantly reduced or control made predictable. At Rothamsted, semi-selective media have been developed to re-isolate *V. chlamydosporium* from field soils (Kerry *et al.*, 1989) and enable quantitative studies to be made on the proliferation and survival of the fungus in field soils. It has proved possible to establish the fungus in soil from applications of actively growing mycelium on cereal grain, in sand-bran mixtures or dried in alginate granules; aqueous suspensions of chlamydospores can also be used without additional carbon sources. The proliferation and survival of two isolates of *V. chlamydosporium* were examined in three soils to which chlamydospores had been added in aqueous suspensions. In general, proliferation was best in the organic soil (Table 7.3) but the response to soil type depended on the isolate; isolate 13 was less affected than isolate 2 which grew less well in all soils. Isolate 13 proliferated rapidly and after 8 weeks was present in soil in

significantly larger numbers than were introduced. To study the colonization and growth of *V. chlamydosporium* in soil, granular formulations of
the fungus were prepared in alginate/wheat bran mixtures (Kerry *et al.*,
1989). An energy source (wheat bran) was essential for the establishment
of the fungus in soil but the extent of colonization also depended on the
fungal isolate introduced and the soil type. There are approximately
5×10^3 colony forming units g^{-1} soil in naturally suppressive soils (Kerry
et al., 1989). However, rates of application of 0·1% w/w required to establish this amount of inoculum are equivalent to 2·5 t ha^{-1} and are
impractical.

To measure hyphal growth, alginate granules were placed on nylon
meshes buried in soil. The hyphae grow through the mesh which can be
removed and stained (Lumsden, 1981; Kerry *et al.*, 1989). Hyphal growth
of several isolates of *V. chlamydosporium* was limited (Table 7.4) and it
seems likely that reductions in the numbers of granules applied will result
in insufficient soil colonization and inadequate nematode control. However, growth and survival are markedly affected by formulation and
production methods and different formulations to alginate/wheat bran
may support more extensive growth in soil.

There are a number of ways in which rates of application may be reduced. *Verticillium chlamydosporium* is known to colonise the rhizosphere
of several crop plants without producing lesions in the roots or affecting
plant growth (Kerry *et al.*, 1984). It may, therefore, be possible to increase
the amount of inoculum in soil on the root surface precisely where it is required to control nematodes. The rhizosphere was important in
prolonging the survival of the fungus in soil (Kerry, 1988b) and may allow
more extensive spread than was reported in non-rhizosphere soil. However, even the most aggressive rhizosphere colonizer, when applied as a
seed treatment, was not able to prevent the multiplication of the beet cyst
nematode. Application of the fungus as a 10 cm band along the row of a
sugar beet crop would reduce rates of application by 67%. Coosemans
(1988) was able to obtain further reductions in application rates by applying the fungus to a compost/soil mixture in paper cups which were
incubated for one month before being planted with sugar beet. The seedlings were transplanted six weeks later into a field infested with beet-cyst
nematode and populations of the pest were significantly reduced and
yields increased. At present it is not considered economical to use this
technique on sugar beet but similar methods could be used to protect
higher value crops such as cabbages. Improvements in production
methods and formulation should result in increased survival and growth
in soil and with better methods of application it should be possible to apply
these fungi at rates at least equivalent to those for fertilizers.

Table 7.4. Hyphal growth of *Verticillium chlamydosporium* introduced into field soil in alginate granules.

	Area (cm^2)	Extent (mm)
Mean*	1·25 ±0·20	5·4 ±0·47
Range	0·52 – 3·17	3·9 – 10·7

*entries show means of 15 replicates±SE.

Control strategies

Natural control has allowed farmers to shorten rotations of susceptible cereals in the presence of *Heterodera avenae*. Cereal crops are frequently grown in monocultures in many soils throughout Northern Europe in which nematode damage is rarely observed. Significant but not sufficient natural control has been observed to occur in soils infested with other cyst and root-knot nematodes but it remains to be demonstrated whether such suppressive soils can enhance chemical control or delay the selection of resistance breaking pathotypes (Kerry, 1986).

Manipulation of the residual soil microflora through the addition of selective soil amendments may enable the activity of nematophagous fungi to be selectively enhanced. Fungi parasitic on eggs produce chitinase and this is thought to be important in the penetration of nematode egg shells (Dackman, Chet & Nordbring-Hertz, 1988). Addition to soil of large amounts (2·5 t ha^{-1}) of bleached crustacean shells containing chitin significantly reduced the populations of root-knot nematodes and enhanced chitinase activity in soil (Godoy *et al.*, 1983b). In general, the large quantities of organic matter required to reduce nematode multiplication significantly limit this strategy to high value crops and those where small areas of soil need treatment. Manipulation of fungi for the biological control of nematode pests in arable crops may be more readily achieved through the addition of selected agents to induce biological control in soils where they are absent.

Strategies for the use of endoparasitic fungi that attack female nematodes and eggs are governed by the fact that these fungi do not prevent nematode invasion or the likelihood of substantial yield losses in the season following application. Their effect is similar to that of a nematode resistant cultivar in that they reduce the post-cropping nematode population left in soil. Hence, if these fungi are to be applied to a soil heavily infested with nematodes then a nematicide would also be necessary. Application of the nematicide, carbofuran, with the nematode-trapping fungus, *Arthrobotrys irregularis* gave better control of *Meloidogyne* spp. on cucumber than when either treatment was applied alone (B'Chir, Horrigue & Verlodt, 1983). Also, the nematicide, ethoprop, at sublethal doses increased infection of second-stage juveniles of *M. incognita* by the weakly pathogenic fungus, *Catenaria anguillulae* (Roy, 1982). However, the combined use of nematicides and the fungal parasites that attack nematode females and eggs have not been tested.

Although fungal parasites may not prevent nematode populations from multiplying, an agent that could reduce post-cropping populations by 50 or 70% would result in 20 or 60% reductions respectively in the length of rotations required before infestations were at non-damaging levels. Facultative parasites, such as *Verticillium chlamydosporium*, could be applied to small nematode infestations to prevent them building up to damaging levels. However, such a strategy would require sophisticated management and regular monitoring. *V. chlamydosporium* may survive in soil for 2 to 3 months from a single application (Kerry, 1988b) but it seems unlikely that sufficient inoculum would persist for more than a year to give long term nematode control. This would be particularly true for in row applications of the fungus in which subsequent cultivations would mix in untreated soil and dilute the inoculum. Hence, fungi such as *V. chlamydosporium* would need to be used as microbial pesticides applied to soil each time the crop is sown or planted.

Biological agents are not a replacement for nematicides and are unlikely to give such rapid and effective control. However, combined with resistant and tolerant cultivars, they could make an important contribution to integrated pest management systems for several important plant nematode pests and greatly decrease the current dependence on nematicides.

Conclusions

There is a need to make surveys for nematophagous fungi, preferably in soils where the pest nematode is indigenous and in soils where it causes less of a problem than expected. Such surveys should result in the isolation of new potential biological control agents (Cook & Baker, 1983) and in increasing knowledge of the factors affecting natural control. More examples of the natural control of cyst and root-knot nematodes are likely to be described and suppressive soils could be exploited in the development of integrated control programmes. However, current evidence suggests that opportunities for the manipulation of natural control are limited. The enhancement of the activities of nematophagous fungi in soils by the addition of even selected amendments such as chitin, require the application of too much material to make this strategy acceptable except in subsistence agriculture and some horticultural/amenity situations. However, more information is required on factors affecting the activity of indigenous nematophagous fungi in suppressive soils if only to ensure that agricultural practices do not remove the beneficial effects that have taken several years to establish.

Much interest is likely to continue in the development of biological control agents for application to soils where they are scarce or absent in order to induce biological control. Such research could lead to a commercial product and is likely to attract support from industry. However, there are still many problems to be resolved if predictable control is to be achieved at practical rates of application. Improved methods of production and formulation are essential and require close collaboration between biologists, microbiologists and production and formulation chemists. Research on the modes of action of nematophagous fungi is also required and may lead to the identification of toxins or enzymes essential for the infection process. Hence, novel nematicides might be identified such as the avermectins produced by *Streptomyces* spp. (Putter *et al.*, 1981) and a detailed understanding of the infection process could lead to opportunities for genetic manipulation, and the production of organisms more effective as biological control agents.

Biological control requires a multidisciplinary approach and, because success is likely to depend upon a detailed knowledge of the pest, agent, and soil interactions, problems will rarely be solved rapidly. However, continuing demands for a reduction in the use of nematicides and progress in understanding the factors affecting the activity of nematophagous fungi in soil should lead to such agents making significant contributions to future nematode management programmes.

References

Baker, R., Elad, Y. & Chet, I. (1984). The controlled experiment in the scientific method with special emphasis on biological control. *Phytopathology*, **74**, 1019-1021.

B'Chir, M. M., Horrigue, N. & Verlodt, H. (1983). Mise au point d'une méthode de lutte integrée, associant un agent biologique et une substance chimique, pour combattre les *Meloidogyne* sous-abris plastiques en Tunisie. *Mededelingen van de faculteit Landbouwwetenschappen Rijksuniversiteit, Gent*, **48**, 421-432.

Bursnall, L. A. & Tribe, H. T. (1974). Fungal parasitism in cysts of *Heterodera*. II. Egg parasites of *H. schachtii*. *Transactions of the British Mycological Society*, **62**, 595-601.

Canto-Saenz, M. & Kaltenbach, R. (1984). Effect of some fungicides on *Paecilomyces lilacinus* (Thom) Samson. In *Abstracts, First International Congress of Nematology*, p. 16. Guelph, Canada.

Cook, R. J. & Baker, K. F. (1983). *The Nature and Practice of Biological Control of Plant Pathogens*. American Phytopathological Society: St. Paul, Minnesota.

Coosemans J. (1988). Rhizosphere inoculation with the parasitising fungus *Verticillium chlamydosporium* against *Heterodera schachtii*. In Abstracts, European Society of Nematologists Symposium, Uppsala, Sweden, 1988, p. 22.

Crump, D. H. & Kerry, B. R. (1977). Maturation of females of the cereal cyst-nematode on oat roots and infection by an *Entomophthora*-like fungus in observation chambers. *Nematologica*, **23**, 398-402.

Crump, D. H. & Kerry, B. R. (1981). A quantitative method for extracting resting spores of two nematode parasitic fungi, *Nematophthora gynophila* and *Verticillium chlamydosporium*, from soil. *Nematologica*, **27**, 330-339.

Dackman, C., Chet, I. & Nordbring-Hertz, B. (1988) Fungal parasitism of the cyst nematode *Heterodera schachtii*: infection and enzymatic activity. *FEMS Microbiological Ecology* (in press).

Davide, R. G. & Zorilla, R. A. (1983). Evaluation of a fungus, *Paecilomyces lilacinus* (Thom.) Samson, for the biological control of the potato cyst nematode *Globodera rostochiensis* Woll. as compared with some nematicides. *Philippine Agriculturist*, **66**, 397-404.

Gintis, B. O., Morgan-Jones, G. & Rodriguez-Kabana, R. (1983). Fungi associated with several developmental stages of *Heterodera glycines*, from an Alabama soybean field soil. *Nematropica*, **13**, 181-200.

Godoy, G., Rodriguez-Kabana, R. & Morgan-Jones, G. (1983a). Fungal parasites of *Meloidogyne arenaria* eggs in an Alabama soil. A mycological survey and greenhouse studies. *Nematropica*, **13**, 201-213.

Godoy, G., Rodriguez-Kabana, R., Shelby, R. A. & Morgan-Jones, G. (1983b). Chitin amendments for control of *Meloidogyne arenaria* in infested soil. II. Effects on microbial population. *Nematropica*, **13**, 63-74.

Hewlett, T. E., Dickson, D. W., Mitchell, D. J. & Kannwischer-Mitchell, M. E. (1988). Evaluation of *Paecilomyces lilacinus* as a biocontrol agent of *Meloidogyne javanica* on tobacco. *Journal of Nematology*, **20**, 578-584.

Irving, F. & Kerry, B. R. (1986). Variation between strains of the nematophagous fungus, *Verticillium chlamydosporium* Goddard. II. Factors affecting parasitism of cyst nematode eggs. *Nematologica*, **32**, 474-485.

Jaffee, B. A. & Zehr, E. I. (1982). Parasitism of the nematode *Criconemella xenoplax* by the fungus *Hirsutella rhossiliensis*. *Phytopathology*, **72**, 1378-1381.

Jaffee, B. A. & Zehr, E. I. (1984). Parasitic and saprophytic potentials of the nematode-attacking fungus, *Hirsutella rhossiliensis*. In *Abstracts, First International Congress of Nematology*, p. 46. Guelph, Canada.

Jatala, P. (1986). Biological control of plant-parasitic nematodes. *Annual Review of Phytopathology*, **24**, 453-489.

Jones, F. G. W. (1978). Plant nematodes: a neglected group of pests. *University of Leeds Review 1978*, **21**, 89-116.

Kerry, B. R. (1980). Biocontrol: Fungal parasites of female cyst nematodes. *Journal of Nematology*, **12**, 253-259.

Kerry, B. R. (1984). Nematophagous fungi and the regulation of nematode populations in soil. *Helminthological Abstracts Series B*, **53**, 1-14.

Kerry, B. R. (1986). An assessment of the role of parasites and predators in the regulation of cyst nematode populations. In *Cyst Nematodes*, NATO ASI Series, ed. F. Lamberti & C. E. Taylor, pp. 433-450. Plenum Press: London.

Kerry, B. R. (1987). Biological Control. In *Principles and Practice of Nematode Control in Crops*, ed. R. H. Brown &. B. R. Kerry, pp. 233-263. Academic Press: Sydney.

Kerry, B. R. (1988a). Fungal parasites of cyst nematodes. *Agriculture, Ecosystems and Environment*, **24**, 293-305.

Kerry, B. R. (1988b). Two micro-organisms for the biological control of plant-parasitic nematodes. In *Proceedings of the British Crop Protection Council Meeting, Brighton, 1988*, pp. 603-607.

Kerry, B. R. & Andersson, S. (1983). *Nematophthora gynophila* och *Verticillium chlamydosporium*, svampparasiter pa cystnematoder, vanliga i svenska jordar med frkornst au strasadescystnematoder. *Vaxtskyddsnotiser*, **47**, 79-80.

Kerry, B. R. & Crump, D. H. (1977). Observations on fungal parasites of females and eggs of the cereal cyst-nematode, *Heterodera avenae*, and other cyst nematodes. *Nematologica*, **23**, 193-201.

Kerry, B. R. & Crump, D. H. (1980). Two fungi parasitic on females of cyst-nematodes (*Heterodera* spp.). *Transactions of the British Mycological Society*, **74**, 119-125.

Kerry, B. R., Crump, D. H. & Mullen, L. A. (1980). Parasitic fungi, soil moisture and multiplication of the cereal-cyst nematode, *Heterodera avenae*. *Nematologica*, **26**, 57-68.

Kerry, B. R., Crump, D. H. & Mullen, L. A. (1982). Studies of the cereal-cyst nematode, *Heterodera avenae* under continuous cereals, 1975-1978. II. Fungal parasitism of nematode females and eggs. *Annals of Applied Biology*, **100**, 489-499.

Kerry, B. R., Irving, F. & Hornsey, J. C. (1984). Variation between strains of the nematophagous fungus, *Verticillium chlamydosporium* Goddard. II. Factors affecting growth *in vitro*. *Nematologica*, **32**, 461-473.

Kerry, B. R., Kirkwood, I. A., Barba, J. & de Leij, F. (1990). Establishment and survival of *Verticillium chlamydosporium* Goddard, a parasite of nematodes, in soil. *Plant Pathology*, in press.

Kerry, B. R., Simon, A. & Rovira, A. D. (1984). Observations on the introduction of *Verticillium chlamydosporium* and other parasitic fungi into soil for control of the cereal-cyst nematode *Heterodera avenae*. *Annals of Applied Biology*, **105**, 509-516.

Lumsden, R. D. (1981). A nylon fabric technique for studying the ecology of *Pythium aphanidermatum* and other fungi in soil. *Phytopathology*, **71**, 282-285.

Minogue, M. J., Francis, I. C., Quatermass, P. Kappagoda, M. B., Bradbury, R., Walls, R. S. & Motum, P. I. (1984). Successful treatment of fungal keratitis caused by *Paecilomyces lilacinus*. *American Journal of Ophthalmology*, **98**, 625-626.

Morgan-Jones, G. & Rodriguez-Kabana, R. (1987). Fungal biocontrol for the management of nematodes. In *Vistas on Nematology: A Commemoration of the 25th Anniversary of the Society of Nematologists*, eds J. A. Veech & D. W. Dickson, pp. 94-99. Society of Nematologists Inc.: Hyattsville, Maryland.

Papavizas, G. C. & Lewis, J. A. (1981). Introduction and augmentation of microbial antagonists for the control of soilborne plant pathogens. In *Biological Control in Crop Production*, BARC Symposium no. 5, ed. G. C. Papavizas, pp. 305-322. Allenheld & Osmum: Totowa, New Jersey.

Putter, I., MacConnell, J. G., Preiser, F. A., Haidri, A. A., Ristich, S. S. & Dybas, R. A. (1981). Avermectins: novel insecticides, acaricides and nematicides from a soil micro-organism. *Experientia*, **37**, 963-964.

Roessner, J. (1987). Pilze als antogonisten von *Globodera rostochiensis*. *Nematologica*, **33**, 106-118.

Roy, A. K. (1982). Effect of ethoprop on the parasitism of *Catenaria anguillulae* on *Meloidogyne incognita. Revue de Nematologie*, **5**, 335-336.

Sayre, R. M. & Starr, M. P. (1988). Bacterial diseases and antagonisms of nematodes. In *Diseases of Nematodes*, vol. I, ed. G. O. Poinar & H. B. Jansson, pp. 70-101. CRC Press Inc.: Boca Raton, Florida.

Seinhorst, J. W. (1986). Effects of nematode attack on the growth and yield of crop plants. In *Cyst Nematodes*, NATO ASI Series, ed. F. Lamberti & C. E. Taylor, pp. 191-209. Plenum Press: Sydney.

Southey, J. F. (1978). *Plant Nematology*. HMSO: London.

Stirling, G. R. (1984). Biological control of *Meloidogyne javanica* with *Bacillus penetrans. Phytopathology*, **74**, 55-60.

Stirling, G. R. (1988). Biological control of plant-parasitic nematodes. In *Diseases of Nematodes*, vol. II., ed. G. O. Poinar & H. B. Jansson, pp. 93-139. CRC Press Inc.: Boca Raton, Florida.

Stirling, G. R. & Kerry, B. R. (1983). Antagonists of the cereal cyst-nematode, *Heterodera avenae* Woll. in Australian soils. *Australian Journal of Experimental Agriculture and Animal Husbandry*, **23**, 318-324.

Stirling, G. R., McKenry, M. V. & Mankau, R. (1979). Biological control of root-knot nematodes (*Meloidogyne* spp.) on peach. *Phytopathology*, **69**, 806-809.

Tribe, H. T. (1977a). Pathology of cyst-nematodes. *Biological Reviews*, **52**, 477-507.

Tribe, H. T. (1977b). A parasite of white cysts of *Heterodera*: *Catenaria auxiliaris. Transactions of the British Mycological Society*, **69**, 367-376.

Chapter 8

Selection, production, formulation and commercial use of plant disease biocontrol fungi: problems and progress

R. D. Lumsden & J. A. Lewis

Biocontrol of Plant Diseases Laboratory, US Department of Agriculture, Agricultural Research Service, Beltsville, Maryland 20705, USA

Introduction

Most plant diseases are caused by fungi belonging to about 50 genera. It is estimated (Papavizas, 1985b) that in the United States alone, losses amount to at least $4 billion annually due to soil-borne fungal pathogens. Current and future legislative regulation, especially in the US, to restrict pesticide (including fungicide) use will compound these conservatively estimated losses even further (Acuff, 1988). Although restrictions are being imposed to protect food quality and the environment, chemicals are still our only recourse at present to prevent diseases of food and fibre crops. In recent years, the need to develop disease control measures as alternatives to chemicals has become a priority of scientists worldwide. Biological control, especially using fungal antagonists against fungal plant pathogens, has gained considerable attention and appears to be promising as a viable supplement or alternative to chemical control (Cook & Baker, 1983; Papavizas, 1981, 1985a).

In the broad sense of biological control (Cook & Baker, 1983), crop resistance to disease is the ideal means of controlling diseases. However, many crops have little or no resistance to certain plant pathogens, so that biological control using microorganisms against plant pathogens is an attractive method for disease control. Considerable research has gone into the study of biological control of plant diseases over the past 25 years, but to date, there are no fungal microbial agents for plant disease control registered in the US, and only two in Europe and the UK [*Peniophora gigantea* against *Heterobasidion annosum* of conifers (Rishbeth, 1975); *Trichoderma viride* against *Chondrostereum purpureum* of plum trees (Ricard, 1981)]. Of the 16 biocontrol microorganisms registered by the US Environmental Protection Agency (EPA), two are fungi for use against weeds and one a fungus against mites; two bacteria are registered against plant diseases; and the rest are registered for use against insects (Betz, Rispin & Schneider, 1987; W. Schneider, personal communication).

Considering the thousands of chemicals registered for agricultural pest control, biocontrol agents at present are insignificant components of pest management systems. Biological control agents have great potential for solving pest problems, but they are much like any other living organism useful to society. In their indigenous or wild state they are uncontrollable to a great extent; and, much as with native wild grasses, maize and wild animals, they do not perform for maximum benefit to humans until they are domesticated, trained and nurtured to improve performance. It is well recognized that biological control agents do not usually perform spectacularly and they often do not work as quickly, as efficiently or as consistently as broad spectrum chemicals (Baker, 1986; Cook & Baker, 1983). In order to establish biocontrol efficacy there are many inherent obstacles to overcome. These include basic aspects of strain selection, efficient production of biomass, formulation, storage ability, and methods of application. Also, a better understanding of mechanisms of action, nutrition and ecology of biological control agents is needed. Further, the potential for genetic manipulation of fungi to create genetically superior strains or hybrids that can perform better than wild types empirically selected from the environment needs to be examined. This review is intended to address these difficulties which we have faced in biological control studies and suggest means for improvement of efficacy.

Selection, screening and improvement of strains

Many past efforts, most of which were unsuccessful, to select biological control fungi have been done empirically, ranging from random selection from soil and plant surfaces to chance encounters on culture media where zones of inhibition are noticed. Emphasis should be placed on searching for antagonists in environmental niches directly related to the incidence of disease or lack thereof (Campbell, 1986). Disease suppressive soils are a case in point, and are defined as those soils in which disease development is suppressed when the pathogen is present or is introduced in the presence of a susceptible host (Cook, 1981; Schneider, 1982). Although disease suppressive soils are unfortunately difficult to recognize, several examples are known (Linderman et al., 1983; Schneider, 1982). For instance, in the Rhône valley in France, *Fusarium*-suppressive soils specific against *Fusarium* vascular wilts were found (Alabouvette, Rouxel & Louvet, 1979). Here, saprotrophic isolates of *F. oxysporum* and *F. solani* in soil repressed activity of pathogenic strains of *F. oxysporum* on melons. In recent studies in our laboratory (Lumsden et al., 1987), *Pythium* damping-off was suppressed in soils from a traditional chinampa agroecosystem in the valley of Mexico compared with soil from a nearby modern agroecosystem. A general type of disease suppression was found in the suppressive soil. Isolates of antagonists from the suppressive soil were capable of reducing disease caused by *Pythium* spp. in the greenhouse. In the absence of identifiable disease suppressive soils, an alternative is to examine healthy plants in otherwise heavily·infested agricultural sites.

All selection approaches involve subsequent isolation of antagonistic fungi in pure culture and identification by established biosystematic approaches (Knutson, 1981). Baiting techniques, using survival structures of the pathogen either caught on nylon as in the case of *Pythium* spp. (Lumsden *et al.*, 1987), or direct burial of sclerotia of *Sclerotinia* spp. (Ayers & Adams, 1981) or *Phymatotrichum omnivorum* (Kenerly *et al.*, 1987) in soil have resulted in the isolation of prospective antagonists. Mycoparasites are favoured in these baiting techniques. This approach associates the potential antagonist with actual survival structures. Triple agar plate methods have been used to identify prospective antagonistic fungi in soil with antibiotic activity toward plant pathogens (Dennis & Webster, 1971).

Most examples of biological control have dealt with indigenous, naturally occurring antagonists. These are the most readily obtainable and are the most easily registered according to the current regulations as formulated by the EPA (Betz *et al.*, 1987). Naturally occurring antagonists will continue to be a major source of new isolates in the future for which we will continue to screen for more efficient strains. However, we believe that genetic improvement also has an important place in biological control. This will involve conventional genetic crossing of strains, when applicable, and inducement of mutations by mutagenic agents. Perhaps more important is the promise of molecular biology and genetic engineering for the creation of new and superior strains of biocontrol agents. This area of research is in its infancy but beginnings have been made. Papavizas, Lewis & Abd-El Moity (1982) induced mutations in *Trichoderma* spp. with ultraviolet light and selected strains tolerant to methyl benzimidazole carbamate fungicides. Some mutants were more efficient than the wild types in disease control. Mutation and selection of strains of *Trichoderma harzianum* for ability to colonize plant rhizospheres has also been accomplished (Ahmad & Baker, 1987). Furthermore, a prelude to the improvement of *Trichoderma harzianum* by gene transfer has been done by Stasz, Harman & Weeden (1988) with protoplast fusion of strains requiring specific growth factors (auxotrophs) or those sensitive to specific antibiotics. Genetic characterization of the strains is underway and some show improved biocontrol ability (G. Harman, personal communication).

Once fungal antagonists have been isolated, selection for maximum biocontrol ability is difficult and tedious. Isolation from natural environments and induction of improved efficiency must be followed by quick, reliable bioassay methods to determine which few isolates are superior for biocontrol ability and which of the vast majority are to be discarded (Spurr, 1985). At present, we do not have such methods. It is generally recognized that Petri plate challenge techniques whereby a potential antagonist is paired opposite the pathogen of interest is not a desirable technique (Linderman et al., 1983; Papavizas & Lewis, 1981). Also, in our experience, this method can be very misleading. Isolates that often show strong zones of inhibition in challenge culture do not perform when used to treat a plant for protection from a pathogen, especially in the field (Linderman et al., 1983). Conversely, some of the more promising isolates used for plant protection do not demonstrate obvious interactions in vitro.

Ideally, assessment of biocontrol ability should be done in an environment as similar as possible to the environment where the antagonist will be applied. If the target is a pathogen of annual crops or potted plants in greenhouse production, then assessment in greenhouses or growth chambers is appropriate (Lumsden & Locke, 1989). If the target is a pathogen of annual or perennial crops in field production then outdoor beds or microplots are best used (Adams & Ayers, 1982). Unfortunately, screening for isolate performance is usually not practically or efficiently done in an outdoor environment. Moreover, if screening is done in a greenhouse or growth chamber for ultimate field application, the performance of the antagonist may be poor in the field (Lumsden et al., 1987). When greenhouse screening assays are performed, attempts should be made to simulate as closely as possible the environment in which biocontrol is expected to work. Natural soils should be used (Linderman et al., 1983) or if appropriate, commercially used soil mixes or soilless potting products (Lumsden & Locke, 1989). The assays should be standardized with a known amount of antagonist inoculant and an amount and type of pathogen inoculum to give a level of disease which does not overwhelm the antagonist. Pathogen inoculum should be as 'natural' as possible.

In some cases, indirect screening assays may yield useful information. Most of these methods are in vitro assays which may give possible insights into mechanisms of action. These assays include studies on mycoparasitism; competition for space and nutrients; production of antibiotics, metabolites or lytic enzymes; induction of resistance; prevention of pathogen propagule germination; and ability of antagonist to proliferate in soil, colonize organic substrates, and affect pathogen survival (Cook & Baker, 1983; Papavizas & Lumsden, 1980). Whipps (1987) devised a bioassay in vitro for assessing the ability of several fungal antagonists of Sclerotinia sclerotiorum to prevent colonization of host tissue and subsequent sclerotium formation. This method successfully demonstrated with living host tissue the inability of Pythium oligandrum to inhibit sclero-

Table 8.1. Names and addresses of fungal culture collection repositories

American Type Culture Collection
12301 Parklawn Drive
Rockville, Maryland 20852, USA

Centraalbureau voor Schimmelcultures
P.0. Box 273 3740 AG Baarn,
The Netherlands

Commonwealth Agricultural Bureau
International Mycological Institute
Ferry Lane, Kew, Richmond
Surrey TW9 3AF, UK

Northern Regional Research Laboratory
USDA, ARS Patent Collection
1815 N. University Street
Peoria, Illinois 61604, USA

tium formation and the ability of *Gliocladium* and *Trichoderma* strains to do so. Another assessment method, carried out in natural soil, was described by Kenerly & Stack (1987) for selection of antagonists against sclerotia of *Phymatotrichum omnivorum*. The following attributes of the potential antagonists isolated from sclerotia of *P. omnivorum* were used: the ability to affect sclerotium germination and survival, and range of temperature for effectiveness. With the three methods of assessment, *Gliocladium roseum* was singled out from among many species of fungi to best colonize sclerotia over a range of temperature from 19 to 31°C. Spurr (1985) devised a quantitative laboratory bioassay to determine the ability of conidia of nonpathogenic strains of *Alternaria alternata* applied to leaf surfaces for control of leafspot caused by pathogenic *A. alternata*. This method employed leaves of the host treated with ethanol for improved assessment of the activity of the protective, saprotrophic *A. alternata*. Each individual assessment method requires careful consideration of which parameters are likely to be important to biological control and which can be accurately and rapidly assessed in a scaled-down system. However, it is important to remember that these methods may only imply *potential* for biocontrol. To date, nothing can replace the results obtained from pathogen-antagonist-host interactions in a natural ecosystem.

Once isolates of biocontrol fungi have been identified it is extremely important to deposit them in culture collection repositories for preservation. Examples of culture collections which, among others, accept biocontrol fungi are listed in Table 8.1.

Equally important, working cultures derived from single spore isolations should be preserved in liquid nitrogen, low-temperature freezers or in silica gel for ready access to reliable, genetically stable, pure cultures for future use in biocontrol studies. Genetic purity and stability are extremely important in biological control because of chance mutations and contamination (Bowers, 1982; Kenney & Couch, 1981; Scher & Castagno, 1986). 'Reputations can be lost upon wide-scale testing or marketing of an ineffective product' (Scher & Castagno, 1986).

Production

Once a biocontrol fungus has shown potential for disease control based on laboratory, greenhouse, and field tests, production of an effective biomass becomes a major concern. Both liquid and semi-solid fermentation systems are used for this purpose.

New advances in fermentation technology have produced bacterial and fungal biomass for use as biocontrol insecticides and herbicides (Bowers, 1982; Churchill, 1982). However, similar technology for the production of biocontrol microorganisms effective against plant pathogens is in its infancy. This deficiency is clearly an obstacle to the advancement of biocontrol research. Since fermentation facilities are already in place, especially in the United States, liquid fermentation for the production of biomass will be the preferred approach (Churchill, 1982). Fermenters of up to 150,000 litre capacity currently provide a wide range of control for pH, agitator speeds, impeller designs, aeration rates, choice of incoming gases, variations in baffling and backpressures, and temperatures (Knight, 1988).

A first step in the production of biocontrol fungi is development of a suitable medium using inexpensive, readily available agricultural by-products with the appropriate nutrient balance (Latgé & Soper, 1977). Acceptable materials include molasses, brewer's yeast, corn steep liquor, sulphite waste liquor, and cottonseed and soy flours (Lisansky, 1985). Although these natural substrates are generally consistent in composition, there may be problems in biomass formation if there are major disparities in batches of similar substrates.

For a successful fermentation, not only must appropriate substrates be used, but sufficient biomass containing adequate amounts of effective propagules must be obtained. Isolates of *Trichoderma* and *Gliocladium* performed best for biocontrol if preparations contained the resistant, survival propagules of the fungus, chlamydospores. Not only were chlamydospore preparations more efficacious, but they also allowed greater proliferation of the fungi in soil than preparations containing conidia (Lewis & Papavizas, 1984; Papavizas *et al.*, 1984). Of various liquid media tested, small-scale fermentation in molasses-brewer's yeast resulted in abundant chlamydospore production (Papavizas *et al.* 1984). Deep-tank fermentation for the formation of conidia of *Trichoderma* has recently been reported (Tabachnik, 1988). One of the advantages of working with isolates of *Trichoderma* and *Gliocladium* is their apparent lack of specific nutritional or cultural requirements during fermentation for production of biomass abundant in chlamydospores. In some cases, it may not be desirable to maintain optimum nutrient, aeration, and pH levels since unfavourable conditions promote spore production. We observed this with chlamydospore production by isolates of *Trichoderma* and *Gliocladium* (Lewis & Papavizas, 1983).

There are several additional factors to consider in liquid fermentation. For example, the rate at which an effective biomass is produced affects cost of production as well as the chance of contamination and viability (Lisansky, 1985). It is desirable to obtain the optimum amount of biomass in the shortest time. With isolates of *Trichoderma*, *Gliocladium*, and *Talaromyces*, satisfactory quantities of biomass were obtained in 6 to 7 days, but this time period is still long compared with that for bacteria. Increased fermentation time also reduces viability of some fungi and increases the risk of contamination (Churchill, 1982).

Since the importance of deep-tank fermentation has become apparent, solid state fermentation facilities have become limited in much of the world. In fact, solid state fermentation has rarely been used in North America because of insufficient consumer demand for the products formed (Cannel & Moo-Young, 1980). In spite of the rapid advances in liquid fermentations demonstrated by industrialized nations, most of the research on mass production of biocontrol fungi continues to emphasise solid (semi-solid) substrate fermentation for the formation of inoculum (Aidoo, Hendry & Wood, 1982). Substrates for the production of biocontrol inoculum of fungi in the genera *Trichoderma*, *Gliocladium*, *Coniothyrium*, *Chaetomium*, *Laetisaria* and *Penicillium* include various grain seeds and meals, bagasse, straws, wheat bran, sawdust, and peat individually or in combination (Papavizas, 1985b; Papavizas & Lewis, 1981). The system is especially useful for small-scale laboratory, greenhouse, and field tests which require minimal facilities for implementation. Solid fermentations are also suitable for the production of fungi which do not sporulate in liquid cultures (Lisansky, 1985). In our laboratory, solid fermentation is used successfully to produce large amounts of ascospores of *Talaromyces flavus* and for large-scale production of spores of the biocontrol fungus *Sporidesmium sclerotivorum* (Adams & Ayers, 1982; Fravel *et al.*, 1985). In the latter case, vermiculite moistened with a liquid medium is infested with *S. sclerotivorum* in a large twin-shell blender, aseptically bagged, and incubated so that the mycoparasite grows and sporulates (Adams & Ayers, 1982). However, there are several inherent problems with solid fermentations which put the system at a disadvantage. Generally, the preparations are bulky; may be subject to a greater risk of contamination; and may require extensive space for processing, incubation, and storage. Furthermore, they may have to be dried and milled with the undesirable formation of dusts containing spores, may require costly packaging and transportation, and special equipment may be necessary for their application. Finally, as with all natural agricultural materials, substrate constituents may be inconsistent and variable.

Formulation

The problems to consider in the formulation of an acceptable product are as important as those involved in growth. A biocontrol formulation with agricultural potential should possess several desirable characteristics such as ease of preparation and application, stability, adequate shelf life, abundant viable propagules, and low cost (Churchill, 1982; Lisansky, 1985).

Fermentation biomass is separated from spent medium by various types of filtration (pressure filtration, rotary vacuum drum filtration), centrifugation, or, in some cases, flocculation (Schmit-Kastner & Gölker, 1987). The biomass can then be dried and milled prior to its incorporation into dusts, granules, pellets, wettable powders, or emulsifiable liquids. The drying process may influence the viability of the propagules in the biomass. For example, we have pan dried, spray dried, and lyophilized biomass of isolates of *Trichoderma*, the propagules of which have survived the processes differently (Lewis & Lumsden, unpublished). Generally, viability was the highest after pan drying, the most cumbersome and inefficient of the three methods. Unfortunately, viability was the least after lyophilization, a process used successfully in many other applications. Similar results were reported for other biocontrol fungi (Churchill, 1982). Dust formulations prepared from ground biomass of several *Trichoderma* spp. and *Gliocladium virens* isolates mixed with a pyrophyllite clay (Pyrax®) as a carrier and applied to field plots reduced *Rhizoctonia* diseases of potato and cotton (Beagle-Ristaino & Papavizas, 1985). These dusts also have the potential for use as seed treatment.

An innovative approach to formulation of biocontrol fungi, which has captured the interest and attention of many scientists, involves the immobilization or entrapment of biomass within an organic matrix. The system used most extensively in our laboratory, and being adapted by industry, is biomass immobilization within an alginate gel pellet or prill (Table 8.2), a process used first for immobilization of chemical herbicides and mycoherbicides (Walker & Connick, 1983). Briefly, the method consists of mixing fermenter biomass (wet or dry) and a carrier (bulking agent) with a sodium alginate solution. Chlamydospores form the bulk of the biomass

Table 8.2. Biocontrol microorganisms incorporated into alginate prill for the reduction of diseases caused by soilborne plant pathogenic fungi

Biocontrol agent	Target disease pathogen	Crop	Reference
Gliocladium virens	Rhizoctonia solani	cotton	Lewis & Papavizas, 1987
Gliocladium virens	Rhizoctonia solani	ornamentals	Lumsden & Locke, 1989
Gliocladium virens	Pythium ultimum	ornamentals	Lumsden & Locke, 1989
Gliocladium virens	Sclerotium rolfsii	snapbeans	Papavizas & Lewis, 1989
Laetisaria arvalis	Rhizoctonia solani	cotton	Lewis & Papavizas, 1988
Pseudomonas cepacia	Pythium ultimum	soybean	Fravel et al., 1985
Pythium oligandrum	various fungi	cress	McQuilken et al., 1989
Talaromyces flavus	Verticillium dahliae	potato, eggplant	Fravel et al., 1985
Trichoderma spp.	Rhizoctonia solani	cotton, sugarbeet	Lewis & Papavizas, 1987

of *Trichoderma* and *Gliocladium* (Fig. 8.1). The carrier may be inert (e.g. Pyrax®), a food base (e.g. powdered wheat bran), or a combination. The mixture is then dripped into a calcium chloride solution to form insoluble gelatinous beads which dry into stable, hard prill of uniform size (Fig. 8.2). The process has been patented by our laboratory (US Patents nos. 4668512 and 4724147). We have prepared alginate prill with bacterial cells, spores of *Talaromyces flavus*, and fermenter biomass of *Trichoderma* and *Gliocladium* isolates and of other fungi (Fravel *et al.*, 1985; Lewis & Papavizas, 1985, 1987)(Table 8.2). Addition of prill containing several biocontrol fungi to soils resulted in the reduction of various diseases (Lewis & Papavizas, 1987; Lumsden & Locke, 1989) and allowing proliferation of the biocontrol fungi (Lewis & Papavizas, 1985; Papavizas, Fravel & Lewis, 1987)(Fig. 8.3). Prill formation may also be accomplished with other polymers such as carrageenan and polyacrylamide, various clay carriers, and other food bases.

Efficacy

Formulated biocontrol products, whether prepared from liquid or solid fermentation biomass, must be delivered effectively into the agricultural

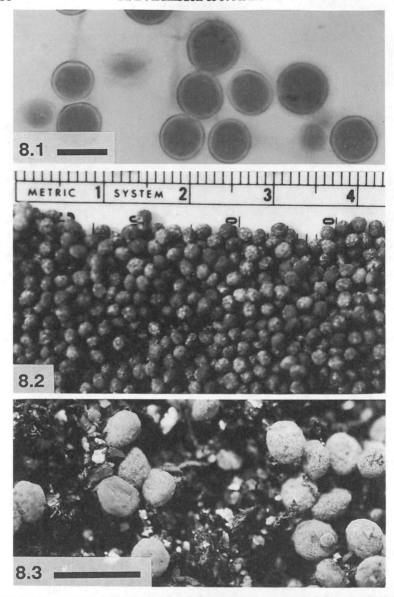

Figs. 8.1–8.3. Sodium alginate-bran formulation of *Gliocladium virens*.
Fig. 8.1. Biomass produced by liquid fermentation on molasses-yeast medium showing chlamydospores and autolysing mycelium. Scale bar = 15 μm.
Fig. 8.2. Dried alginate pellets made from biomass, wheat bran and Na-alginate dripped into $CaCl_2$ solution (centimetre scale).
Fig. 8.3. Pellets with sporulation of *G. virens* on the surface of peat-vermiculite medium after moistening and incubation for 4 days. Scale bar = 5 mm.

system. But how effective are the delivered products in controlling disease? Historically, deliberate introduction of antagonists into the environment, especially soil, for control of disease was considered futile (Garrett, 1956). Attempts to increase the population of antagonists were thought to be contrary to the ecological principle that the population reflects the habitat, and induced changes are only transient (Papavizas & Lewis, 1981). Consequently, the biological control literature abounds with cases of variability in formulation performance from highly effective results in one experiment to poor results in another similar experiment (Baker, 1986; Papavizas & Lewis, 1981; Scher & Castagno, 1986; Spurr, 1985). In addition, many failures are not reported in the literature. If the basis for these failures were known, they too could contribute to improvements in biocontrol (Boudreau & Andrews, 1987).

In the light of this lack of notable successes in biocontrol of fungal plant pathogens, a new focus is required to reduce variability in performance and increase the usefulness of biocontrol systems for eventual development by industry. Several factors emerge as being important in improving efficacy. These include:

- the most likely system in which the applications of biocontrol will be successful;
- the methods of formulation and delivery;
- reliable methods of monitoring activity and testing efficacy in agricultural applications;
- consideration of environmental fluctuations and variation in pathogen inoculum potential;
- providing a formulation with reasonable shelf-life and marketing potential.

Which agricultural environments and crops are the most amenable to application of biocontrol? This question has been addressed several times (Lumsden & Locke, 1989; Lynch & Ebben, 1986; Papavizas, 1985b). Crops and environments in glasshouses where a degree of control of physical factors can be maintained are more amenable to biocontrol than fluctuating field environments (Lumsden & Locke, 1989; Locke, Marois & Papavizas, 1985). Simplicity in an ecosystem also favours biocontrol by reducing competition of the soil microbiota with antagonists for available nutrients (Papavizas, 1985b).

In addition to the environment where applied, it is also important to select carefully the method of formulation delivery and to give the biocontrol agent best advantage over the pathogen. As previously mentioned in this chapter, formulation of biocontrol agents into alginate prill offers a unique advantage to biological control agents (Fravel *et al.*, 1985; Lewis

& Papavizas, 1987; Lumsden & Locke, 1989). With this system, biological control fungi such as *Gliocladium virens*, *Trichoderma harzianum*, or *Talaromyces flavus* can be incorporated into a bead-like package containing appropriate food base, selective nutrients, or stimulants for the antagonist. In addition, pH may be adjusted to favour the biocontrol agent of interest. The physical, chemical and biological integrity of the preparation must be maintained from formulation until final use (Bowers, 1982). The stability must be for six months to two years, if possible, and ideally the preparation should be able to withstand high temperatures (*ca* 50°C) for several hours during shipping. All stages in the production, formulation and application of biocontrol agents should be carefully monitored with a bioassay system to determine the ability of the antagonist to control disease.

Seed treatments with biocontrol fungi have recently been improved with amendments to stimulate antagonists (Nelson, Harman & Nash, 1988). *Trichoderma harzianum* was favoured in pea seed treatments by certain carbohydrates and *T. koningii* was stimulated by fatty or organic acids. With this system, as with alginate pellets, combinations and concentrations can be adjusted to favour the biological agent over pathogens. Recently, improvement in efficacy was reported by Harman & Taylor (1988) with a solid matrix priming system for seed application. With this method, seeds were hydrated to a controlled level with moistened finely-ground organic matter to allow pre-germinative metabolic activity to occur, but without allowing emergence of the radicle. Performance of *T. harzianum* against damping-off was markedly enhanced with solid matrix priming when applied onto cucumber and tomato seeds with coal dust. Tomato seeds were additionally protected because they were strongly acidic, thus favouring the action of *Trichoderma* to the detriment of competing bacteria. Another method for treatment of seed involves incorporation of the biocontrol fungus into semi-liquid gels used for fluid-drilling (Conway, 1986).

Granular preparations have been applied successfully in furrow or broadcast. Backman & Rodriguez-Kabana (1975) used a two-component carrier system consisting of diatomaceous earth impregnated with a molasses food base. Similarly, Jones, Pettit & Taber (1984) described a lignite and stillage carrier formulated with *Trichoderma* and *Gliocladium* into granules for use against *Rhizoctonia solani*. With the use of in-furrow or broadcast materials the cost, rate and ease of application with existing farm machinery, must all be considered. Often the dosage levels of biocontrol formulations are out of reason when considered as broadcast applications. For example, incorporation of a material at a rate of $0 \cdot 5\%$ to a depth of 15 cm would require an exorbitant amount of 12 t ha^{-1}. Applications in furrow would require much less. At a rate of 1 g per 30 cm of row, a total of 14 kg ha^{-1} (100 cm rows) would be required and would thus be a more realistic method of application compared to broadcast application.

Successful field testing requires careful consideration of several factors (Ayers & Adams, 1981; Scher & Castagno, 1986). Several representative locations should be used under standard conditions with the anticipated delivery system and formulation. Ideally, the soil should be naturally infested with the pathogen at sufficient inoculum density to ensure uniform disease development. Since this situation is usually not available, artificial infestation is often necessary. Adams & Ayers (1982) successfully infested a field plot previously devoid of the pathogen of interest, *Sclerotinia minor*. They grew the pathogen on autoclaved oats and inoculated mature lettuce plants in the field plots to induce a high level of disease. These plants were allowed to overwinter in order to allow the sclerotia of the pathogen to mature. The mycoparasitic fungus, *Sporidesmium sclerotivorum*, was applied the following spring in a solid preparation at $0 \cdot 02$ to $20 \cdot 82$ kg ha^{-1}, which delivered 1 to 10^5 macroconidia of the antagonist per g of soil. The mycoparasite reduced the population of sclerotia by 95% in 6 months and provided 40 to 83% control of lettuce drop in four successive crops over a 2-year period when compared with untreated plots. The degree of biocontrol success depended on the rate of mycoparasite application and presence of adequate moisture. Also important for the success of this field test was consideration of the level of pathogen inoculum, which was carefully 'naturalized' to simulate natural infestation. The mycoparasite was allowed a period of time to affect the pathogen inoculum before planting the host crop.

Consideration should be given in field testing to proper choice of the responses and conditions to measure (Kommedahl & Windels, 1981). Evaluation of seedling emergence, plant stand, disease incidence and severity, plant weight, yield, flowering time, and weight of marketable crop will depend on the host-pathogen combination being tested. Additionally, care must be taken in expressing the amount of disease control attributed to the biocontrol agent. According to Bowers (1982), the EPA has defined 'control' as 70-100% reduction in a pest population, or as in the case of fungal pathogens, in the level of disease development. Suppression, aids in control, or other such terms should be used for less than 70-100% reduction in the target pathogen.

Commercialization

Commercialization of biocontrol agents may begin at several stages. Industry may prefer to obtain biocontrol agents on its own and carry out the process independently. However, a second, more effective system is to develop close liaison between industry, public research institutions, extension or advisory services and eventually growers, forming a network of experimentation to develop a product useful to agriculture. A third method, in which research organizations carry the full responsibility for development, is unreasonable.

What are the commercial advantages of biocontrol agents over conventional chemical control? Such agents are usually considered nontoxic to animals, including humans, and to be harmless to crops and the environment. They may have unique modes of action in contrast to chemicals, and consequently may not affect nontarget microorganisms. There are no known examples of plant pathogens becoming resistant to the action of fungal biological control agents, unlike chemicals which may lose their effectiveness through this cause. Biocontrol agents may thus become available to replace chemicals that have been eliminated by stringent regulations to protect the environment, or they may be used in conjunction with minimal amounts of chemicals in an integrated system (Locke *et al.*, 1985).

What are the disadvantages of biocontrol agents? They are living organisms which are subject to deleterious effects of environmental extremes. Their shelf life may be limited or they may require special handling to maintain viability. Unlike many chemical controls, biological control is usually slow acting and not expected to eliminate pathogens. Therefore, biological control may require re-education of the user not to expect complete control in one growing season. Biocontrol agents are probably specific to the types of pathogen or diseases they affect. Also, they may control a pathogen on one crop but not another. Finally, market size may be limited because of specific applications.

For industry to become interested in developing a biocontrol agent, the above factors must be weighed. One of the first steps is for industry to conduct a preliminary economic analysis to determine if an eventual product might be profitable. Results of these surveys are usually not made public, but one example has been published. Heim *et al.* (1986) assessed the economic benefits of biological seed treatment for control of take-all of wheat to Pacific Northwest farmers. More assessments of economic benefits of this type would be useful.

Once profit potential is established, production and formulation processes must be determined. As pointed out previously, it is essential to monitor each step in this process for quality control with emphasis on:

- quantity of effective propagules of the antagonist in the formulation;
- purity of the preparation in terms of genetically stable cultures of the biocontrol agent free from contaminating microorganisms;
- efficacy of the formulation for biocontrol of the target pathogen.

At some point in the research process, a formulation is considered optimal for the intended use. At this time the formulation is 'frozen', or development halted, for testing of its safety. In the USA the EPA requires information on the potential risks to human health or the environment (Betz et al., 1987). A tier system is used to evaluate toxicological and nontarget risks. Tier 1 requires a protocol to establish acute pulmonary, intravenous, dermal, and oral toxicity and pathogenicity to rats. Primary eye irritation and infectivity tests are also required. Nontarget avian, fish, beneficial insect, crustacean and plant tests are included in tier 1. Certain requirements may be exempted depending on intended use and exposure in the environment. If no problems are encountered at this tier level further testing at subsequent levels is not required. Further tier tests are more stringent as justified.

Toxicological data are submitted to the EPA for registration of the biocontrol agent. At the present time the EPA is charging a fee for review of biochemical and microbial registrations. The review process may require six months to two years. Although registration by the EPA may be time consuming, this process is essential to assure environmental safety of newly introduced organisms or those placed in their native environment but in large numbers. The current regulations of the EPA indicate evaluation of microbial pest control agents on an individual basis. This will reduce extensive testing on organisms that are indigenous to North America and that are to be applied on a limited scale (under 4 ha).

Similar registration and testing procedures are in effect in Europe and the UK (Ricard, 1981). In France, the Plant Protection Service of the Ministry of Agriculture handles registration. In the UK, registration is in two parts. Clearance under the Pesticide Safety Precautions Scheme must be obtained first, then claims for efficacy are considered under the Agricultural Chemicals Approval Scheme.

Simultaneous to or after registration, extensive field testing should be done at several representative locations under representative field conditions. The delivery system being proposed should be used. This portion of the development process should be carefully done, because too often products or potential products have been tested on a large scale or even sold to the public before total confidence in the product is gained. Biological control products must be efficacious, economical, safe, and of high quality before they will be accepted by users. It is therefore essential that quality control and extensive testing are carefully carried out to assure a quality product. Besides the assurance of a high quality, efficacious product, users should be assured that biocontrol products are safer to the environment, to users, and to workers than the chemical control products biologicals are intended to replace.

Biocontrol agents, processes for their manufacture, or their use for disease control are more attractive to industrial firms if they are protected by patents. Patenting of microorganisms or biotechnological processes for biocontrol agents are now allowable for novel forms of living organisms or for unique processes of production and use. Exclusive licensing of patents is also a means by which industry can achieve an advantage for marketing a potential biocontrol agent. Patents are a useful mechanism for protecting commercial interests in biocontrol agents and should be a consideration during development of a potential product. Additionally, industry is concerned with protecting their interests in microbial products by developing the ability to identify strains of microorganisms by 'fingerprinting' methods.

Conclusions

Although biological control of fungal plant pathogens with fungal antagonists has not dramatically solved agricultural problems so far, we are at a turning point in technology to make significant advances. It may be several years before microbial biocontrol products are accepted for agricultural use but it seems likely that the time is coming. Several important considerations must be implemented in order to expedite future use of microbial agents for disease control. Foremost among these is greater cooperation between industrial and nonindustrial scientists. This interchange is an essential key to transferring basic scientific information into research and development activities. Cooperative efforts are now formally encouraged by recent legislation passed in the US and the UK to implement technology transfer from government agencies to private industry. New approaches must be introduced to develop improved production, formulation and shelf-life of biocontrol agents. Increased understanding is essential concerning persistent problems related to dependability and reliability of microbial agents. Models need to be developed for analysis of epidemiological, economic and biological data to determine maximum benefits and how they might be introduced into integrated pest management systems. We need to determine mechanisms of action of biological control agents, including factors responsible for naturally occurring suppressive soils; to understand better the ways of improving microbial action against pathogens both through manipulation of their biology and their genetic constitution. Knowledge is needed of the factors that determine ability of microbial agents to adhere to and colonize host plant surfaces, especially leaf surfaces and the rhizosphere. Investigation of new technologies is needed, especially genetic manipulation and hybridization; to induce new biotypes for improved performance, adaptation of microbial agents to new disease control requirements, and improved compatibility with chemical fungicides. Financial support of research to understand and develop biological control of fungal pathogens has been limited in the past and must be improved for significant advances to be made.

Even in a competitive environment, which is healthy for both a scientific and industrial atmosphere, we must be mutually supportive of efforts to develop biocontrol agents. It is imperative that successful cases of biological control be introduced to agriculture to set precedents for industry and crop producers to show that this approach is worthy of efforts to register and commercialize these products. These successes should also demonstrate that microbial biocontrol agents are economically feasible, safe and above all, effective.

References

Acuff, G. (1988). New rules will affect pesticide use. *American Vegetable Grower*, **36**, 11-14.

Adams, P. B. & Ayers, W. A. (1982). Biological control of *Sclerotinia* lettuce drop in the field by *Sporidesmium sclerotivorum*. *Phytopathology*, **72**, 485-488.

Ahmad, J. S. & Baker, R. (1987). Rhizosphere competence of *Trichoderma harzianum*. *Phytopathology*, **77**, 182-189.

Aidoo, K. E., Hendry, R. & Wood, B. J. B. (1982). Solid substrate fermentations. *Advances in Applied Microbiology*, **28**, 201-235.

Alabouvette, C., Rouxel, F. & Louvet, J. (1979). Characteristics of *Fusarium* wilt-suppressive soils and prospects for their utilization in biological control. In *Soilborne Plant Pathogens*, ed. B. Shippers & W. Gams, pp. 165-182. Academic Press: New York.

Ayers, W. A. & Adams, P. B. (1981). Mycoparasitism and its application to biological control of plant diseases. In *Biological Control in Crop Production*, Beltsville Agricultural Research Center Symposium 5, ed. G. C. Papavizas, pp. 91-103. Allanheld & Osmun: Totowa, New Jersey.

Backman, P. A. & Rodriguez-Kabana, R. (1975). A system for the growth and delivery of biological control agents to the soil. *Phytopathology*, **65**, 819-821.

Baker, R. (1986). Biological control: an overview. *Canadian Journal of Plant Pathology*, **8**, 218-221.

Beagle-Ristaino, J. E. & Papavizas, G. C. (1985). Biological control of *Rhizoctonia* stem canker and black scurf of potato. *Phytopathology*, **75**, 560-564.

Betz, F., Rispin, A. & Schneider, W. (1987). Biotechnology products related to agriculture. Overview of regulatory decisions at the US Environmental Protection Agency. *ACS Symposium series* 334, pp. 316-327. American Chemical Society: Washington, D.C.

Boudreau, M. A. & Andrews, J. H. (1987). Factors influencing antagonism of *Chaetomium globosum* to *Venturia inaequalis*: A case study in failed biocontrol. *Phytopathology*, **77**, 1470-1475.

Bowers, R. C. (1982). Commercialization of microbial biological control agents. In *Biological Control of Weeds with Plant Pathogens*, ed. R. Charudattan & H. L. Walker, pp. 157-173. John Wiley & Sons: New York.

Campbell, R. (1986). The search for biological control agents against plant pathogens: A pragmatic approach. *Biological Agriculture and Horticulture*, **3**, 317-327.

Cannel, E. & Moo-Young, M. (1980). Solid-state fermentation systems. *Process Biochemistry*, **9**, 24-28.

Churchill, B. W. (1982). Mass production of microorganisms for biological control. In *Biological Control of Weeds with Plant Pathogens*, ed. R. Charudattan & H. L. Walker, pp. 139-156. John Wiley & Sons: New York.

Conway, K. E. (1986). Use of fluid-drilling gels to deliver biological control agents to soil. *Plant Disease*, **70**, 835-839.

Cook, R. J. (1981). Biocontrol of plant pathogens: Overview. In *Biological Control in Crop Production*, Beltsville Agricultural Research Center Symposium 5, ed. G. C. Papavizas, pp. 23-44. Allanheld & Osmun: Totowa, New Jersey.

Cook, R. J. & Baker, K. F. (1983). *The Nature and Practice of Biological Control of Plant Pathogens*. American Phytopathological Society: St. Paul, Minnesota.

Dennis, C. & Webster, J. (1971). Antagonistic properties of species-groups of *Trichoderma*. I. Production of non-volatile antibiotics. *Transactions of the British Mycological Society*, **57**, 25-39.

Fravel, D. R., Marois, J. J., Lumsden, R. D. & Connick, W. J., Jr. (1985). Encapsulation of potential biocontrol agents in an alginate-clay matrix. *Phytopathology*, **75**, 774-777.

Garrett, S. D. (1956). *Biology of Root-infecting Fungi*. Cambridge University Press: Cambridge.

Harman, G. E. & Taylor, A. G. (1988). Improved seedling performance by integration of biological control agents at favorable pH levels with solid matrix priming. *Phytopathology*, **78**, 520-525.

Heim, M., Folwell, R. J., Cook, R. J. & Kirpes, D. J. (1986). Economic benefits and costs of biological control of take-all to the Pacific Northwest wheat industry. *Washington State University Research Bulletin* 988. Washington State University: Pullman, Washington.

Jones, R. W., Pettit, R. E. & Taber, R. A. (1984). Lignite and stillage: Carrier and substrate for application of fungal biocontrol agents to the soil. *Phytopathology*, **74**, 1167-1170.

Kenerly, C. M., Jeger, M. J., Zuberer, D. A. & Jones, R. W. (1987). Populations of fungi associated with sclerotia of *Phymatotrichum omnivorum* buried in Houston black clay. *Transactions of the British Mycological Society*, **89**, 437-445.

Kenerly, C. M. & Stack, J. P. (1987). Influence of assessment methods on selection of fungal antagonists of the sclerotium-forming fungus *Phymatotrichum omnivorum*. *Canadian Journal of Microbiology*, **33**, 632-635.

Kenney, D. S. & Couch, T. L. (1981). Mass production of biological agents for plant disease, weed and insect control. In *Biological Control in Crop Production*, Beltsville Agricultural Research Center Symposium 5, ed. G. C. Papavizas, pp. 143-150. Allanheld & Osmun: Totowa, New Jersey.

Kommedahl, T. & Windels, C. E. (1981). Introduction of microbial antagonists to specific courts of infection: Seeds, seedlings, and wounds. In *Biological Control in Crop Production*, Beltsville Agricultural Research Center Symposium 5, ed. G. C. Papavizas, pp. 227-248. Allanheld & Osmun: Totowa, New Jersey.

Knight, P. (1988). Fermentation special report. *Biotechnology*, **6**, 505-516.

Knutson, L. V. (1981). Symbiosis of biosystematics and biological control. In *Biological Control in Crop Production*, Beltsville Agricultural Research Center Symposium 5, ed. G. C. Papavizas, pp. 61-78. Allanheld & Osmun: Totowa, New Jersey.

Latgé, J. P. & Soper, R. S. (1977). Media suitable for industrial production of *Entomophthora virulenta* zygospores. *Biotechnology and Bioengineering*, **19**, 1269-1284.

Lewis, J. A. & Papavizas, G. C. (1983). Production of chlamydospores and conidia by *Trichoderma* spp. in liquid and solid growth media. *Soil Biology and Biochemistry*, **15**, 351-357.

Lewis, J. A. & Papavizas, G. C. (1984). A new approach to stimulate population proliferation of *Trichoderma* spp. and other potential biocontrol fungi introduced into natural soils. *Phytopathology*, **74**, 1240-1244.

Lewis, J. A. & Papavizas, G. C. (1985). Characteristics of alginate pellets formulated with *Trichoderma* and *Gliocladium* and their effect on the proliferation of the fungi in soil. *Plant Pathology*, **34**, 517-577.

Lewis, J. A. & Papavizas, G. C. (1987). Application of *Trichoderma* and *Gliocladium* in alginate pellets for control of *Rhizoctonia* damping-off. *Plant Pathology*, **36**, 438-446.

Lewis, J. A. & Papavizas, G. C. (1988). Biocontrol of *Rhizoctonia solani* (Rs) by some novel soil fungi. *Phytopathology*, **78**, 862.

Linderman, R. G., Moore, L. W., Baker, K. F. & Cooksey, D. A. (1983). Strategies for detecting and characterizing systems for biological control of plant pathogens. *Plant Disease*, **67**, 1058-1064.

Lisansky, S. G. (1985). Production and commercialization of pathogens. In *Biological Pest Control*, ed. N. W. Hussey & N. Scopes, pp. 210-218. Blandford Press: Poole, England.

Locke, J. C., Marois, J. J. & Papavizas, G. C. (1985). Biological control of *Fusarium* wilt of greenhouse-grown chrysanthemums. *Plant Disease*, **69**, 167-169.

Lumsden, R. D., García-E. R., Lewis, J. A. & Frías-T., G. A. (1987). Suppression of damping-off caused by *Pythium* spp. in soil from the indigenous Mexican chinampa agricultural system. *Soil Biology and Biochemistry*, **19**, 501-508.

Lumsden, R. D. & Locke, J. C. (1989). Biological control of *Pythium ultimum* and *Rhizoctonia solani* damping-off with *Gliocladium virens* in soilless mix. *Phytopathology*, **79**, 361-366.

Lynch, J. M. & Ebben, M. H. (1986). The use of microorganisms to control plant disease. *Journal of Applied Bacteriology Symposium Supplement*, 115S-126S.

McQuilken, M. P., Cooke, R. C. & Whipps, J. M. (1989). Production and formulation of *Pythium oligandrum* inocula. In *New Directions In Biological Control*, UCLA Symposium on Molecular and Cellular Biology, eds. R. Baker & P. Dunn, p. 175. A. R. Liss: New York.

Nelson, E. B., Harman, G. E. & Nash, G. T. (1988). Enhancement of *Trichoderma*-induced biological control of *Pythium* seed rot and preemergence damping-off of peas. *Soil Biology and Biochemistry*, **20**, 145-150.

Papavizas, G. C. (1981). *Biological Control in Crop Production*. Beltsville Agricultural Research Center Symposium 5. Allanheld & Osmun: Totowa, New Jersey.

Papavizas, G. C. (1985a). *Trichoderma* and *Gliocladium*: Biology, ecology, and potential for biocontrol. *Annual Review of Phytopathology*, **23**, 23-54.

Papavizas, G. C. (1985b). Soilborne plant pathogens: new opportunities for biological control. In *Proceedings of the 1984 British Crop Protection Conference–Pests and Diseases, Brighton, UK,* pp. 371-378. British Crop Protection Council: Thornton Heath, Surrey, UK.

Papavizas, G. C., Dunn, M. T., Lewis, J. A. & Beagle-Ristaino, J. (1984). Liquid fermentation technology for experimental production of biocontrol fungi. *Phytopathology,* **74,** 1171-1175.

Papavizas, G. C., Fravel, D. R. & Lewis, J. A. (1987). Proliferation of *Talaromyces flavus* in soil and survival in alginate pellets. *Phytopathology,* **77,** 131-136.

Papavizas, G. C. & Lewis, J. A. (1981). Introduction and augmentation of microbial antagonists for the control of soilborne plant pathogens. In *Biological Control in Crop Production,* BARC Symposium 5, ed. G. C. Papavizas, pp. 305-322. Allanheld & Osmun: Totowa, New Jersey.

Papavizas, G. C. & Lewis, J. A. (1989). Effect of *Gliocladium* and *Trichoderma* on damping-off and blight of snapbean caused by *Sclerotium rolfsii. Plant Pathology,* **38,** 277-286.

Papavizas, G. C., Lewis, J. A. & Abd-El Moity, T. H. (1982). Evaluation of new biotypes of *Trichoderma harzianum* for tolerance to benomyl and enhanced biocontrol capabilities. *Phytopathology,* **72,** 126-132.

Papavizas, G. C. & Lumsden, R. D. (1980). Biological control of soilborne fungal propagules. *Annual Review of Phytopathology,* **18,** 389-413.

Ricard, J. L. (1981). Commercialization of a *Trichoderma* based mycofungicide: some problems and solutions. *Biocontrol News and Information,* **2,** 95-98.

Rishbeth, J. (1975). Stump inoculation: a biological control of *Fomes annosus.* In *Biology and Control of Soilborne Plant Pathogens,* ed. G. W. Bruehl, pp. 158-162. *American Phytopathological Society:* St. Paul, Minnesota.

Scher, F. M. & Castagno, J. R. (1986). Biocontrol: a view from industry. *Canadian Journal of Plant Pathology,* **8,** 222-224.

Schmidt-Kastner, G. & Gölker, C. F. (1987). Downstream processing in biotechnology. In *Basic Biotechnology,* ed. J. D. Bu'Lock & B. Kristiansen, pp. 173-196. Academic Press: Orlando, Florida.

Schneider, R. W. (1982). *Suppressive Soils and Plant Disease.* American Phytopathological Society: St. Paul, Minnesota.

Spurr, H. W., Jr. (1985). Bioassays critical to biocontrol of plant disease. *Journal of Agricultural Entomology,* **2,** 117-122.

Stasz, T. E., Harman, G. E. & Weeden, N. F. (1988). Protoplast preparation and fusion in two biocontrol strains of *Trichoderma harzianum. Mycologia,* **80,** 141-150.

Tabachnik, M. (1988). *Trichoderma:* Production and formulation. *Trichoderma Newsletter,* **4,** 7.

Walker, H. L. & Connick, W. J., Jr. (1983). Sodium alginate for production and formulation of mycoherbicides. *Weed Science,* **31,** 331-338.

Whipps, J. M. (1987). Behaviour of fungi antagonistic to *Sclerotinia sclerotiorum* on plant tissue segments. *Journal of General Microbiology,* **133,** 1495-1501.

Chapter 9

Mechanisms of biological disease control with special reference to the case study of *Pythium oligandrum* as an antagonist

Karen Lewis[1], J. M. Whipps[2] & R. C. Cooke[1]

[1]*Department of Animal and Plant Sciences, University of Sheffield, Sheffield S10 2TN, UK and* [2]*Microbiology Department, AFRC Institute of Horticultural Research, Littlehampton, West Sussex, BN17 6LP, UK*

Introduction

There has been increasing interest in biological control of plant diseases in recent years but little commercial success in the field. This failure may reside in the lack of understanding of the mode of action of the antagonists, their ecology as well as inadequate knowledge of methods of production and formulation for commercial use (Whipps, 1986; Whipps, Lewis & Cooke, 1988; Lumsden & Lewis, Chapter 8). Integrated studies of applied eco-physiology are required rather than any single empirical approach. Nevertheless, the use of model systems and critical *in vivo* studies are an essential prerequisite to successful biological control. Such systems can provide a detailed knowledge of mode of action and factors that affect a control system. They also highlight future possibilities of genetic manipulation and the *de novo* selection of desirable traits between isolates.

Biocontrol agents may utilize several modes of action, therefore it is important to know the proportion and timing of each mode of action that may occur. Information of this type can be obtained from *in vitro* studies or using plants grown under gnotobiotic conditions where the potential activity of biocontrol agents can be assessed. However, such studies do not provide information on their modes of action *in vivo*, particularly within plants where separation of plant response or antagonist activity is not always possible or in soil where direct observation and chemical analyses are difficult. These limitations must be borne in mind when extrapolating results obtained in the laboratory to the natural environment.

Pythium oligandrum has shown biological control activity against a number of important soilborne plant pathogens both in the laboratory and in the field. Notably, oospore preparations of *P. oligandrum* applied as a seed treatment have been shown to reduce the incidence of *Pythium ultimum* induced damping-off on cress and sugar beet, *Mycocentrospora acerina* on carrot and *Phoma betae* on sugar beet (Vesely, 1977, 1979; Al-Hamdani, Lutchmeah & Cooke, 1983; Vesely & Hejdanek, 1984; Lutchmeah & Cooke, 1985; Martin & Hancock, 1987; Walther & Gindrat-1987). Nevertheless, the specific modes of action involved with the biological control of plant diseases by *P. oligandrum* are unclear and are considered in detail later.

General mechanisms of biological disease control

Induced resistance and cross protection

Induced resistance is a plant response to challenge by micro-organisms or abiotic agents such that following the inducing challenge *de novo* resistance to pathogens is shown in normally susceptible plants (de Wit, 1985). Induced resistance can be localized, when it can be detected only in the area immediately adjacent to the inducing challenge or systemic, where resistance occurs subsequently at sites throughout the plant. Both localized and systemically induced resistance are non-specific and can act against a whole range of pathogens, but whereas localized resistance occurs in many different plant species, systemic resistance appears to be restricted to plants such as green bean (*Phaseolus vulgaris*), tobacco (*Nicotiana tabacum*), cucumber (*Cucumis sativus*) and muskmelon (*Cucumis melo*). Cross protection differs from induced resistance in that following inoculation with avirulent strains of pathogens or other microorganisms both inducing microorganisms and challenge pathogens occur on or within the protected tissues (de Wit, 1985). Consequently, competition, antibiosis and hyperparasitism are all theoretically possible in addition to induced resistance *per se* but often induced resistance has been indicated as the most important feature of cross protection. Nevertheless, the relative proportion of these modes of action occurring within plant tissues have not been well characterized in most cases.

Several possible mechanisms for localized resistance have been proposed. These include phytoalexin production and alterations to plant cell walls including increased production of suberin, hydroxyproline-rich glycoproteins and lignification (Hammerschmidt, Lamport & Muldoon, 1984) and correlations between resistance and lignin formation, peroxidase activity and protease inhibitors have been found (Dean & Kuc, 1987; Roby *et al.*, 1987). In systemically protected tobacco or cucumber, increases in newly formed pathogenesis-related (also termed b-) proteins have also been recorded and these may be chitinase or β-glucanase (Gianinazzi *et al.*, 1980; Fritig *et al.*, 1987; Metraux, Streit & Staub, 1988).

The most commonly reported cases of cross protection involving fungi are probably those used against vascular wilts. Inoculation with non-pathogenic strains or weakly virulent strains of pathogenic formae speciales of *Fusarium* and *Verticillium* species or with other fungi or bacteria have all given levels of cross protection (e.g. Matta & Garibaldi, 1977; Ogawa & Komada, 1985; Sneh, Agami & Baker, 1985; Hillocks, 1986). In most of these experiments, cut or wounded stems and roots or intact bare roots were dipped or sprayed with suspensions of the putative cross protecting agent and this may well bypass resistance systems operating in intact, uninjured roots and stems. However, cross protection is also thought to occur in intact root systems against other pathogens. For instance, the avirulent fungus *Phialophora graminicola* limited the mycelial spread of the take-all fungus, *Gaeumannomyces graminis* var *tritici* on wheat (Deacon, 1973) and precolonization of wheat roots by non-pathogenic isolates of *G. graminis* var. *graminis* protected against *G. graminis* var. *tritici* and *G. graminis* var. *avenae* (Wong, 1975).

Mycorrhizal fungi can also act as cross protection agents decreasing the incidence of root disease. With ectomycorrhizas, antibiosis against the pathogen, physical protection by the mantle, competition with the pathogen for nutrients coming from the roots, stimulation of antagonistic microflora associated with the mantle and induction of host plant resistance have all been suggested as possible mechanisms involved in the protection of roots (Marx, 1972; Chakravarty & Unestam, 1987). Similarly, plants with endomycorrhizal associations can be more resistant to pathogens than non-mycorrhizal plants of similar size and developmental stage (Dehne, 1982). In some cases, the effect may be partially systemic (Rosendahl, 1985). More recently, interactions between the extramatrical mycelium of a vesicular arbuscular mycorrhizal fungus and *F. oxysporum* f. sp. *radicis-lycopersici* also appeared involved in disease reduction on tomatoes but the mechanisms were not investigated (Caron, Richard & Fortin, 1986).

Hypovirulence

Hypovirulence is a term used to describe reduced virulence found in some strains of pathogens. This phenomenon was first observed in *Endothia parasitica* on European chestnut (*Castanea sativa*) in Italy where naturally occurring hypovirulent strains were able to reduce the effect of virulent strains (Grente & Sauret, 1969b). These slower growing hypovirulent strains were found to contain a single cytoplasmic element of double stranded RNA (dsRNA), similar to that found in mycoviruses, which was transmitted via anastomoses in compatible strains through natural virulent populations of *E. parasitica* (Grente & Sauret, 1969a; Van Alfen *et al.*, 1975). Consequently, hypovirulent strains of *E. parasitica* have been used as biological control agents of chestnut blight (Anagnostakis, 1982). This may be considered a specialized form of cross protection which is limited to the control of established compatible strains only (van Alfen & Hansen, 1984).

Hypovirulence has also been reported in many other pathogens including *Rhizoctonia solani*, *Gaeumannomyces gramini* var *tritici* and *Ophiostoma ulmi* but the transmissible elements responsible for hypovirulence and or reduced vigour of the fungi are subject to debate and may be due to dsRNAs, plasmids or viruses (Rogers, Buck & Brasier, 1986; Koltin, Finkler & Ben-Zvi, 1987).

Competition

Competition occurs between microorganisms when space or nutrients are limiting. Implicit in this definition is the understanding that combative interactions such as antibiotic production or mycoparasitism or the occurrence of induced resistance in the host are not included even though these mechanisms may form an important part of the overall processes occurring in the interaction.

For instance, organic amendments or artificial infestations of soils with different amounts of pathogen and biocontrol agent have commonly been used to investigate 'competition' and biological control in the soil (Martin & Hancock, 1986; Paulitz & Baker, 1987; Paulitz, Park & Baker, 1987). Resulting populations of pathogen and biocontrol agent in the soil or on the organic amendment are then related to disease occurring in test plants. Here, however, only the overall outcome of the interactions is observed, not competition for materials or space *per se*. Consequently, although the term is commonly used to describe interactions between microorganisms, direct evidence for the occurrence of competition is sparse, the majority of examples coming from circumstantial evidence.

The rhizosphere is of major concern where competition for space as well as nutrients occurs. For instance, one attribute of a successful rhizosphere biocontrol agent would be the ability to maintain a high population on the root surface providing protection of the whole root for the duration of its life. Rhizosphere competent isolates of *Trichoderma* spp. have been obtained by mutagenesis and a correlation between rhizosphere competence and the ability to utilize cellulose substrates associated with the root has been observed (Ahmad & Baker, 1987). Similarly, both *Microdochium bolleyi* and *Phialophora graminicola* utilize cortical cells that senesce in the normal course of cereal root development and it has been suggested that these fungi compete with the take-all pathogen *Gaeumannomyces graminis* var. *graminis* for colonization of the senescing root cortex thus diminishing take-all (Kirk & Deacon, 1987). Competition for substrates also occurs naturally, external to the root, particularly in regions enriched by organic amendments. In some soils pathogens can be suppressed from causing disease, albeit in part, by this mechanism. For example, *Fusarium* species were suppressed in the Chateaurenard soil in France (Alabouvette, Couteadier & Louvet, 1985). A greater microbial biomass was present in this suppressive soil compared with a conducive soil and it was suggested that this led to greater nutrient competition and consequent inhibition of *Fusarium* spp. Specific antagonists showing combative behaviour were also active in the suppressive soil but the relative importance of the mechanisms involved was unclear.

Antibiosis

The production of antibiotics by actinomycetes, bacteria and fungi is very simply demonstrated *in vitro*. Numerous agar plate tests have been developed to detect volatile and non-volatile antibiotic production by putative biological control agents and to quantify their effects on pathogens (Whipps, 1987). In general, however, the role of antibiotic production in biological control *in vivo* remains largely unproven (but see Fravel, 1988).

Secondary metabolite production is influenced by cultural conditions and although many micro-organisms produce antibiotics in culture there is little evidence that antibiotics are produced in natural environments, except after input of organic materials. Even so, it is possible that detection techniques are insensitive, that antibiotics are rapidly degraded or that they are bound to the substrate such as clay particles in soil preventing detection (Howell & Stipanovic, 1980; Papavizas & Lumsden, 1980; Williams & Vickers, 1986). Unusually, there is evidence that the protection from seed rots occurring after treatment of seeds with ascospore suspensions of *Chaetomium globosum* is due to the presence of a non-diffusible antibiotic which remains bound to the surface of the hyphae (Hubbard, Harman & Eckenrode, 1982). Hence, even though the antibiotic could be extracted from the mycelium of the fungus or from inoculated seed coats, no zones of inhibition were produced against pathogenic *Pythium* species *in vitro*.

Species of *Gliocladium* and *Trichoderma* are well known biological control agents that produce a range of antibiotics which are active against pathogens *in vitro* (Dennis & Webster, 1971a, b; Bell, Wells & Markham, 1982; Claydon *et al.*, 1987) and, consequently, antibiotic production has commonly been suggested as a mode of action for these fungi. For example, cotton seeds treated with *Gliocladium virens* were protected from damping-off caused by both *Pythium ultimum* and *Rhizoctonia solani* (Howell, 1982). *In vitro*, *G. virens* parasitized hyphae of *R. solani* but not those of *P. ultimum* and in soil it decreased the number of sclerotia of *R. solani* but did not affect the number of oospores of *P. ultimum*. Using mutants of *G. virens* which were deficient in, or over-producers of, the antibiotic gliovirin, a correlation was found between protection from *P. ultimum* damping-off and production of gliovirin (Howell & Stipanovic, 1983). *R. solani* was unaffected by this antibiotic. This suggested that gliovirin production was the important mode of action against *P. ultimum* but not against *R. solani*. It also illustrates that one antagonist may have different modes of action dependent on the pathogen examined.

Speed of control has also been used to distinguish between modes of action of *Trichoderma* species against *Pythium* species on seed coats of pea (Lifshitz, Windham & Baker, 1986). As germination of sporangia of pathogenic strains of *Pythium* spp. took place within 10 h and embryo infection of pea occurred within 24-48 h under the environmental conditions used, for success, any biological control agent must have a fast action. Inoculation of *Trichoderma* species onto pea seeds gave control of *Pythium* rot but *Trichoderma* spores required 10-14 h for over 50% germination making mycoparasitism unlikely. Competition for nutrients was also considered unlikely as addition of nutrients (glucose and asparagine) to the seeds did not affect disease control and the presence of *Trichoderma* did not affect germination of *Pythium* sporangia. However, antibiotic production with activity against *P. ultimum* was detected in culture filtrates of *Trichoderma* and microscopic observation showed growth abnormalities in *Pythium* hyphae when *Trichoderma* hyphae were in close proximity but before any mycoparasitism occurred. This effect was not apparent at a macroscopic level. Consequently with the isolates of *Trichoderma* tested in this study, antibiosis was considered the key factor in the biological control of *Pythium* on seed coats of pea.

Mycoparasitism

Mycoparasitism occurs when one fungus exists in intimate association with another from which it derives some or all of its nutrients while conferring no benefit in return. Biotrophic mycoparasites have a persistent contact with or occupation of living cells whereas necrotrophic mycoparasites kill the host cells, often in advance of contact and penetration. Mycoparasitism is a commonly observed phenomenon *in vitro* and *in vivo* and this mode of action and its involvement in biological disease control has been reviewed extensively in the last few years (Ayers & Adams, 1981; Lumsden, 1981; Whipps *et al.*, 1988).

The first phase of mycoparasitism involves location of host mycelium or spores by the mycoparasite. For instance, hyphae of *Trichoderma harzianum* have been found to grow towards hyphae of susceptible fungi before contact is made (Chet, Harman & Baker, 1981), presumably due to chemical signals originating from the host. Subsequently, extensive hyphal coiling and short branching on the host occurs, which in some cases may be lectin mediated (Elad, Barak & Chet, 1983a; Barak *et al.*, 1985). Cytoplasmic degradation may occur before contact or penetration (De Oliveira, Bellei & Borges, 1984). But often, penetration takes place immediately, arising directly from parent hyphae or from appressoria, followed by cytoplasmic breakdown and host exploitation (Elad *et al.*, 1983b). Cell wall degrading enzymes such as chitinases, β-1,3-D-glucanases and proteases which enable penetration of host hyphae are probably involved here (Ridout, Coley-Smith & Lynch, 1988).

In contrast to the host cell degradation which occurs due to direct action by necrotrophic mycoparasites, *Sporidesmium sclerotivorum* possesses some biotrophic characteristics. For instance, haustoria are produced within hyphae of sclerotia of *Sclerotinia minor* (Bullock *et al.*, 1986) and rather than producing its own β-1,3-D-glucanase to degrade the sclerotia, the activity of the host enzyme is stimulated (Adams & Ayers, 1983).

Considerable evidence now exists that specific sclerotial mycoparasites can be successfully introduced into the field and give control of disease. For example, *Coniothyrium minitans* grown on a barley-rye-sunflower seed mixture and applied to sunflower seed rows reduced *Sclerotinia*-wilt by 30% over a two year period (Huang, 1980). Similarly, a single application of *Sporidesmium sclerotivorum* grown on non-sterile sand containing 1% w/v live sclerotia of *Sclerotinia minor*, reduced lettuce drop in the field by 40-80% in four successive crops over two years (Adams & Ayers, 1982; Adams, Marois & Ayers, 1984). Macroconidia of *Sporidesmium sclerotivorum* germinated near sclerotia, and after colonization, its hyphae grew from sclerotium to sclerotium through soil, utilizing them as protected nutrient sources (Ayers & Adams, 1979; 1981). Massive production of macroconidia from infected sclerotia maintained a large inoculum level within the soil. Self-maintenance by a mycoparasite was also found in glasshouse trials using *Coniothyrium minitans* against *Sclerotinia sclerotiorum* disease of lettuce and celery (Whipps, Budge & Ebben, 1989).

Modes of action of *Pythium oligandrum* - a case study

P. oligandrum was first isolated from diseased roots of pea (*Pisum sativum*) by Drechsler in 1930. The frequent occurrence of *P. oligandrum* among isolates of phytopathogenic *Pythium* species led Drechsler (1946) to conclude that the fungus was a mycoparasite and only a secondary invader of plant tissue. More recent studies have confirmed *P. oligandrum* to be an aggressive mycoparasite with a wide host range (Deacon & Henry, 1978; Vesely, 1978; Vesely & Hejdanek, 1984; Lutchmeah & Cooke, 1984; Foley & Deacon, 1986; Whipps, 1987). Laboratory studies using co-inoculations of *P. oligandrum* and a range of cellulolytic fungi, with cellulose as the main carbon source, have shown that both mycoparasitism and nutrient competition can occur but the relative activity of each mechanism was dependent on the experimental conditions and fungal iso-

Table 9.1. Types of hyphal interactions observed between *Pythium oligandrum* and host fungi on the surface of cellulose film overlaying 2% distilled water agar

Fast lysis	Fast granulation	Slow, no lysis
Alternaria dauci	*Rhizoctonia solani*	*Pythium ultimum*
Alternaria radicina		*Pythium intermedium*
Botrytis cinerea		*Pythium debaryanum*
Fusarium oxysporum		
Phoma betae		
Pleospora betae		
Sclerotinia sclerotiorum		
Sclerotium cepivorum		
Stemphylium radicinum		
Pyrenochaeta terrestris		

lates used (Tribe, 1966; Deacon, 1976; Al-Hamdani & Cooke, 1983). Some workers have suggested a role for substrate competition in disease control in the soil and spermosphere (Martin & Hancock, 1986, 1987) but mycoparasitism is strongly implicated as a major mode of action of *P. oligandrum* as a biocontrol agent.

Hyphal interactions

Mycoparasitism in the soil will involve encounters between sparse hyphal systems and more often specifically, individual hyphae. Consequently, single hyphal interactions between *P. oligandrum* and a range of fourteen important plant pathogenic fungi from all major taxonomic groups have been examined on cellulose film overlaying distilled water agar with a view to ascertaining modes of action and identifying possible target species (Table 9.1).

Figs. 9.1-9.4. *(Facing page)* Interactions between *P. oligandrum* (E) and *B. cinerea* (P) hyphae showing fast lysis (Scale bars = 20 μm). Fig. 9.1, Hyphae in close proximity; Fig. 9.2, five minutes later showing formation of lateral branches (arrowed) and directed growth of *P. oligandrum*; Fig. 9.3, five minutes later showing contact; Fig. 9.4, five minutes after contact with further lateral branches (arrowed) effecting penetration.

Figs. 9.5-9.6. Interactions between *P. oligandrum* (P) and *Pleospora betae* (L) hyphae showing stages of fast lysis (scale bars = 20 μm). Fig. 9.5, hyphae in close proximity; Fig. 9.6, ten minutes later, showing formation of lateral branches which effect contact with the target hypha of *P. betae*.

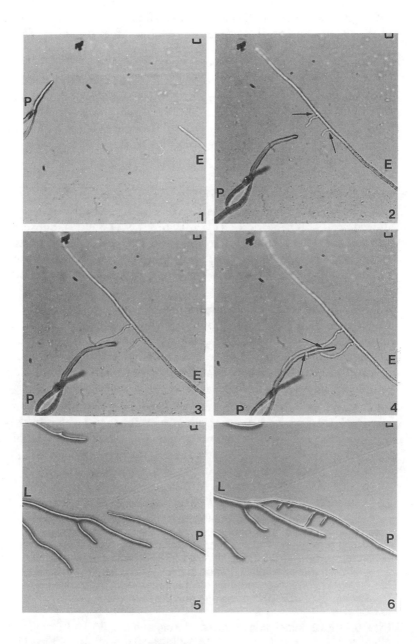

P. oligandrum consistently parasitised all fourteen fungi examined, thus indicating the considerable range of potential target organisms for control by the antagonist. In all cases there was no alteration in host mycelia prior to contact with *P. oligandrum*. When the hyphae came into close proximity, lateral branches formed on the *P. oligandrum* hypha and these grew towards the host over distances up to 100 μm (Figs. 9.1 to 9.3 & 9.5 to 9.6). Subsequently, penetration of the host cell walls was effected by fine invasive hyphae (Figs. 9.4, 9.6 & 9.9). Responses of the host to penetration by the mycoparasite were found to be dependent on the target species involved. Three distinct patterns of hyphal interaction were observed.

Fast lysis: Hyphal interactions of this type were characterised by the complete lysis of host hyphae and extrusion of hyphal contents within seconds following cell wall penetration (Figs. 9.7 to 9.9, 9.11 & 9.12). Subsequently, susceptible hyphae ceased to elongate and lost opacity whilst the mycoparasite continued to grow, branching profusely in the area of the interaction (Fig. 9.10). Fast lysis occurred with most of the pathogens examined but not with *Pythium* spp. or *Rhizoctonia solani*.

Fast granulation: This second hyphal interaction was also rapid, taking place within minutes of contact but rarely involved lysis. Cell wall penetration was followed most commonly by granulation and disorganization of host hyphal contents leading to loss of opacity and cessation of hyphal growth (Figs. 9.13 & 9.14). *R. solani* was the only pathogen examined exhibiting this interaction (Table 9.1).

Slow, no lysis: In contrast to the other hyphal interaction types, interactions in this group were neither rapid nor did they involve host lysis. Disorganization of hyphal contents and loss of opacity occurred in susceptible hyphae 1 to 8 h following contact with *P. oligandrum* (Figs. 9.15 & 9.16). Only *Pythium* spp. gave this reaction (Table 9.1).

Figs. 9.7-9.10. *(Facing page)* Interactions between *P. oligandrum* (P) and *Sclerotium cepivorum* (V) hyphae showing fast lysis (scale bar = 20 μm). Fig. 9.7, a growing hyphal tip of *P. oligandrum* approaches the side of a *S. cepivorum* hypha; Fig. 9.8, five minutes later, contact occurs; Fig. 9.9, after a further ten minutes, a lateral branch of *P. oligandrum* (arrowed) effects lysis of the target hypha; Fig. 9.10, thirty-five minutes later, intracellular growth of *P. oligandrum* (arrowed) is occurring and the hypha of *S. cepivorum* ceases to elongate.

Figs. 9.11-9.12. Interactions between *P. oligandrum* (P) and *Stemphylium radicinum* hyphae (S) showing fast lysis (scale bar = 20 μm). Fig. 9.11, the tips of two *S. radicinum* hyphae (S1 and S2) approaching the side of a *P. oligandrum* hypha; Fig. 9.12, lysis of the *S. radicinum* (S1) hyphal tip on contact with *P. oligandrum*.

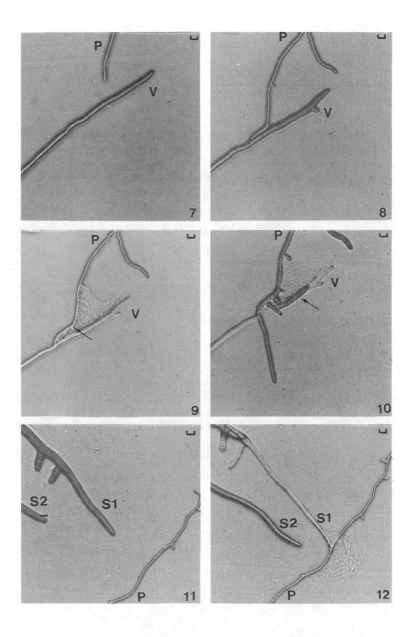

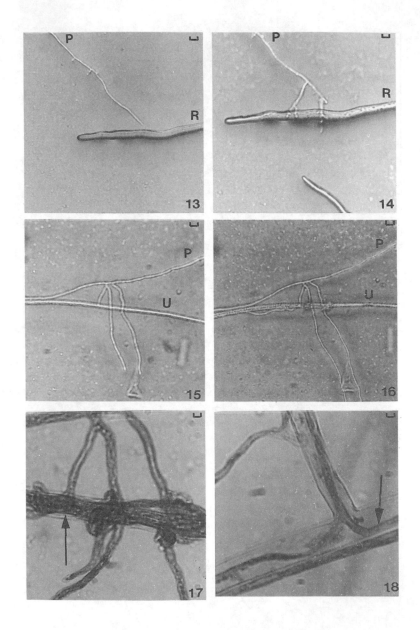

The later stages of all interaction types resulted in the intracellular growth of *P. oligandrum* throughout the host mycelium. The mycoparasite often exited via fine branches to cause further lysis and penetration of other hyphae, thus parasitizing the whole mycelial network (Figs. 9.17 to 9.23).

P. oligandrum is therefore an aggressive mycoparasite with a wide range of hosts including genera in the Mastigomycotina, Asco-Deuteromycotina and Basidiomycotina, with the precise nature and rapidity of an interaction being dependent upon the host species involved. Moreover, similar mycoparasitism by *P. oligandrum* has been confirmed on glass slides buried in sand during biocontrol experiments against *R. solani* (Lewis, 1988). Nevertheless, some caution must be observed in extrapolating these observations to the natural soil environment.

Antibiosis

The effect of non-volatile and volatile metabolites of *P. oligandrum* on the growth of the fourteen plant pathogenic fungi previously employed was examined using standard agar plate techniques (Lewis, 1988). These included 'face-to-face' tests for volatile antibiotics, and cellophane replacement and dual culture techniques for total inhibitory metabolites (Whipps, 1987). In addition, comparison of pathogen growth on agar plates on which *P. oligandrum* or *P. ultimum* had been grown on cellophane and then removed were carried out. A number of host fungal colonies, especially those exhibiting fast lysis interactions showed significantly reduced growth on *P. oligandrum* precolonized plates when compared to controls (Fig. 9.24). As there was no evidence of volatile antibiotic production by *P. oligandrum*, the effects on precolonized plates may have been due to the production of non-volatile growth inhibiting compounds, although more efficient nutrient depletion of the agar by *P. oligandrum* compared with *P. ultimum* cannot be ruled out.

These findings confirm those of other workers (Deacon, 1976; Whipps, 1987) that *P. oligandrum*, although not producing vast quantities of anti-

Figs. 9.13-9.14. *(Facing page)* Interactions between *P. oligandrum* (P) and *Rhizoctonia solani* (R) hyphae showing fast granulation (scale bar = 20 μm). Fig. 9.13, hyphal tip of *P. oligandrum* approaching the side of a hypha of *R. solani*; Fig. 9.14, penetration by *P. oligandrum* (P) causing granulation of the *R. solani* cytoplasm.

Fig. 9.15-9.18. Interactions between *P. oligandrum* (P) and *Pythium ultimum* (U) of slow, no lysis type. Fig. 9.15, *P. oligandrum* and *P. ultimum* in contact; Fig. 9.16, three hours later showing disorganisation of the *P. ultimum* hyphal contents; Fig. 9.17, the same hypha six hours after contact with *P. oligandrum* growing within a *P. ultimum* hypha; Fig. 9.18, *P. oligandrum* (arrowed) growing within a *P. ultimum* hypha (scale bars = 20 μm and 5 μm for Figs. 9.15-9.16 and Figs. 9.17-9.18 respectively).

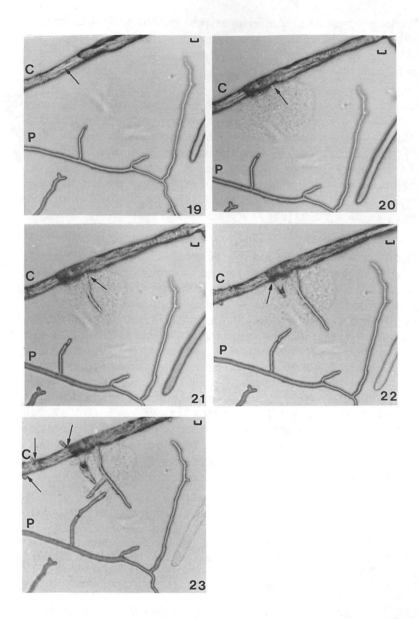

Table 9.2. Extracellular cell wall hydrolysing enzymes produced by *Pythium oligandrum*

Enzyme	Substrate(s)	Production Solid media	Liquid media
Cellulase	Bose cellulose	–	nt
	Bose cellulose+glucose	–	nt
	Methyl cellulose	–	–
	Methyl cellulose+glucose	–	–
β-1,3-D-glucanase	Laminarin	nt	+
	Laminarin+glucose	nt	–
Chitinase	Colloidal chitin	nt	–
Lipase	Tween 80	+	nt
Protease	Hide powder azure	+	+

Key: '+' indicates production of the enzyme, '–' lack of production, nt = not tested.

biotics, does secrete metabolites that are inhibitory to certain host species *in vitro*. Antibiosis in this context is distinct from that classically defined as involving gross morphological changes over long distances. Here the process can be considered an integral part of mycoparasitism leading to inhibited growth of the host and increased susceptibility to invasion. Also, localized antibiotic production could be involved in some fast lysis interactions because of the rapidity of the lytic effect.

Enzyme production

P. oligandrum penetrates host cell walls and consequently enzymatic hydrolysis may be involved. Enzymes capable of hydrolysing the major cell wall polymers of fungi have been shown to include an array of polysaccharidases such as cellulase, β-1,3-D-glucanase, chitinase, as well as lipases and proteases. Initially, the ability of *P. oligandrum* to produce

Figs. 9.19-9.23. *(Facing page)* Growth of *P. oligandrum* (P) following a fast lysis interaction with a *Sclerotinia sclerotiorum* (C) hypha (scale bars 20 μm); Fig. 9.19, *P. oligandrum* (arrowed) growing within a *S. sclerotiorum* hypha; Fig. 9.20, five minutes later, the exit of *P. oligandrum* (arrowed) effects lysis of the *S. sclerotiorum* hypha in which it has grown; Fig. 9.21, five minutes later, continuing extension of the *P. oligandrum* exit branch (arrowed); Fig. 9.22, ten minutes later, *P. oligandrum* exhibiting further branching within the *S. sclerotiorum* hypha (arrowed); Fig. 9.23, *P. oligandrum* (arrowed) continuing to grow intracellularly with frequent exits from hypha (C).

these hydrolases was tested using solid assay media (Table 9.2). Extracellular lipase and protease were produced but not chitinase. There was no indication of cellulase production with either Bose cellulose or methyl cellulose as substrates whether in the presence or absence of glucose. This confirms previous reports of the non-cellulolytic nature of *P. oligandrum* (Deacon, 1976). More extensive studies in defined liquid culture further corroborated these findings and, in addition, β-1,3-D-glucanase was found to be consistently produced. The production of these enzymes by

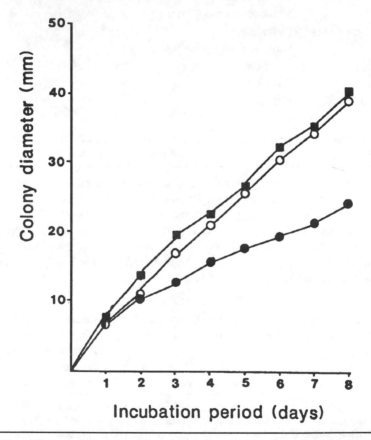

Fig. 9.24. Diameters of *Alternaria dauci* colonies (mm) growing on cornmeal agar precolonised by *Pythium oligandrum* (closed circles), *Pythium ultimum* (open circles) or uninoculated cellophane (closed squares). Significant differences occurred between *Pythium oligandrum* precolonised plates and *Pythium ultimum* or uninoculated controls on days 3-8 ($P = 0.01$).

Table 9.3. Extracellular β-1,3-D-glucanase production by *Pythium oligandrum* in various liquid media containing different sources of carbon

Medium[c]	β-1,3-D-glucanase activity[a]	
	EU mg^{-1} protein[b]	EU g^{-1} dry weight
laminarin	6·00 ±0·9	15·6 ±1·6
Pythium ultimum cell walls	4·63 ±1·3	18·4 ±1·5
Rhizoctonia solani cell walls	2·83 ±1·0	17·3 ±1·5
Stemphylium radicinum cell walls	5·58 ±1·4	21·6 ±2·5
glucose	0·00 ±0·0	0·00 ±0·0
glucose+laminarin	0·00 ±0·0	0·00 ±0·0
no glucose+no laminarin	0·00 ±0·0	0·00 ±0·0
LSD (P = 0·05)	1·01	1·77

[a]assayed by measuring glucose release from laminarin from a 5-day-old culture; [b]EU = enzyme units; [c]basal medium used throughout was a modified Czapek Dox medium containing: (g l^{-1} distilled water) NaNO3 (1·0), KH2PO4 (1·0), MgSO4.7H2O (1·0), KCl (0·5), FeSO4 (0·01), cholesterol (0·03; prepared as a solution 1·5% w/v in 95% v/v ethanol), thiamine HCl (0·125). Entries are the means of four replicates ±SEM.

P. oligandrum indicates a potential for their involvement in penetration of host fungi in all three main fungal divisions.

Extracellular β-1,3-D-glucanase and protease production by *P. oligandrum* was inducible. β-1,3-D-glucanase was induced by growth on laminarin, cell walls of *Pythium ultimum*, *Stemphylium radicinum* and *Rhizoctonia solani* but was repressed in the presence of glucose, indicating possible catabolite repression (Table 9.3). Protease activity was also induced by the presence of protein (bovine serum albumin) or isolated host cell walls (Table 9.4). In addition, levels of activity were decreased when bovine serum albumin was supplemented with carbon, nitrogen and sulphur nutrients, supplied as glucose, asparagine and MgSO4 suggesting that catabolite repression may also occur. The fact that induction of both these enzymes can occur in the presence of host cell walls indicates the potential importance of these enzymes in hyphal wall penetration. The stimulation of protease activity which in this case is of a neutral/alkaline type with maximal activity at pH values between 5·5 and 10 (Fig. 9.25) has commonly been ignored in the majority of mycoparasitic research. In addition, subsequent to wall penetration protease enzymes may also aid in the degradation of host hyphal contents.

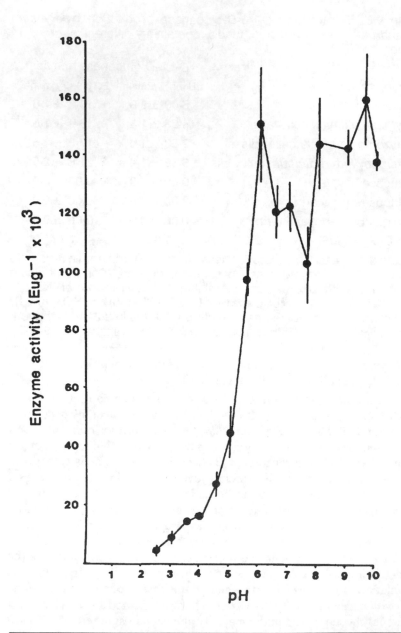

Fig. 9.25. The effect of pH on the protease activity of a cell free culture filtrate of *Pythium oligandrum*. Each point is the mean of four replicates shown with 95% confidence limits.

Table 9.4. Extracellular protease production by *Pythium oligandrum* in various liquid media

Culture medium[3]	Protease activity[1] (EU g^{-1} dry weight)[2]
No protein[4]; no CNS[5]	0.00 ± 0.00^a
No protein + CNS	0.00 ± 0.00^a
+ protein + CNS	0.08 ± 0.00^b
+ protein; no CNS	0.23 ± 0.03^d
+ *Pythium ultimum* cell walls	0.17 ± 0.02^c
+ *Rhizoctonia solani* cell walls	0.16 ± 0.01^c
+ *Stemphylium radicinum* cell walls	0.12 ± 0.07^c
LSD ($P = 0.05$)	0.05

[1]assayed colorimetrically using Hide Powder Azure as substrate from a 2-day-old culture.
[2]EU = enzyme units. [3]The minimal medium used throughout contained (g l^{-1} distilled water): KH$_2$PO$_4$ (1.0), MgCl$_2$ (0.05), NaCl (0.2), CaCl$_2$ (0.2), cholesterol (0.03; prepared as a 1.5% w/v solution in 95% v/v ethanol), thiamine-HCl (0.125).
[4]Protein, as bovine serum albumin, was added at 1 mg l^{-1}.
[5]CNS = carbon, nitrogen and sulphur, added as D-glucose, L-asparagine and MgSO$_4$.7H$_2$O at 1.0 g l^{-1}. Each entry is the mean for four replicates $\pm$SEM. Figures followed by different letters are significantly different ($P = 0.05$).

Environmental factors

Environmental factors, particularly temperature, pH, and water potential are known to have significant effects on the ability of antagonists to control disease in the field. *P. oligandrum* is a widespread fungus occurring in various soil types, but most commonly in those of neutral pH (Foley & Deacon, 1985; Dick, Ali-Shtayeh & Mohammed, 1986). Laboratory experiments show that optimum mycelial growth of *P. oligandrum* takes place between pH 6 and 7 with no growth occurring below pH 3 or above pH 8 (Fig. 9.26). Soil pH would therefore be an important consideration prior to the application of *P. oligandrum* as a biocontrol agent. Similarly, *in vitro* agar plate studies show *P. oligandrum* to have a very rapid growth rate (optimum 30°C). There is however little growth at temperatures below 7°C or above 37°C (Fig. 9.27). All new soil isolates are as a result screened for low temperature growth. Water potentials below -3 to -3.5 MPa also prevent mycelial growth (M. P. McQuilken, unpublished results). The effects of environmental parameters on hyphal interactions have been demonstrated before (Whipps, 1987; Whipps & Magan, 1987;

Magan & Whipps, 1988) and clearly have important implications for bio-control in general.

Future utilization of Pythium oligandrum

Mode of action studies have shown *P. oligandrum* to be a highly successful mycoparasite of several plant pathogens *in vitro* with the ability to exhibit directed growth toward potential hosts, secrete cell wall degrading enzymes and produce some inhibitory metabolites. The relative importance and precise operation of these factors in natural soils is not well defined and is clearly an area for future research. The findings of these fundamental studies of interactions are important for the design of

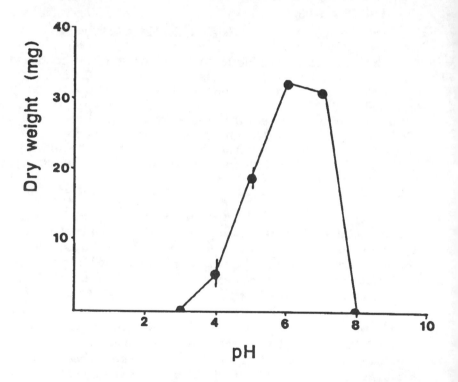

Fig. 9.26. The effect of pH on mycelial dry weight (mg) of 5-day-old colonies of *Pythium oligandrum* grown in V8-Vegetable Juice liquid medium. Each point is the mean of four replicates shown with 95% confidence limits.

future biocontrol studies of *P. oligandrum* in more natural environments. On the basis of these results and earlier studies using seed pelleting, further experiments relating to ecophysiology, inoculum production and oospore biology are also being undertaken, to enable sensible strategies for practical use to be defined.

Oospores are the main survival structures of *P. oligandrum* and are the active starting material used for biocontrol. It is well established that oospores are long-lived and able to withstand stress, including that imposed during the commercial pelleting process (Lutchmeah & Cooke, 1985). Current research is aimed towards mass inoculum production in solid and

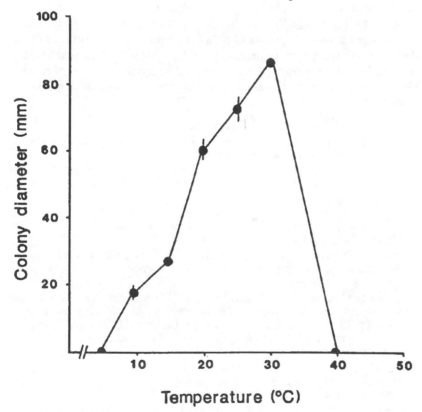

Fig. 9.27. Mean colony diameter of *Pythium oligandrum* following three days incubation on cornmeal agar at different temperatures. Each point is the mean of four replicates shown with 95% confidence limits.

liquid fermentations on various waste products including molasses, whey, bran, cereal grains and straw. The use of new formulations is also being investigated including the application of novel seed pelleting technology and alginate pellets. Through prolonged exposure to increasingly higher concentrations of the fungicide Benomyl, an isolate of *P. oligandrum* with a LD50 twice that of untreated isolates has been obtained. Integrated biological and chemical control experiments using this isolate have also been assessed with some success. It is now possible to envisage genetic manipulation and *de novo* selection of *P. oligandrum* isolates for desirable factors. Evidently there is now a mass of information relating to the mode of action, biology, and physiology of *P. oligandrum* establishing the need for current large scale field trials. The work on *P. oligandrum* presented in this paper provides a good example as to nature and importance of fundamental studies prior to, or in conjunction with, field application of a potential biocontrol agent. A more rational approach of this type will inevitably lead to greater predictability and commercial success of biocontrol systems.

References

Adams, P. B. & Ayers, W. A. (1982). Biological control of sclerotinia drop in the field by *Sporidesmium sclerotivorum*. *Phytopathology*, **72**, 485-488.

Adams, P. B. & Ayers, W. A. (1983). Histological and physiological aspects of infection of sclerotia of *Sclerotinia* species by two mycoparasites. *Phytopathology*, **73**, 1072-1076.

Adams, P. B., Marois, J. J. & Ayers, W. A. (1984). Population dynamics of the mycoparasite, *Sporidesmium sclerotivorum*, and its host, *Sclerotinia minor*, in soil. *Soil Biology & Biochemistry*, **16**, 627-633.

Ahmad, J. S. & Baker, R. (1987). Competitive saprophytic ability and cellulolytic activity of rhizosphere-competent mutants of *Trichoderma harzianum*. *Phytopathology*, **77**, 358-362.

Alabouvette, C., Couteaudier, Y. & Louvet, J. (1985). Soils suppressive to *Fusarium* wilt: mechanisms and management of suppressiveness. In *Ecology and Management of Soilborne Plant Pathogens*, ed. C. A. Parker, A. D. Rovira, K. J. Moore, P. T. W. Wong & J. F. Kollmorgen, pp. 101-106. American Phytopathological Society: St Paul, Minnesota .

Al-Hamdani, A. M. & Cooke, R. C. (1983). Effects of the mycoparasite *Pythium oligandrum* on cellulolysis and sclerotium production by *Rhizoctonia solani*. *Transactions of the British Mycological Society*, **81**, 619-621.

Al-Hamdani, A. M., Lutchmeah, R. S. & Cooke, R. C. (1983). Biological control of *Pythium ultimum*-induced damping-off by treating cress seed with the mycoparasite *Pythium oligandrum*. *Plant Pathology*, **32**, 449-454.

Anagnostakis, S. L. (1982). Biological control of chestnut blight. *Science*, **215**, 466-471.

Ayers, W. A. & Adams, P. B. (1979). Mycoparasitism of sclerotia of *Sclerotinia* and *Sclerotium* species by *Sporidesmium sclerotivorum*. *Canadian Journal of Microbiology*, **25**, 17-23.

Ayers, W. A. & Adams, P. B. (1981). Mycoparasitism and its application to biological control of plant diseases. In *Biological Control in Crop Production*, ed. G. C. Papavizas, pp. 91-103. Allanheld Osmun: Totowa, N.J.

Barak, R., Elad, Y., Mirelman, D. & Chet, I. (1985). Lectins: a possible basis for specific recognition in the interaction of *Trichoderma* and *Sclerotium rolfsii*. *Phytopathology*, **75**, 458-462.

Bell, D. K., Wells, H. D. & Markham, C. R. (1982). *In vitro* antagonism of *Trichoderma* species against six fungal pathogens. *Phytopathology*, **72**, 379-382.

Bullock, S., Adams, P. B., Willetts, H. J. & Ayers, W.A. (1986). Production of haustoria by *Sporidesmium sclerotivorum* in sclerotia of *Sclerotinia minor*. *Phytopathology*, **76**, 101-103.

Caron, C., Richard, C. & Fortin, J. A. (1986). Effect of preinfestation of the soil by a vesicular-arbuscular mycorrhizal fungus, *Glomus intraradices*, on *Fusarium* crown and root rot of tomatoes. *Phytoprotection*, **67**, 15-19.

Chakravarty, P. & Unestam, T. (1987). Differential influence of ectomycorrhizae on plant growth and disease resistance in *Pinus sylvestris* seedlings. *Journal of Phytopathology*, **120**, 104-120.

Chet, I., Harman, G. E. & Baker, R. (1981). *Trichoderma hamatum*: its hyphal interactions with *Rhizoctonia solani* and *Pythium* spp. *Microbial Ecology*, **7**, 29-38.

Claydon, N., Allan, M., Hanson, J. R. & Avent, A. G. (1987). Antifungal alkyl pyrones of *Trichoderma harzianum*. *Transactions of the British Mycological Society*, **88**, 503-513.

Deacon, J. W. (1973). Control of the take-all fungus by grass leys in intensive cereal cropping. *Plant Pathology*, **22**, 88-94.

Deacon, J. W. (1976). Studies on *Pythium oligandrum*, an aggressive parasite of other fungi. *Transactions of the British Mycological Society*, **66**, 383-391.

Deacon, J. W. & Henry, C. M. (1978). Mycoparasitism by *Pythium oligandrum* and *P. acanthicum*. *Soil Biology & Biochemistry*, **10**, 409-415.

Dean, R. A. & Kuc, J. (1987). Rapid lignification in response to wounding and infection as a mechanism for induced systemic protection in cucumber. *Physiological and Molecular Plant Pathology*, **31**, 69-81.

Dehne, H. W. (1982). Interaction between vesicular-arbuscular mycorrhizal fungi and plant pathogens. *Phytopathology*, **72**, 1115-1119.

Dennis, C. & Webster, J. (1971a). Antagonistic properties of species-groups of *Trichoderma*. I. Production of non-volatile antibiotics. *Transactions of the British Mycological Society*, **57**, 25-39.

Dennis, C. & Webster, J. (1971b). Antagonistic properties of species-groups of *Trichoderma*. II. Production of volatile antibiotics. *Transactions of the British Mycological Society*, **57**, 41-48.

De Oliveira, V., Bellei, M. de M. & Borges, A. C. (1984). Control of white rot of garlic by antagonistic fungi under controlled environmental conditions. *Canadian Journal of Microbiology*, **30**, 884-889.

de Wit, P. J. G. M. (1985). Induced resistance to fungal and bacterial diseases. In *Mechanisms of Resistance to Plant Diseases*, ed. R. S. S. Fraser, pp. 405-424. Nijhoff/Junk: Dordrecht.

Dick, M. W., Ali-Shtayeh & Mohamed, S. (1986). Distribution and frequency of *Pythium* species in parkland and farmland soils. *Transactions of the British Mycological Society*, **86**, 49-62.

Drechsler, C. (1930). Some new species of *Pythium*. *Journal of the Washington Academy of Sciences*, **20**, 398-418.

Drechsler, C. (1946). Several species of *Pythium* peculiar in their sexual development. *Phytopathology*, **36**, 781-864.

Elad, Y., Barak, R. & Chet, I. (1983a). Possible role of lectins in mycoparasitism. *Journal of Bacteriology*, **154**, 1431-1435.

Elad, Y., Chet, I., Boyle, P. & Henis, Y. (1983b). Parasitism of *Trichoderma* spp. on *Rhizoctonia solani* and *Sclerotium rolfsii* - scanning electron microscopy and fluorescence microscopy. *Phytopathology*, **73**, 85-88.

Foley, M. F. & Deacon, J. W. (1985). Isolation of *Pythium oligandrum* and other necrotrophic mycoparasites from soil. *Transactions of the British Mycological Society*, **85**, 631-639.

Foley, M. F. & Deacon, J. W. (1986). Susceptibility of *Pythium* spp. and other fungi to antagonism by the mycoparasite *Pythium oligandrum*. *Soil Biology & Biochemistry*, **18**, 91-95.

Fravel, D. R. (1988). Role of antibiosis in the biocontrol of plant diseases. *Annual Review of Phytopathology*, **26**, 75-91.

Fritig, B., Kauffmann, S. Dumas, B., Geoffroy, P., Kopp, M. & Legrand, M. (1987). Mechanism of the hypersensitivity reaction of plants. In *Plant Resistance to Viruses*, Ciba Foundation Symposium 133, pp. 92-108. Wiley: Chichester.

Gianinazzi, S., Ahl, P., Cornu, A. & Scalla, R. (1980). First report of host b-protein appearance in response to a fungal infection in tobacco. *Physiological Plant Pathology*, **16**, 337-342.

Grente, J. & Sauret, S. (1969a). L'"hypovirulence exclusive' est-elle contrôlée par des determinants cytoplasmiques? *Comptes Rendus Hebdomadaires des Séances de l'Academie des Sciences, Séries D*, **268**, 3173-3176.

Grente, J. & Sauret, S. (1969b). L'hypovirulence exclusive phénomène original en pathologie végétale. *Comptes Rendus Hebdomadaires des Séances de l'Academie des Sciences, Séries D*, **268**, 2347-2350.

Hammerschmidt, R., Lamport, D. T. A. & Muldoon, E. (1984). Cell wall hydroxyproline enhancement and lignin deposition as an early event in the resistance of cucumber to *Cladosporium cucumerinum*. *Physiological Plant Pathology*, **24**, 43-47.

Hillocks, R. J. (1986). Cross protection between strains of *Fusarium oxysporum* f. sp. *vasinfectum* and its effect on vascular resistance mechanisms. *Journal of Phytopathology*, **117**, 216-225.

Howell, C. R. (1982). Effect of *Gliocladium virens* on *Pythium ultimum*, *Rhizoctonia solani*, and damping-off of cotton seedlings. *Phytopathology*, **72**, 496-498.

Howell, C. R. & Stipanovic, R. D. (1980). Suppression of *Pythium ultimum* induced damping-off of cotton seedlings by *Pseudomonas fluorescens* and its antibiotic pyoluteorin. *Phytopathology*, **70**, 712-715.

Howell, C. R. & Stipanovic, R. D. (1983). Gliovirin, a new antibiotic from *Gliocladium virens*, and its role in the biological control of *Pythium ultimum*. *Canadian Journal of Microbiology*, **29**, 321-324.

Huang, H. C. (1980). Control of sclerotinia wilt of sunflower by hyperparasites. *Canadian Journal of Plant Pathology*, **12**, 26-32.

Hubbard, J. P., Harman, G. E. & Eckenrode, C. J. (1982). Interaction of a biological control agent, *Chaetomium globosum*, with seed coat microflora. *Canadian Journal of Microbiology*, **28**, 431-437.

Kirk, J. J. & Deacon, J. W. (1987). Control of the take-all fungus by *Microdochium bolleyi*, and interactions involving *M. bolleyi*, *Phialophora graminicola* and *Periconia macrospinosa* on cereal roots. *Plant & Soil*, **98**, 231-237.

Koltin, Y., Finkler, A. & Ben-Zvi, B. -S. (1987). Double-stranded RNA viruses of pathogenic fungi: virulence and plant protection. In *Fungal Infection of Plants*, ed. G. F. Pegg & P. G. Ayres, pp. 334-348. Cambridge University Press.

Lewis, K. (1988). Biological control mechanisms of the mycoparasite *Pythium oligandrum* Drechsler. Ph.D. thesis, University of Sheffield.

Lifshitz, R., Windham, M. T. & Baker, R. (1986). Mechanism of biological control of preemergence damping-off of pea by seed treatment with *Trichoderma* spp. *Phytopathology*, **76**, 720-725.

Lumsden, R. D. (1981). Hyperparasitism for control of plant pathogens. In *CRC Handbook of Pest Management in Agriculture Vol. I*, ed. D. Pimentel, pp. 475-484. CRC Press: Boca Raton, Florida.

Lutchmeah, R. S. & Cooke, R. C. (1984). Aspects of antagonism by the mycoparasite *Pythium oligandrum*. *Transactions of the British Mycological Society*, **83**, 696-700.

Lutchmeah, R. S. & Cooke, R. C. (1985). Pelleting of seed with the antagonist *Pythium oligandrum* for biological control of damping-off. *Plant Pathology*, **34**, 528-531.

Magan, N. & Whipps, J. M. (1988). Growth of *Coniothyrium minitans*, *Gliocladium roseum*, *Trichoderma harzianum* and *T. viride* from alginate pellets and interaction with water availability. *EPPO Bulletin*, **18**, 37-45.

Martin, F. N. & Hancock, J. G. (1986). Association of chemical and biological factors in soils suppressive to *Pythium ultimum*. *Phytopathology*, **76**, 1221-1231.

Martin, F. N. & Hancock, J. G. (1987). The use of *Pythium oligandrum* for biological control of preemergence damping-off caused by *P. ultimum*. *Phytopathology*, **77**, 1013-1020.

Marx, D. H. (1972). Ectomycorrhizae as biological deterrents to pathogenic root infection. *Annual Review of Phytopathology*, **10**, 429-454.

Matta, A. & Garibaldi, A. (1977). Control of *Verticillium* wilt of tomato by preinoculation with avirulent fungi. *Netherlands Journal of Plant Pathology*, **83** (Suppl. 1), 457-462.

Metraux, J. P., Streit, L. & Staub, T. (1988). A pathogenesis-related protein in cucumber is a chitinase. *Physiological and Molecular Plant Pathology*, **33**, 1-9.

Ogawa, K. & Komada, H. (1985). Biological control of fusarium wilt of sweet potato with cross-protection by nonpathogenic *Fusarium oxysporum*. In *Ecology and Management of Soilborne Plant Pathogens*, ed. C. A. Parker, A. D. Rovira, K. J. Moore, P. T. W. Wong & J. F. Kollmorgen, pp. 121-123. American Phytopathological Society: St. Paul, Minnesota.

Papavizas, G. C. & Lumsden, R. D. (1980). Biological control of soilborne fungal propagules. *Annual Review of Phytopathology*, **18**, 389-413.

Paulitz, T. C. & Baker, R. (1987). Biological control of *Pythium* damping-off of cucumbers with *Pythium nunn*: influence of soil environment and organic amendments. *Phytopathology*, **77**, 341-346.

Paulitz, T. C., Park, C. S. & Baker, R. (1987). Biological control of *Fusarium* wilt of cucumber with nonpathogenic isolates of *Fusarium oxysporum*. *Canadian Journal of Microbiology*, 33, 349-353.

Ridout, C. J., Coley-Smith, J. R. & Lynch, J. M. (1988). Fractionation of extracellular enzymes from a mycoparasitic strain of *Trichoderma harzianum*. *Enzyme & Microbial Technology*, 10, 180-187.

Roby, D., Toppan, A., Esquerre-Tugaye, M. -T. (1987). Cell surfaces in plant microorganism interactions. VIII. Increased proteinase inhibitor activity in melon plants in response to infection by *Colletotrichum lagenarium* or to treatment with an elicitor fraction from this fungus. *Physiological and Molecular Plant Pathology*, 30, 453-460.

Rogers, H. J., Buck, K. W. & Brasier, C. M. (1986). Transmission of double-stranded RNA and a disease factor in *Ophiostoma ulmi*. *Plant Pathology*, 35, 277-287.

Rosendahl, S. (1985). Interactions between the vesicular-arbuscular mycorrhizal fungus *Glomus fasciculatum* and *Aphanomyces euteiches* root rot of peas. *Phytopathologische Zeitschrift*, 114, 31-40.

Sneh, B., Agami, O. & Baker, R. (1985). Biological control of *Fusarium*-wilt in carnation with *Serratia liquefaciens* and *Hafnia alvei* isolated from rhizosphere of carnation. *Phytopathologische Zeitschrift*, 113, 271-276.

Tribe, H. T. (1966). Interactions of soil fungi on cellulose film. *Transactions of the British Mycological Society*, 49, 457-466.

Van Alfen, N. K. & Hansen, D. R. (1984). Hypovirulence. In *Plant Microbe Interactions. Molecular and Genetic Perspectives Vol. 1*, ed. T. Kosuge & E. W. Nester, pp. 400-419. New York: MacMillan.

Van Alfen, N. K., Jaynes, R. A., Anagnostakis, S. L. & Day, P. R. (1975). Chestnut blight: biological control by transmissible hypovirulence in *Endothia parasitica*. *Science*, 189, 890-891.

Veseley, D. (1977). Potential biological control of damping-off pathogens in emerging sugar beet by *Pythium oligandrum*. *Phytopathologische Zeitschrift*, 90, 113-115.

Veseley, D. (1978). Parasitic relationships between *Pythium oligandrum* Drechsler and some other species of the Oomycetes class. *Zentraalblatt für Bakteriologie und Parasitkunde*, II Abt. Bd, 133, 341-349.

Veseley, D. (1979). Use of *Pythium oligandrum* to protect emerging sugar-beet. In *Soil-Borne Plant Pathogens*, ed. B. Schippers & W. Gams, pp. 593-595. Academic Press: London.

Veseley, D. & Hejdanek, S. (1984). Microbial relations of *Pythium oligandrum* and problems in the use of this organism for the biological control of damping-off in sugar beet. *Zentralblaat für Mikrobiologie*, 139, 257-265.

Walther, D. & Gindrat, D. (1987). Biological control of *Phoma* and *Pythium* damping-off of sugar-beet with *Pythium oligandrum*. *Journal of Phytopathology*, 119, 167-174.

Whipps, J. M. (1986). Use of microorganisms for biological control of vegetable diseases. *Aspects of Applied Biology*, 12, 75-94.

Whipps, J. M. (1987). Effect of media on growth and interactions between a range of soil-borne glasshouse pathogens and antagonistic fungi. *New Phytologist*, 107, 127-142.

Whipps, J. M. & Magan, N. (1987). Effects of nutrient status and water potential of media on fungal growth and antagonist-pathogen interactions. *EPPO Bulletin*, **17**, 581-991.

Whipps, J. M., Budge, S. P. & Ebben, M. H. (1989). Effect of *Coniothyrium minitans* and *Trichoderma harzianum* on *Sclerotinia* disease of celery and lettuce in the glasshouse at a range of humidities. In *Proceedings of CEC Joint Experts Meeting, Cabrils, Spain 27-29 May 1987; Integrated Pest Management in Protected Crops*, in press.

Whipps, J. M., Lewis, K. & Cooke, R. C. (1988). Mycoparasitism and plant disease control. In *Fungi in Biological Control Systems*, ed. M. N. Burge, pp. 161-187. Manchester University Press.

Williams, S. T. & Vickers, J. C. (1986). The ecology of antibiotic production. *Microbial Ecology*, **12**, 43-52.

Wong, P. T. W. (1975). Cross-protection against the wheat and oat take-all fungi by *Gaeumannomyces graminis* var. *graminis*. *Soil Biology and Biochemistry*, **71**, 189-194.

Chapter 10

Some perspectives on the application of molecular approaches to biocontrol problems

Ralph Baker

Plant Pathology and Weed Science, Colorado State University, Fort Collins, Colorado 80523, USA

Introduction

Cook & Baker (1983) list 44 agents used in biocontrol, 25 being fungi that have been used in control of plant diseases other than those induced by nematodes. The mechanisms of action of these agents (Baker, 1984) range from those that operate intimately within a host, such as by inducing resistance and hypovirulence, to those acting antagonistically to pathogens before or during penetration of a host (antibiosis, competition, exploitation).

Considerable attention has been directed toward the use of bacterial agents not only in biocontrol but also in induction of increased growth responses (Kloepper, Lifshitz & Schroth, 1988); however, fungi may possess superior beneficial attributes. For example, inconsistent control by bacteria is primarily related to variable rhizosphere colonization (Weller, 1988). Indeed, only a proportion of the roots of a plant may have detectable rhizosphere populations of a bacterium introduced by seed treatment (Bahme & Schroth, 1987). Further, those that are colonized may harbour the agent at only a fraction of the population density of the total bacteria and densities may decrease drastically at the extremities of roots. Weller (1988) has reviewed traits associated with rhizosphere competence, especially in bacteria. These include attachment to roots, possession of flagella, chemotaxis and osmotolerance. The importance of these attributes to rhizosphere competence in bacteria is still conjectural and it seems less likely they would be key factors for compctcnce in fungi. However, rhizosphere-competent mutants of *Trichoderma* colonize all roots from a seed treatment and maintain a high population density to the tip of the root (Ahmad & Baker, 1987a & b, 1988a & c).

There is also the possibility that fungi possess attributes useful in involving a wider variety of mechanisms for biocontrol than is characteristic of bacteria (Baker, 1984). For instance, many avirulent strains of fungi possess the ability to breech intact host barriers and invade tissues, thereby potentially inducing resistance. Bacteria may operate efficiently only by activity in wounds (e.g. Kerr, 1980). Parasitic bacteria, like *Bdellovibrio bacterovorus*, are rarely thought of as biocontrol agents whereas mycoparasitism is the basis for biocontrol in many systems (Baker, 1988a). Perhaps the most favourable attribute possessed by fungi is hyphal extension. This permits rapid colonization, utilization and subsequent antagonistic activity in possessing substrates (resource capture, see Rayner & Webber, 1984) which can be vital in rhizosphere colonization (Ahmad & Baker, 1987a) or in colonization of soil organic material leading to decreased inoculum potential of soilborne pathogens (Paulitz & Baker, 1987a & b, 1988a & b). Such saprotrophic activity can provide the relatively large amounts of carbon substrate necessary for production of antibiotics.

Both fungi (Baker, 1988b) and bacteria (Kloepper *et al.*, 1988) may induce increased growth responses in plants. Although the evidence for production of plant growth stimulating hormones by bacteria is not convincing, a diffusible factor inducing increased growth produced by *Trichoderma* spp. has been demonstrated (Windham, Elad & Baker, 1986). Finally, frequently used bacterial biocontrol agents, such as *Pseudomonas* spp., usually do not have specialized resting structures whereas fungi produce such propagules and are better equipped to persist for long periods in soil.

Thus, fungi exhibit many desirable attributes which might be exploited for use in biocontrol programmes. The challenge is to establish the molecular basis for these fungal activities so that they can be manipulated. The most important factor limiting progress is our ignorance of genetic determinants associated with biocontrol and growth promotion by fungi. Progress has been made in research on qualitative and even quantitative aspects of some of the mechanisms involved in biocontrol (e.g. Baker, 1968, 1984, 1988a & b, Baker & Scher, 1987; Cook & Baker 1983; Papavizas, 1987; Salt, 1978) and this can be used to accent the attributes which remain to be clearly identified and described.

Genetic manipulations

Biocontrol agents must obviously be manipulated and improved in order to be effectively used for control of plant diseases. In many cases the biotechnologist will be attempting to adapt to his own use genetic information which has evolved to serve totally different purposes. This principle is often overlooked in biological control. For example, fungi which are seen as potential biocontrol agents have developed numerous mechanisms for activity and survival in their appropriate ecological niches. Only some of these characteristics may be considered to be beneficial in growth promotion and/or biocontrol, so the challenge is to manipulate the organism so that its performance matches more closely the purpose for which it is to be employed.

The prime strategy for improvement is by genetic manipulation. Thus, fungal biocontrol agents could acquire desirable traits by sexual recombination, but of the 25 fungal biocontrol agents listed by Cook & Baker (1983), 13 are members of the Deuteromycotina where sexual stages are rarely or never observed. The remainder may produce sexual propagules but there are difficulties in producing these in culture; in any case, genetic modification by classical sexual recombination has not been used to any great extent in biocontrol for acquisition of desirable traits.

There are three methods that have the most potential for genetic manipulation of fungi without functional sexual stages. These methods include conventional mutagenesis, protoplast fusion and transformation. Papavizas (1987) provided a good review of the basic approaches used in such manipulations.

Ultraviolet light (UV) mutation of *Trichoderma* spp. to induce benomyl resistance improved the efficiency of these antagonists in biocontrol of *Rhizoctonia solani*, *Pythium ultimum*, *Sclerotium cepivorum*, and *S. rolfsii* (Papavizas, 1987). Several UV-induced mutants of *T. harzianum*, unlike wild type strains, produced two antibiotic metabolites in fermentation medium. The strain producing the highest amounts of these metabolites was the most effective biocontrol agent against *S. cepivorum*. Besides induction of antibiotic production by species not previously producing such metabolites (Papavizas, 1987), it is possible to increase the yield of antimicrobial compounds. Howell & Sipanovic (1983) used UV to induce a mutant of *Gliocladium virens* that had enhanced ability to produce gliovirin and which provided good biocontrol.

The successful application of the technique of protoplast fusion has potential value in genetic recombination of species in which sexual or parasexual mechanisms are not present or are difficult to exploit. Techniques and applications have been reviewed extensively in the literature (e.g. see Chapter 11 and Peberdy, 1980).

While the isolation and culture of protoplasts from a wide range of microorganisms is now possible, there are few studies related to fungal biocontrol agents. Seh & Kenerley (1988) fused protoplasts and regenerated three *Gliocladium* spp., and Harman and associates (Stasz, Harman & Weeden, 1988) fused a rhizosphere-competent mutant (T-95) with a wild type strain (T-12), both members of the *T. harzianum* group. Personal communication with these scientists indicates that their experience with fusants parallels observations reviewed by Peberdy (1980) – namely, that there is a complex pattern of chromosome segregation that defies traditional criteria used in the recovery of segregants. Clearly, fused protoplasts must be screened for stability of altered genetic characteristics and for potential improvement in biocontrol capabilities. In soils infested with *Pythium ultimum* and planted with various crops, seed treatment with either parent or fusant strains of *T. harzianum* provided increased stands compared with untreated controls and were as effective as treatment with thiram (Harman, Taylor & Stasz, 1989). Of most interest was the observation that stands were improved when seeds were treated with the two progeny strains compared with those treated with either parent. The progeny also induced more robust cucumber seedlings than the parent strains as measured by greater root volumes.

Transformation of fungi is also possible (Hynes, 1986). A variety of vectors with markers is available to insert genes stably into the fungal genome. Applications of this technology to biocontrol with fungal agents is just beginning. Henson, Blake & Pilgeram (1988) transformed *Gaeumannomyces graminis* var. *graminis* and var. *tritici* to achieve benomyl resistance by using *p*BT6 (the plasmid encoding fungicide-resistant *β*-tubulin). Such transformations enable study of pathogenic determinants since *G. graminis* var. *graminis* is not virulent to wheat and transfer of virulence genes from *G. graminis* var. *tritici* to *G. graminis* var. *graminis* may be possible. In addition, *G. graminis* var. *graminis* induced increased resistance to *G. graminis* var. *tritici* in wheat (Wong & Southwell, 1980) and could potentially be combined with benomyl seed treatments to achieve integrated chemical and biological control.

In another potential application, cloning technology could be used for increased efficiencies of biocontrol agents. For example, Turgeon & Yoder (1985) speculate that genes promoting toxin production could be transferred to a mycoherbicide. Since mycoherbicides currently under development are host specific, construction of pathogens which produce toxins could result in enhanced virulence of these biological control agents of weeds. This should ensure reduction in a weed population in a way that is environmentally safe.

Plasmids have also been found in strains of antagonists belonging to the *Gliocladium virens* complex. The majority of the strains examined contained one or more plasmids in the mitochondrial DNA fraction (Mischke, 1988a & b). Cloning and mapping are underway to determine relationships among the plasmids.

Usefulness of genetic modifications in research and practical applications

Attempts to enhance the activity of biocontrol agents have not been initiated systematically to any great extent in the past (Baker & Scher, 1987) although advances in biotechnology have provided many new tools and opportunities.

A recurrent problem has been the inability to detect, monitor, and recover microorganisms purposely released into the environment. This became even more of a concern when regulatory agencies required monitoring of released microorganisms—especially those that were genetically modified. Benomyl resistant mutants of *Trichoderma* spp. (Ahmad & Baker, 1987b; Papavizas 1985, 1987) are useful in such studies and applications. If applied to soil, the mutant strains can be isolated in pure culture on media supplemented with benomyl which eliminates contamination by many fungi commonly found in soil.

Other strategies for detection of labelled fungi have been developed. Schneider (1984) exposed *Fusarium óxysporum* f.sp. *apii* to UV and selected orange mutants that were used to aid quantification of root infection. Hadar, Katan & Katan (1989) grew pathogenic strains on a medium containing chlorate and selected chlorate-resistant nitrate-non-utilizing (*nit*) mutants of *F. oxysporum*. In these cases, not only is pathogenicity retained but survival in soil is comparable to wild type parents. Beyond use for detection of released strains in the environment, labelling is a powerful tool in ecological studies.

The mitochondrial DNA of *Talaromyces flavus*, an effective antagonist of *Verticillium dahliae* (Marois *et al.*, 1982), has been analyzed (Mischke & Papavizas, 1987). The cytochrome oxidase subunit III gene was located on a specific fragment of mitochondrial DNA by Southern hybridization experiments. This technique may be potentially useful for strain identification in field experiments.

Considerable attention in the past has centered on the significance of antibiosis as a mechanism of biocontrol (Baker, 1984). Substrates suitable for inducing antibiotic production in soil are necessary if this mechanism is to be employed in biocontrol. Soils typically lack such substrates unless they are present in added organic matter, dead seed coats or lesions on roots. Even when they are produced, antibiotics may be inactivated by adsorption on clay colloids. In spite of these factors, though, the role of antibiosis in certain systems with microniches favourable for production of antibiotics is being elucidated. Howell & Stipanovic (1983) used a gliovirin-deficient mutant of *Gliocladium virens* as a seed treatment in biocontrol of *Pythium ultimum*. The mutant did not protect cotton seedlings from damping-off while gliovirin-producing wild types did. In another approach, Howell (1987) eliminated the mycoparasitic attributes of the same antibiotic producing antagonist by mutation and demonstrated that it still induced biocontrol of *Rhizoctonia* damping-off in cotton seedlings.

Rhizosphere competence

A desirable attribute for fungal biocontrol agents is the ability to colonize plant rhizospheres (rhizosphere competence). This attribute is usually a characteristic of only certain bacteria (Weller, 1988) and potentially could insure that the developing roots would be protected by seed treatment application of the biocontrol agent. But neither wild types nor UV-light induced mutants of *Trichoderma* spp. developed by Papavizas (1985) were rhizosphere competent. Mutations to rhizosphere competence, however, were induced by treatment with N-methyl-N'-nitro-N-nitrosoguanidine (NTG)(Ahmad & Baker, 1987b).

The hypothesis originally advanced to account for the induction of rhizosphere competence was that benomyl-resistant mutants of *Trichoderma* could colonize rhizospheres more easily when benomyl was mixed into soil. The expectation being that benomyl-susceptible fungi competing in the rhizosphere would be limited in their activities, thus conferring advantage to the mutants. This strategy resulted in colonization from a seed treatment of bean, cucumber, maize, radish, and tomato at population densities approaching 10^6 colony forming units (cfu) g^{-1} rhizosphere soil 8 cm below the seed (Ahmad & Baker, 1987b). A surprising observation was that rhizosphere colonization of mutants occurred at comparable levels when benomyl was not added to the soil. Mutants colonized rhizospheres at soil reactions from pH 5-7 and temperatures of 19-33°C (Fig. 10.1).

There is no conclusive evidence that benomyl resistance *per se* is necessary for colonization of rhizospheres by mutants of *Trichoderma* spp. Upon exposure of 10^6 conidia to NTG in the mutation process described by Ahmad & Baker (1987b), about 10^4 conidia typically survive (Ahmad & Baker, 1988c) of which approximately 10 can be shown to be benomyl resistant. Depending on species, approximately 95% of the benomyl-resistant mutants may be rhizosphere-competent. Therefore, benomyl resistance is not completely correlated with rhizosphere competence but appears to be associated with this characteristic in high frequency.

While the generation of rhizosphere-competent mutants is easily accomplished with fast-growing *T. harzianum*, the task was particularly challenging with *T. polysporum* since this species has a linear growth rate much lower than that of roots of typical crop plants. Rhizosphere competence was accomplished in this instance, however, by selecting mutants with increased growth rates compared with wild types (Ahmad & Baker, 1988c).

The rhizosphere-competent mutants of *Trichoderma* spp. (Ahmad & Baker, 1987b) afforded an excellent opportunity to study attributes asso-

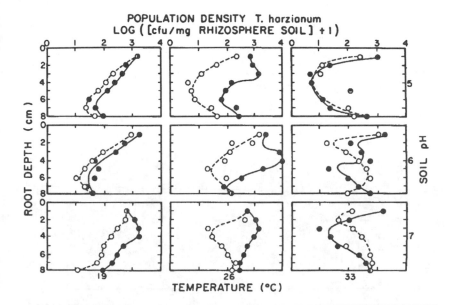

Fig. 10.1 Population densities of a rhizosphere-competent mutant of *Trichoderma harzianum* (T-95) in rhizosphere soil of cucumber plants grown at three pH levels and three temperatures with benomyl mixed in the soil (solid symbols) and without benomyl (open symbols). From Ahmad & Baker (1987b) with permission.

ciated with such competence in fungi. As indicated above, correlation between benomyl resistance of mutants and rhizosphere competence was not perfect, but there was a perfect correlation between increased cellulase production and rhizosphere competence (Ahmad & Baker, 1988c) – even when benomyl resistant, mutants were not rhizosphere-competent unless they also expressed increased cellulase production in comparison with wild type.

The association of the increased ability to utilize cellulose substrates by the mutants with rhizosphere competence prompted a hypothesis explaining mechanisms associated with rhizosphere competence. Foster, Rovira & Cock (1983) identified the main component of the mucigel of roots as the remains of the epidermal cell walls. Such primary components of cells are rich in cellulose. This suggests that enhanced cellulose utilization by mutants, in comparison with wild type strains, enabled efficient colonization of this substrate by the mutants resulting in growth of the fungus on the rhizoplane of roots. Indeed, 30 to 50 mm total hyphae cm^{-1} of root were observed on developing roots from seeds treated with a rhizosphere-competent mutant, whereas 5 to 20 mm total hyphae cm^{-1} of root

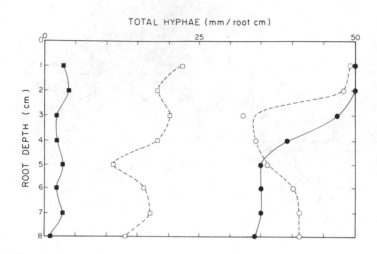

Fig. 10.2. Total length of hyphae observed per root centimetre depth of cucumber plants grown from seeds treated with *Trichoderma harzianum* strain T-95 (circles) and seeds not treated (squares) sown in soil with (closed symbols) or without (open symbols) benomyl. From Ahmad & Baker (1987b) with permission.

were detected with a non-rhizosphere competent wild type strain (Fig. 10.2)

The ability of different strains and their rhizosphere-competent mutant progeny to colonize wheat straw or cellulose in soil was determined by Ahmad & Baker (1987a). A competitive saprotrophic ability (CSA) index was computed for each isolate. CSA indices were correlated directly with production of cellulase and with a rhizosphere competence (RC) index. Given a *Trichoderma* spp. with linear growth rate comparable with that of roots, these observations suggest that mutation to increased cellulase production is a key factor in colonization of the mucigel.

Pure culture studies comparing growth of wild types and mutants on carbon substrates of cotton linters, microcrystalline cellulose, wood cellulose, xylan, or simple carbohydrate derivatives of these materials

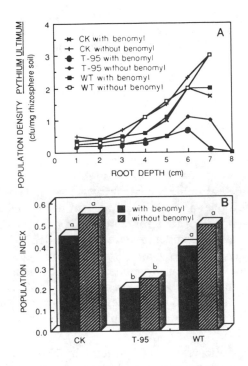

Fig. 10.3. Spacial distribution (A) and population index (B) of *Pythium ultimum* in rhizosphere soil of 8-day old cucumber plants grown from seeds treated with *Trichoderma harzianum* strains T-95 and the wild type parent (WT). Bars labelled with the same letters are not significantly different ($P = 0 \cdot 05$). From Ahmad & Baker (1988b) with permission.

revealed other aspects contributing to development of this theory (Ahmad & Baker, 1988a). Both mutants and wild types grew slowly when glucose, galactose, cellobiose, or xylose were the sole carbon source. Rhizosphere competent mutants, however, developed four to five times the biomass of their wild type parents when grown on complex carbohydrate substrates.

Compared with wild types, the mutants produced significantly greater weights of mycelium on three qualitatively different forms of cellulose and xylan all of which are components of the mucigel (Foster *et al.*, 1983): however, simple sugars are also present in the root exudates (Schroth & Hildebrand, 1964) and these could cause catabolite repression (Montencourt, 1983). One rhizosphere-competent strain (T-95) had significantly higher production of mycelial dry weight on a cellulose substrate when glucose was added at concentrations of $0 \cdot 5$ to 3% by weight (Ahmad & Baker, 1988a) as compared with weights generated on cellulose alone. In contrast, another rhizosphere-competent mutant (T-12B) produced significantly less growth when glucose was added to the cellulose substrate. It was interesting that strain T-95 had a lower RC index than T-12B (Ahmad & Baker, 1988b). This might be predicted from their comparative growth rates on the cellulose substrate amended with simple carbohydrate–an analogue of the mucigel-exudate interaction (Ahmad & Baker, 1988a).

The question arises as to the significance in fungi of mutations to rhizosphere competence. Hypothetically, such an attribute would protect both the seed and developing roots from attack by pathogens (Papavizas, 1985). Indeed, this phenomenon was apparent when conidia of a rhizosphere competent mutant were applied to cucumber seeds and compared with its non-rhizosphere competent parent (Ahmad & Baker, 1988b). The population density index of *Pythium ultimum* was significantly reduced along 8 cm of root when the rhizosphere-competent mutant was applied to the seed in contrast to seed treatment with the wild type parent (Fig 10.3). The same relationships were observed in sporangial germination of *P. ultimum* in the presence of rhizosphere competent or incompetent strains (Fig. 10.4).

The most dramatic responses induced by rhizosphere-competent strains was observed in connection with the increased growth response of plants. *Trichoderma* spp. induced increases in dry weight of many plant species by up to 300% (Chang *et al.*, 1986). The increased growth induced in apparently 'healthy' plants by *Trichoderma* spp. may be due to control of minor pathogens (Salt, 1978) or to production of a growth stimulating factor (Windham *et al.*, 1986). *Pythium* spp. may well be a primary minor pathogen (Salt, 1978) and it is known that *Trichoderma* spp. are antagonists to these parasites (Baker, 1988b; Papavizas, 1985). Apparently, this is the explanation for increased incidence of emergence and resultant plant growth when seed were treated with rhizosphere-competent strains in comparison with non-rhizosphere competent wild type parents (Ahmad & Baker, 1988b).

With the induction of resistance or the inherent tolerance of fungal biocontrol agents (like *Trichoderma* spp.) to fungicides, it logically follows that such agents could be integrated with chemical controls. Thus, control of damping-off of radish by both seed treatment with *T. harzianum* (T-95) and soil incorporation of benodanil was additive (Lifshitz, Lifshitz & Baker, 1985). Papavizas (1987) reviewed similar strategies for manipulating the genomes of biocontrol agents so that they could operate efficiently in a biological-chemical control system. Again, conidia of *T.*

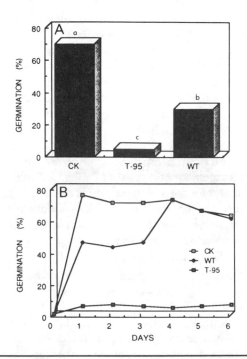

Fig. 10.4. Germination of sporangia of *Pythium ultimum* on a membrane placed on (A) cucumber seeds, and (B) rhizosphere soil of cucumber seedlings treated with *Trichoderma harzianum* strains T-94 and WT (wild type). Bars labelled with the same letters are not significantly different ($P = 0.05$). From Ahmad & Baker (1988b) with permission.

harzianum (T-95) applied in rooting hormone to carnation cuttings reduced incidence of *Rhizoctonia* rot 50% more during propagation (Baker & Scher, 1987). A similar degree of control was noticed by adding benomyl to the hormone solution. But when both were added, 100% control was achieved.

Another strategy for increasing efficiency of biocontrol agents is to combine seed treatment with solid matrix priming at favourable pH levels (Harman & Taylor, 1988). Whereas stands of peas in field tests were not significantly enhanced by seed treatment with *Trichoderma* strains in the absence of priming, they were improved if such priming was applied (Harman *et al.*, 1989). Moreover, strains derived from protoplast fusion were more effective than the parent strains.

Biocontrol: possibilities and constraints

From this brief overview of ongoing research related to the potential for molecular manipulation of biocontrol fungal agents, it is evident that relatively few laboratories are investigating these strategies in comparison with the total effort now underway in the biological sciences. This situation is due, at least in part, to the diminishing support for research in the agricultural sciences – 67% reduction in the last decade in the USA alone according to data released by the Office of Legislative and Public Affairs, National Science Foundation. This occurred even in the face of enormous apprehension over the wide distribution of chemical pesticides. Nevertheless, biological control is a modern, comprehensive, non-polluting approach to the management of diseases (Tauber & Baker, 1988). It is obvious, however, that biocontrol agents must be enhanced for greater efficiency if they are to compete successfully with chemicals. Identification and enhancement of mechanisms involved in biological control is a logical strategy for solution of the problem.

Mechanisms are being identified in some systems (e.g. Baker, 1968, 1984; Baker & Scher, 1987) and there is at least some perception of the potentially transferable genetic characteristics that might be incorporated into biocontrol agents for enhancement; for example, the intricate steps leading to mycoparasitism are now being elucidated (Baker, 1988a). The question remains, however, as to how such mechanisms may be enhanced by inducing increased virulence and/or by expanding host range. Hypothetically, virulence might be increased by enhancing the production of chitinase for improved penetration of fungal pathogen cell walls. This might be accomplished by transfer of genes from chitinolytic microorganisms or by mutation. Similarly, a wider host range could be accomplished by transfer of genes promoting the production of enzymes aiding penetration of previously resistant cell walls (Elad, Lifshitz & Baker, 1985). Again, if the production of cellulase coupled with growth rate comparable to elongation of roots enables colonization of the mucigel of roots, which in turn leads to rhizosphere competence (Baker & Ahmad, 1987a), a fungal biocontrol agent artificially endowed with such properties could be useful if applied as a seed treatment. Other possibilities abound. Basic studies involving recognition and host reactions in host-pathogen relationships could be applied to engineer an agent capable of occupying host tissue without pathogenicity, where it could induce resistant reactions to pathogens but still remain immune itself (Weltring *et al.*, 1988). Antibiotic production could be increased or the molecules altered so that they are more toxic and/or persist for longer periods in soil and/or rhizospheres. Double stranded RNA may be engineered to increase effectiveness of hypovirulent strains of fungi such as *Endothia parasitica*. In the same systems, incompatibility factors could be negated so that anastomosis occurs with any strain of the pathogen.

A speculator, however, can only dream of such possibilities until the present constraints in biotechnology of fungi are removed. The complexity of genes associated with mechanisms make manipulation difficult and vectors for gene transfer in fungi are limited. Few fungal biocontrol agents have sexual stages so that crosses have to be performed by protoplast fusion and in such manipulations there is no predictable genetic outcome or pattern (Peberdy, 1980). Nevertheless, the promising potentials of the new biotechnology have changed the direction of research in the biological sciences. It has been enthusiastically incorporated into existing biological control programmes as a powerful tool in identifying mechanisms and describing ecological interactions.

References

Ahmad, J. S. & Baker, R. (1987a). Competitive saprophytic ability and cellulytic activity of rhizosphere-competent mutants of *Trichoderma harzianum*. *Phytopathology*, 77, 358-362.

Ahmad, J. S. & Baker, R. (1987b). Rhizosphere competence of *Trichoderma harzianum*. *Phytopathology*, 77, 182-189.

Ahmad, J. S. & Baker, R. (1988a). Growth on carbon substrates of rhizosphere-competent mutants of *Trichoderma harzianum*. *Canadian Journal of Microbiology*, 34, 807-814.

Ahmad, J. S. & Baker, R. (1988b). Implications of rhizosphere-competence of *Trichoderma harzianum*. *Canadian Journal of Microbiology*, 34, 229-234.

Ahmad, J. S. & Baker, R. (1988c). Rhizosphere competence of benomyl tolerant mutants of *Trichoderma* spp. *Canadian Journal of Microbiology*, 34, 694-696.

Bahme, J. B. & Schroth, M. N. (1987). Spatial-temporal colonization patterns of a rhizobacterium on underground organs of potatoes. *Phytopathology*, 77, 1093-1100.

Baker, R. (1968). Mechanisms of biological control of soil-borne pathogens. *Annual Review of Phytopathology*, 6, 263-294.

Baker, R. (1984). Biological control of plant pathogens: Definitions. In *Biological Control in Agricultural IPM Systems*, ed. M. A. Hoy & D. C. Herzog, pp. 24-39. Academic Press: New York.

Baker, R. (1988a). Mycoparasitism: Ecology and physiology. *Canadian Journal of Plant Pathology*, 9, 370-379.

Baker, R. (1988b). *Trichoderma* spp. as plant growth stimulants. *CRC Critical Reviews*, 7, 97-106.

Baker, R. & Scher, F. M. (1987). Enhancing the activity of biological control agents. In *Innovative Approaches to Plant Disease Control*, ed. I. Chet, pp. 1-17. John Wiley & Sons: New York.

Chang, Y.-C., Chang, Y.-C., Baker, R., Kleifeld, O. & Chet, I. (1986). Increased growth of plants in the presence of the biological control agent *Trichoderma harzianum*. *Plant Disease*, 70, 145-148.

Cook, R. J. & Baker, K. F. (1983). *The Nature and Practice of Biological Control of Plant Pathogens.* The American Phytopathological Society: St. Paul, Minnesota, USA.

Elad, Y., Lifshitz, R. & Baker, R. (1985). Enzymatic activity of the mycoparasite *Pythium nunn* during interaction with host and non-host fungi. *Physiological Plant Pathology*, **27**, 131-143.

Foster, R. C., Rovira, A. D. & Cock, T. W. (1983). *Ultrastructure of the Root-Soil Interface.* American Phytopathological Society: St. Paul, Minnesota, USA.

Hadar, E., Katan, J. & Katan, T. (1989). The use of nitrate-nonutilizing mutants and a selective medium for studies of pathogenic *Fusarium* strains. *Plant Disease*, **73**, in press.

Harman, G. E. & Taylor, A. G. (1988). Improved seedling performance by integration of biological control agents at favorable pH levels with solid matrix priming. *Phytopathology*, **78**, 520-525.

Harman, G. E., Taylor, A. G., & Stasz, T. E. (1989). Combining effective strains of *Trichoderma harzianum* and solid matrix priming to provide improved biological seed treatment systems. *Plant Disease*, **73**, in press.

Henson, J. M., Blake, N. K & Pilgeram, A. L. (1988). Transformation of *Gaeumannomyces graminis* to benomyl resistance. *Current Genetics*, **14**, 113-117.

Howell, C. R. (1987). Relevance of mycoparasitism in the biological control of *Rhizoctonia solani* by *Gliocladium virens*. *Phytopathology*, **77**, 992-994.

Howell, C. R. & Stipanovic, R. D. (1983). Gliovirin, a new antibiotic from *Gliocladium virens*, and its role in the biological control of *Pythium ultimum*. *Canadian Journal of Microbiology*, **29**, 321-324.

Hynes, M. J. (1986). Transformation of filamentous fungi. *Experimental Mycology*, **10**, 1-8.

Kerr, A. (1980). Biological control of crown gall through production of agrocin 84. *Plant Disease*, **64**, 25-30.

Kloepper, J. W., Lifshitz, R. & Schroth, M. N. (1988). *Pseudomonas* inoculants to benefit plant protection. In *ISI Atlas of Science: Animal and Plant Sciences, 1988*, pp. 60-64.

Lifshitz, R., Lifshitz, S. & Baker, R, (1985). Decrease in incidence of *Rhizoctonia* preemergence damping-off by the use of integrated and chemical controls. *Plant Disease*, **69**, 431-434.

Marois, J. J., Johnson, S. A., Dunn, M. T. & Papavizas, G. C. (1982). Biological control of *Verticillium* wilt of eggplant in the field. *Plant Disease*, **66**, 1166-1168.

Mischke, S. (1988a). The presence of plasmids in the biocontrol fungus *Gliocladium*. *Phytopathology*, **78**, 863 (abstract).

Mischke, S. (1988b). The nature of mitochondrial plasmids in *Gliocladium virens*, a biocontrol fungus. *Phytopathology*, **78**, 1591 (abstract).

Mischke, S. & Papavizas, G. C. (1987). Mitochondrial DNA of the biocontrol agent *Talaromyces flavus*. *Phytopathology*, **77**, 1756 (abstract).

Montencourt, B. S. (1983). *Trichoderma reesei* cellulolysis. *Trends in Biotechnology*, **1**, 156-161.

Papavizas, G. C. (1985). *Trichoderma* and *Gliocladium*: biology, ecology and potential for biocontrol. *Annual Review of Phytopathology*, **23**, 23-54.

Papavizas, G. C. (1987). Genetic manipulation to improve the effectiveness of bio-control fungi for plant disease control. In *Innovative Approaches to Plant Disease Control*, ed. I. Chet, pp. 193-212. John Wiley & Sons: New York.

Paulitz, T. C. & Baker, R. (1987a). Biological control of *Pythium* damping-off of cu-cumbers with *Pythium nunn*: influence of soil environment and organic amend-ments. *Phytopathology*, 77, 341-346.

Paulitz, T. C. & Baker, R. (1987b). Biological control of *Pythium* damping-off of cu-cumbers with *Pythium nunn*: population dynamics and disease suppression. *Phy-topathology*, 77, 335-340.

Paulitz, T. C. & Baker, R. (1988a). The formation of secondary sporangia by *Py-thium ultimum*: the influence of glucose, organic amendments and *Pythium nunn*. *Soil Biology and Biochemistry*, 20, 151-156.

Paulitz, T. C. & Baker, R. (1988b). Colonization of bean leaves by *Pythium nunn* and *Pythium ultimum*. *Canadian Journal of Microbiology*, 34, 947-951.

Peberdy, J. F. (1980). Protoplast fusion–a tool for genetic manipulation and breed-ing in industrial microorganisms. *Enzyme and Microbial Technology*, 2, 23-29.

Rayner, A. D. M. & Webber, J. F. (1984). Interspecific mycelial interactions–an over-view. In *The Ecology and Physiology of the Fungal Mycelium*, British Mycological Society symposium vol. 8, ed. D. H. Jennings & A. D. M. Rayner, pp. 383-417. Cambridge University Press.

Salt, G. H. (1978). The increasing interest in minor pathogens. In *Soilborne Plant Pathogens*, eds. B. Schippers & W. Gams, pp. 289-312. Academic Press: New York.

Schneider, R. W. (1984). Effects of nonpathogenic strains of *Fusarium oxysporum* on celery root infection by *F. oxysporum* f. sp. *apii* and a novel use of the Line-weaver-Burk double reciprocal plot technique. *Phytopathology*, 74, 646-653.

Schroth, M. N. & Hildebrand, D. C. (1964). Influence of plant exudates on root-in-fecting fungi. *Annual Review of Phytopathology*, 2, 101-132.

Seh, M. L. & Kenerley, C. M. (1988). Protoplast formation and regeneration of three *Gliocladium* species. *Journal of Microbiological Methods*, 8, 121-130.

Stasz, T. E., Harman, G. E. & Weeden, N. F. (1988). Protoplast preparation and fu-sion in two biocontrol strains of *Trichoderma harzianum*. *Mycologia*, 80, 141-150.

Tauber, M. J. & Baker, R. (1988). Every other alternative but biological control. *Bio-Science*, 38, 660.

Turgeon, G. & Yoder, O. C. (1985). Genetic engineered fungi for weed control. In *Biotechnology: Applications and Research*, ed. P. W. Cherenisinoff & R. P. Ouel-lette, pp. 221-230. Technomics Publishing Co.: Lancaster, Pennsylvania.

Weller, D. M. (1988). Biological control of soilborne plant pathogens in the rhizos-phere with bacteria. *Annual Review of Phytopathology*, 26, 379-407.

Weltring, K. M., Turgeon, B. C., Yoder, O. C. & Van Etten, H. D. (1988). Isolation of a phytoalexin detoxification gene from the plant pathogenic fungus *Nectria haematococca* by detecting its expression in *Aspergillus nidulans*. *Gene*, 68, 335-344.

Windham, M. T., Elad, Y. & Baker, R. (1986). A mechanism for increased plant growth induced by *Trichoderma* spp. *Phytopathology*, 76, 518-521.

Wong, P. T. W. & Southwell, R. J. (1980). Field control of take-all of wheat by aviru-lent fungi. *Annals of Applied Biology*, 94, 41-49.

Chapter 11

Protoplast technology and strain selection

M. J. Hocart & J. F. Peberdy

*Microbial Biochemistry and Genetics Group, Department of Botany,
School of Biological Sciences, University of Nottingham,
Nottingham, NG7 2RD, UK*

Introduction

The genetic improvement of fungi for industrial purposes has traditionally involved induced mutagenesis and screening survivors for strains with the appropriate enhanced characteristics (Rowlands, 1984). Such mutagenic programmes result in the introduction of deleterious mutations into the strains in addition to the desired genetic changes. The maintenance of fitness characters, allowing the survival of strains in a natural environment, is not normally an important consideration in the improvement of industrial microorganisms. For strains to be used as bioinoculants or for biocontrol purposes however, their ability to proliferate in a competitive microenvironment is crucial. Consequently, mutagenic treatments are largely inappropriate for the improvement of such strains.

The interaction between a biological control fungus and its target organism is complex, involving mycoparasitism, antibiosis and competition. These characters are themselves complex, and likely to be under the control of a large number of genes. The recombination of whole genomes, therefore, rather than the manipulation of single genes, is required for the generation of new strains. Protoplast fusion provides a procedure for promoting recombination of whole genomes, even between incompatible strains. In this chapter we discuss the basic elements of this technology and the potential for its application.

Protoplast isolation

Some species of insect parasitic fungi, belonging to the Entomophthorales, include a protoplast stage as a normal part of their life cycle (Tyrrell, 1977). On certain media these cells can be cultured *in vitro* for use in biochemical or genetic studies (Nolan, 1985, 1988). Protoplast isolation from the majority of fungi, however, requires the enzymatic digestion of at least part of the cell wall.

Mycolytic enzymes

Enzymes suitable for protoplast release from fungal mycelium are now readily available from commercial sources (Hamlyn et al., 1981; Peberdy, 1985). A number of these enzymes are listed in Table 11.1. Most are crude or partially purified preparations with multiple activities, the main components being chitinase, α and β-glucanases and protease. Although in several studies attempts have been made to analyse the various components of these enzyme preparations, little success has been achieved in correlating yields of protoplasts with specific activities. It is thought that unspecified side activities are often crucial in determining the effectiveness of a particular enzyme (Hamlyn et al., 1981; Peberdy, 1985; Hocart, Lucas & Peberdy, 1987).

The choice of mycolytic enzyme for protoplast isolation is largely empirical, the major components of a preparation may or may not be important for protoplast release. Often a mixture of two or more enzymes is more effective than either used independently.

Where commercial enzymes are unavailable or ineffective, 'in house' preparations can be produced. These may be derived from autolytic cultures of the fungus of interest or produced by a suitable microorganism grown on purified fungal cell walls as the sole carbon source (Reyes et al., 1984; Peberdy, 1985; Davis, 1985). In all cases, whether using commercial products or 'in house' preparations, considerable batch variation is to be expected. Consequently when using different enzyme batches the conditions of protoplast production may need to be modified.

The organism

After enzyme composition, the next most important factor affecting protoplast isolation is the nature of the mycelium. The physiological age of the mycelium exerts a profound influence on protoplast yield. In most species the region of the cell wall most susceptible to lysis is near the hyphal tip, where the wall is thinnest and wall synthesis is actively occurring.

Consequently the maximum yield of protoplasts is obtained during the period of growth coinciding with the occurrence of maximum branching and cell wall synthesis. In general, the maximum yields of protoplasts are from mycelium in the exponential phase of growth (Peberdy, 1979). The period of maximum susceptibility to lysis may be relatively narrow with the result that protoplast yields from mycelium cultured for just a few hours longer than optimum can be dramatically reduced (Coudray & Canevascini, 1980). Therefore it is important to establish the conditions for the production of mycelium to obtain consistently high yields.

Mycelium for protoplast isolation is generally produced in shake culture. Alternatively, colonies grown on sterile cellophane discs, laid over

Table 11.1. Commercially available mycolytic enzymes.

Enzyme preparation	Source organism	Manufacturer or Supplier*
Cellulase CP	*Penicillium funiculosum*	John & E. Sturge Ltd., Selby, North Yorkshire
Cellulase 'onozuka' R-10 Cellulase RS	*Trichoderma viride*	Yakult Honsha Co. Ltd, 8-21 Singikancho, Nishinomiya Hyogo T662, Japan.
Cellulase (Mayvil)	*Aspergillus niger*	Mayvil Chemicals Ltd., Rookery Bridge, Sandbach, Cheshire CW11 9QZ
Driselase	*Irpex lacteus*	Kyowa Hakko Kiogo Co. Ltd., Ohtemachi Building, Ohtemachi, Chiyoda-Ku, Tokyo, Japan.
		*Sigma Chemical Co. Ltd., Fancy Road, Poole, Dorset, BH17 7NH
β-Glucanase	*Penicillium emersonii*	*BDH Chemicals Ltd., Fourways, Carlyon Ind. Est. Atherstone, Warwickshire, CV9 1JG.
β-Glucuronidase	*Helix pomatia*	Sigma Chemical Co., St. Louis, Mo., USA.
Helicase	*Helix pomatia*	L'Industrie Biologique Française, Clichy, France.
Novozym 234	*Trichoderma harzianum*	Novo Industri A/S, Enzyme Division, Bagsvaard, Denmark.
Rhozyme HP150	Unknown	Rohm & Haas Co., Independence Mall West, Philadelphia, PA 19105, USA.
		*Pollock & Poole Ltd., Ladbroke Close, Woodley, Reading RG5 4DX.
Zymolyase 5,000 Zymolyase 60,000	*Arthrobacter luteus*	Kirin Breweries Co. Ltd., Takasaki, Japan.

agar medium, can be used. After an appropriate period of incubation, the discs with the accompanying mycelium, are transferred to the enzyme solution. Occasionally chemical treatment of the mycelium, prior to enzymatic digestion, can enhance protoplast isolation (Peberdy, 1979). Preincubation with 2-mercaptoethanol has been shown to increase protoplast yield from *Beauveria bassiana* (Kawamoto & Aizawa, 1986).

Usually protoplasts are released through small pores in the wall from cells nearest the hyphal tip. More rarely the whole mycelium is converted to protoplasts. In both instances the resulting protoplast population will show considerable heterogeneity for size, organelle content and physiological activity. For some purposes (e.g. electrofusion) a more uniform protoplast population is preferable. Protoplasts derived from spores may be expected to be more homogeneous than those from hyphae. Enzyme systems have been developed for the isolation of protoplasts from the spores of several fungi (Toyama, Shinmyo & Okada, 1983; Bos, 1985). Although the conidial wall is more resistant to digestion its susceptibility can be increased by altering the conditions under which they are produced or by preincubation prior to lysis.

Osmotic stabilizer

Isolated protoplasts are osmotically sensitive and therefore require stabilization to prevent lysis. Inorganic salts, such as sodium, magnesium or potassium chloride or magnesium sulphate, are the most frequently used stabilizers for protoplasts from filamentous fungi, although sugars and sugar alcohols have also been employed.

The choice of stabilizer for protoplast isolation will greatly influence the yield obtained. After release, the protoplasts must be separated from the undigested mycelial debris. This is usually accomplished by filtration and the resulting protoplast suspension washed free of the enzyme solution by centrifugation. Problems may be encountered during the harvesting process using certain stabilizers (e.g. sucrose, $MgSO_4$) in that the protoplasts fail to pellet during centrifugation and remain in suspension. Dilution of the suspension with an alternative stabilizer may allow the protoplasts to be collected by centrifugation. The increase in buoyancy of protoplasts maintained in $MgSO_4$ has long been recognised (de Vries & Wessels, 1972; Peberdy, 1979). Thus, after centrifugation at a relatively high speed with concentrations of $MgSO_4$ between $0 \cdot 8$ to $1 \cdot 2$ M, protoplast buoyancy may be sufficiently enhanced to permit the concentration of the protoplasts at the surface of the suspension.

Complete separation of protoplasts from hyphal debris may require more complicated procedures. Hashiba & Yamada (1982) used a two-phase stabilizer system to purify suspensions of *Rhizoctonia solani* protoplasts. The suspension, stabilized with $0 \cdot 6$ M mannitol, was layered onto a 0·6 M sucrose solution. After centrifugation, the protoplasts, free of hyphal fragments, were located at the interface of the two stabilizers.

The optimum stabilizer for protoplast release may not be the most suitable for the subsequent stabilization of the isolated protoplasts. It may be necessary to change the stabilizer after digestion is complete to prevent the gradual loss of protoplasts through lysis. Osmotic stabilization is also required throughout the regeneration process. The most appropriate stabilizer for this purpose need not necessarily be that which is optimum for protoplast isolation and therefore, a number of alternatives should be tested.

Induced protoplast fusion

Protoplast fusion can be induced either chemically, with polyethylene glycol (PEG), or physically by electrofusion.

PEG-induced fusion

The fusogenic activity of PEG was first recognised with plant protoplasts (Kao & Michayluk, 1974; Wallin, Glimelius & Eriksson, 1974) and was soon applied to fungal protoplasts (Anne & Peberdy, 1975; Ferenczy, Kevei & Szegedi, 1975). It is now the most widely used fusogenic treatment. Briefly, the procedure is as follows: protoplasts of the parental strains (typically 10^7 of each strain) are mixed together, pelleted by centrifugation and resuspended in 1 to 2 ml of the PEG solution, buffered at pH 7·5, containing 10 to 100 mM $CaCl_2$. This causes aggregation of the protoplasts and random fusions within each aggregate. After a 10 to 20 minute incubation the PEG is diluted by addition of stabilizer. The fusion mixture is then either plated directly or washed free of PEG by centrifugation prior to plating. The optimum conditions for fusion may vary for each species but typically solutions containing 25 to 40% PEG of molecular weight between 4000 and 6000 are used (Anne & Peberdy, 1975, 1976; Ferenczy et al., 1976).

The mechanism of the fusion process has been studied in animal cells. Although the precise mechanism is unknown, close membrane/membrane contact is established between protoplasts within each aggregate. It is thought that during this contact localised reorganization of the membrane proteins occur and small cytoplasmic bridges are formed which enlarge as the two protoplasts fuse (Ferenczy, 1981). The presence of calcium ions and the dilution of the PEG solution are a crucial part of the fusion process (Ferenczy, 1981).

Electrofusion

In this process, protoplasts of the parental strains are mixed and placed in a fusion chamber which contains two closely placed electrodes. An alternating electric field is applied which causes the protoplasts to form chains of 2 to 20 cells, a process called dielectrophoresis. This ensures the close membrane/membrane contact essential for fusion. The actual fusion event is induced by the application of one to several short duration, direct current pulses, which momentarily destabilise the protoplast membrane. If the conditions are suitable this leads to a high frequency of fusion involving two or more adjacent protoplasts (Zimmerman & Vienken, 1982; Groth et al., 1987; Sonnenberg & Wessels, 1987).

Since fusion can be observed microscopically, and the strength and duration of the dielectrophoretic field and pulses finely controlled, it is possible to investigate the precise effect of these factors on fusion frequency (Broda, Schnettler & Zimmerman, 1987). High frequency fusion depends on (1) the establishment of strong inter-membrane contact and (2) optimum membrane destabilization. Dielectrophoresis and chain formation is dependent on the amplitude and frequency of the applied alternating field. In addition, the size distribution of the protoplasts is also important. A large disparity in protoplast size between the two parental strains results in their dielectrophoretic separation with a corresponding reduction in hybrid fusion frequency (Broda et al., 1987).

Membrane destabilization is also size dependent. As pulse strength and duration are increased membrane breakdown becomes more extensive. Ultimately the area of membrane destabilization becomes so great that the process is irreversible and the protoplasts lyse. The point of irreversible breakdown is attained sooner with large protoplasts than with small; the latter may require much greater pulse strengths even to reach the point at which fusion can occur. It can be seen, therefore, that for high frequency electrofusion the protoplasts of both strains should be relatively uniform in size and similar for size distribution. For the best results the protoplast preparations should not contain a high proportion of small protoplasts or cell debris (Broda et al., 1987).

Selection strategies

Auxotrophic markers

After protoplast fusion it is usually necessary to select against the parental types so that the fusion products can be recognised. The commonest system of selection is the use of complementary auxotrophic mutants. The fused protoplast suspension is plated on an osmotically stabilized minimal medium lacking the nutritional supplements required by the parental strains. Although widely used, this system has the disadvantage that the introduction of suitable auxotrophic markers into the parental isolates is often laborious and time consuming. Furthermore, there is a real risk that the use of mutagenic treatments will induce unwanted and deleterious mutations in addition to the auxotrophic markers.

It is possible to avoid these risks by screening for spontaneous auxotrophic or non-utilization mutants. Resistance to some compounds is often associated with specific auxotrophic requirements or an inability to utilise particular substrates as nitrogen, sulphur or carbon sources (Apirion 1962, 1965; Arst, 1968; Cove 1976a, b; Boeke, La Croute & Fink, 1984; Singh & Sherman 1974). In some cases, however, the nutritional requirements of such mutants may not be sufficiently stringent to give the fusion products the necessary selective advantage over the parental isolates (Hamlyn, 1982).

Pigmentation markers

Direct visual identification of fusion products may be possible if the parental strains carry suitable pigmentation markers. In species with uninucleate conidia where the determination of spore colour is autonomous (i.e. determined by conidial genes rather than by the underlying mycelium), heterokaryons and diploids may be distinguished from parental colonies simply on the basis of colony pigmentation (Fargasova, Sipiczki & Betina, 1985; Ogawa, Brown & Wood, 1987). In species with multinucleate conidia, or in which pigmentation is determined by the nuclear content of the mycelium, fusion products may be distinguished from the parental colonies but the identification of diploids or recombinant types from heterokaryons will require further manipulations and additional markers.

Resistance markers

Resistance to antimicrobial compounds can be used as a method of selection of fusion products. Protoplasts from strains carrying different fungicide resistance markers are fused and plated directly on to regeneration medium containing both fungicidal compounds. Alternatively the fusion mixture can be plated on to fungicide-free regeneration medium and incubated for a period to allow regeneration to occur, before overlayering with fungicide-amended medium. Which approach is adopted, and the suitability of particular resistance phenotypes for this method of selection, will depend on the genetic basis of the resistance and the likely outcome of the fusion event in terms of the parasexual cycle.

Often different selection strategies are combined. The use of pigmentation markers are frequently used in conjunction with other methods of selection since they give a visual indication of the progress of the parasexual cycle. Resistance markers used in conjunction with auxotrophic requirements allow fusion with unmarked, fungicide-sensitive strains, and the selection of recombinant types on minimal medium containing the inhibitor (Bradshaw & Peberdy, 1984; Norman, 1988).

Dead donor strategies

The selection strategies outlined above are applicable equally to protoplast fusion and orthodox crosses through hyphal anastomoses. They all necessitate the introduction of markers into the parental strains. An alternative selection strategy has been devised which exploits the unique features of protoplast fusion and avoids the need for selectable markers in both parental strains. This approach is known as the dead donor strategy. In its simplest form the protoplasts of one strain are chemically or physically inactivated so that they are no longer able to regenerate, while maintaining their membrane integrity. These inactivated protoplasts are fused with protoplasts from an auxotrophic strain and the fusion mixture plated on a minimal medium. Fusion of the 'dead' protoplasts with the viable protoplasts of the auxotroph rescues the inactivated parent by restoring the damaged cellular functions. Under suitable conditions, only the hybrids are able to form colonies.

Alternatively, the use of an auxotrophic strain can be avoided by using, as the inactivated parent, a strain with a fungicide resistance marker. This is fused with untreated, viable protoplasts of a fungicide sensitive strain and the fusants selected on medium containing the inhibitor.

Protoplasts can be inactivated physically by heat treatment or exposure to UV-light (Lhotakova & Vondrejs, 1984), or chemically with compounds such as crystal violet, N-ethyl-maleimide or 8-hydroxyquinoline (Ferenczy et al., 1987). By using different methods of inactivation it should be possible to perform fusions between strains with no additional, selectable markers. As with the use of resistance markers, the success of these approaches depends on the genetic outcome of the fusion event.

Parasexuality and nature of fusants

The parasexual cycle, which made the genetic analysis of imperfect fungi possible, was first recognised and described in *Aspergillus nidulans* (Pontecorvo et al., 1953). Since then many reports of parasexual phenomena in a wide range of fungi have been published, suggesting that the system may be a feature common to all fungal groups (Pontecorvo, 1956; Pontecorvo & Kafer, 1958; Bradley, 1962; Tinline & MacNeill, 1969; Caten, 1981).

The sequence of events occurring in *A. nidulans*, and shown to occur in several other species is considered the standard parasexual cycle (Fig. 11.1; Pontecorvo, 1956; Fincham, Day & Radford, 1979; Hastie, 1981; Caten, 1981). The first part of this process is the formation of a heterokaryon, with the presence of unlike nuclei in the same mycelium. In many fungi this can be achieved by hyphal anastomosis. Protoplast fusion provides a direct procedure by which heterokaryons can be generated at high

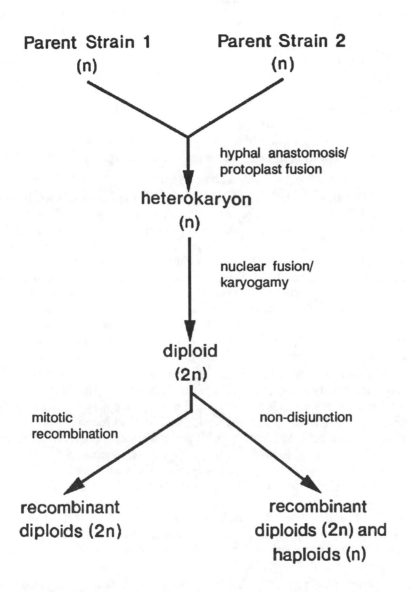

Fig. 11.1. Sequence of events in the standard parasexual cycle.

frequency. In species where heterokaryon formation via hyphal anastomosis is limited, such as *Cephalosporium acremonium* (Nuesch, Treichler & Liersch, 1973; Elander, 1974), protoplast fusion is the predominant means by which the parasexual cycle can be manipulated (Hamlyn *et al.*, 1985). Chance nuclear fusions, or karyogamy, may occur to produce diploid nuclei. If karyogamy occurs between unlike nuclei the resulting diploid nucleus will be heterozygous at all the loci for which the parental strains differed. Two processes may take place in such nuclei to yield novel genotypes, mitotic recombination and non-disjunction. As might be expected, not all fungi conform exactly to this sequence of events. Variations in the extent and duration of each stage occur, which have consequences for the manipulation of the process in the laboratory.

Heterokaryosis

The first stage of the parasexual cycle, heterokaryon formation, is considerably influenced by the cytology of the fungus. Species in which the mycelium is composed primarily of multinucleate cells, like *Aspergillus nidulans*, can readily accommodate the heterokaryotic condition. The majority of the vegetative cells will contain nuclei of both parents, and the relative proportions of the different nuclei will generally be stable and characteristic of the heterokaryon. Where conidia produced by such heterokaryons are multinucleate as in *Aspergillus sojae*, the heterokaryon may be transmitted via the spores (Bradley, 1962). In species with uninucleate conidia, unless karyogamy generates diploid nuclei, plating spores from a heterokaryon will produce only parental types.

Species in which uninucleate vegetative cells predominate may still show heterokaryotic growth, as a result of hyphal anastomosis forming bridges between individual cells. Nuclear migration may occur along the bridge to yield isolated, heterokaryotic cells embedded in a mycelium of monokaryotic cells. The term 'mosaic heterokaryon' has been used to describe this type of growth, typified by heterokaryons in *Verticillium* (Puhalla & Mayfield, 1974; Typas & Heale, 1976) and *Fusarium* (Puhalla & Speith, 1983). Where the parental strains carry complementary nutritional requirements, this limited degree of heterokaryosis is often sufficient, under selective conditions, to support the growth of the surrounding homokaryotic mycelium. Conidia from such heterokaryons will generally have been derived from uninucleate cells and hence of be parental type (Hastie & Heale, 1984).

In other species a sustainable heterokaryotic phase may not occur. Attempts to isolate heterokaryons in *Cephalosporium acremonium* were unsuccessful (Elander, 1974), but recombinant progeny could be isolated following protoplast fusion (Hamlyn & Ball, 1979).

Diploidization

The generation of novel genotypes through parasexuality requires the formation of at least a transient, heterozygous diploid phase. The frequency of karyogamy, and the stability of the resulting diploids, can vary considerably between different species and even between strains. Diploids are usually obtained from *Aspergillus*-type heterokaryons by plating dense conidial suspensions on selective medium. The recovery of diploid conidia depends not only on the frequency of karyogamy but also on the relative rates of division of haploid and diploid nuclei and the degree of sporulation by haploid and diploid hyphae (Caten & Day, 1977; Caten, 1981). In some instances a higher frequency of diploids may be obtained among protoplasts prepared from a heterokaryon than from conidia of the same heterokaryon (F. Kevei personal communication).

Rare diploid conidia produced by mosaic heterokaryons can be detected in the same way, by high density plating of spores on selective medium (Jackson & Heale, 1987). Mosaic heterokaryons are maintained by the formation of new hyphal bridges and their growth is dependent on the rate at which this occurs. In *Verticillium dahliae* hyphal anastomoses are formed relatively frequently at 21°C, but rarely at 30°C (Puhalla & Mayfield, 1974). As a result, heterokaryotic growth ceases when colonies are transferred to the higher temperature. Growth of diploid mycelium, produced by the fusion of nuclei in anastomosed cells, is not temperature dependent. Consequently, diploid mycelium may be isolated directly, as faster growing sectors, from the edge of heterokaryotic colonies which are incubated at the restrictive temperature.

In certain species the heterokaryotic phase is very restricted. De Bertoldi & Caten (1975) demonstrated the isolation of somatic diploids in a *Humicola* species, directly from mixed cultures of auxotrophic strains. The heterokaryotic phase was assumed to be confined to the original fused cell, further growth being dependent upon nuclear fusion and the formation of a heterozygous diploid.

The frequency of diploid formation in heterokaryons may be affected by various factors including heat treatment, UV-light and (+)camphor (Alikhanian, Kameneva & Krylov, 1960; Day & Day, 1974; Ogawa *et al.*, 1987). Cell cycle can also determine the frequency of karyogamy. Yeast cells treated with 8-hydroxyquinoline prior to protoplast isolation yielded a far higher proportion of heterozygous diploid colonies after electrofusion (Emeis, 1987; Forster & Emeis, 1986). It was thought that the pre-treatment caused G1 arrest of the cells (ie. prevented their undergoing DNA synthesis prior to mitosis) and this resulted in a higher frequency of karyogamy following fusion.

Mitotic recombination

In mitotic recombination, crossing-over in diploid nuclei occurs between homologous chromosomes. The subsequent migration of the recombinant chromatids to the same or opposite poles at mitosis, leads to diploid daughter nuclei that are either homozygous for all loci distal to the point of crossover, or heterozygous but with altered linkage relationships for the parental markers (Fig.11.2). In the former case, all recessive markers distal to the point of recombination will be expressed in the diploid, whereas in the latter recombination will only be apparent after haploidization and analysis of the haploid genotypes. A number of agents have been shown to increase the frequency of mitotic crossing-over. These include UV-light (Holliday, 1961; Day & Jones, 1968; Zimmerman, 1971), heat treatment (Tanabe & Garber, 1980) and chemical 'recombinogens' such as 5-fluorodeoxyuridine, 5-fluorouracil, nitrogen mustard (methyl-*bis*- (β-chloroethyl)amine), ethyl methane sulphonate and mitomycin C (Holliday, 1964; Esposito & Holliday, 1964; Beccari, Modigliani & Morpurgo, 1967; Yost, Chaleff & Finerty, 1967; Tanabe & Garber, 1980; Kokontis & Garber, 1983). Since many of these agents are likely to cause DNA damage it is thought that they induce crossing-over by stimulating the DNA repair mechanisms of the cell.

Non-disjunction and haploidization

Non-disjunction involves the uneven segregation of whole chromosomes at mitosis. Sister chromatids, of one or more chromosomes, both pass to the same pole and are included in the same nucleus during nuclear division (Fig.11.3). As a result, one daughter nucleus contains three copies of the non-disjunctional (trisomic) chromosome $(2n+1)$ while the other contains only one $(2n-1)$. In subsequent nuclear divisions the additional chromosome in hyperdiploid nuclei may be lost, restoring the diploid chromosome complement; which is generally more stable than the aneuploid state. Depending on which two of the three copies of the trisomic chromosome are retained the resulting diploid nucleus may be identical to the original diploid, and heterozygous for all chromosomes, or homozygous for the non-disjunctional linkage group. In this latter case any recessive markers carried on this chromosome would be expressed.

Hypodiploid nuclei ($2n-1$, $2n-2$ etc.) may yield haploid nuclei as a result of the progressive loss of additional chromosomes: a process called haploidization. Since this loss is generally a random process the resultant haploid nuclei will contain genetic material derived from either parent.

Non-disjunction and haploidization can also be artificially induced by chemical and physical agents (Barron, 1962; Bignami *et al*., 1974; Kafer, Scott & Kappas, 1986). The haploidizing effect of *p*-fluorophenylalanine and the benzimidazole fungicide benomyl have long been recognised

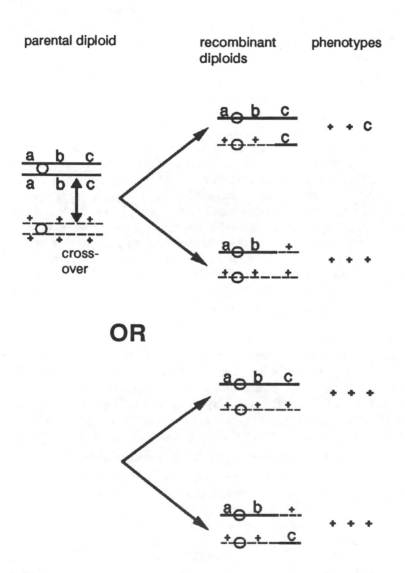

Fig. 11.2. Genetic outcome of mitotic recombination between homologous non-sister chromatids in a heterozygous diploid.

(Morpurgo, 1961; Lhoas, 1961; Hastie, 1970) and are routinely used to induce haploidization of diploids for genetic analysis. In *Metarhizium anisopliae* these two compounds proved ineffective for the breakdown of diploids but the fungicide chloroneb was used successfully to obtain haploid recombinants (Messias & Azevedo, 1980).

Variation in the parasexual cycle and selection strategies

In the standard parasexual cycle mitotic recombination and non-disjunction occur at a relatively low frequency (Caten, 1981), and diploids can be maintained almost indefinitely. In some species, however, the diploid phase is much less stable. In these cases, the existence of diploid nuclei being inferred from the occurrence of parasexual recombinants in the heterokaryon spore population rather than from the direct isolation of the diploid itself (Molnar, Pesti & Hornok, 1985; Silveira & Azevedo, 1987). In *Cephalosporium acremonium* for example, the diploid stage is presumed to be transient: recombinant haploids and unstable heterozygotes are obtained by plating heterokaryons, produced by protoplast fusion, directly onto selective media (Nuesch, Treichler & Liersch, 1973; Hamlyn & Ball, 1979; Hamlyn *et al.*, 1985).

Variations in the parasexual cycle have important consequences for the selection procedure employed. In species with a relatively stable heterokaryotic phase, in which nuclear fusions occur only rarely, it is necessary to maintain the selection over a long period to permit the detection of diploids or recombinants. For this purpose the most appropriate selection strategies may involve the use of auxotrophic or dominant resistance markers. If a stable diploid phase does not occur, recombinant types may be selected directly from the heterokaryon and it is sufficient simply to eliminate the parental types. In this case recessive resistance markers or dead donor strategies can be successfully applied.

Overcoming vegetative incompatibility

Vegetative incompatibility, restricting parasexual gene exchange between unrelated strains via hyphal anastomosis, is widespread in fungi (Caten & Jinks, 1966; Leach & Yoder, 1983; Hastie & Heale, 1984; Typas, 1983) and constitutes a major barrier to strain improvement. Incompatibility can occasionally be overcome through protoplast fusion (Zhemchuzhina *et al.*, 1985; Jackson & Heale, 1987; Silveira & Azevedo, 1987). Where this is possible it has been interpreted as indicating that the expression of incompatibility occurs primarily at the level of the cell wall, rather than at the cytoplasmic or nuclear level (Dales & Croft, 1977).

The nature of the fusion products produced between vegetatively incompatible strains often differs considerably from that of hybrids produced between compatible strains. In some, the heterokaryons may be

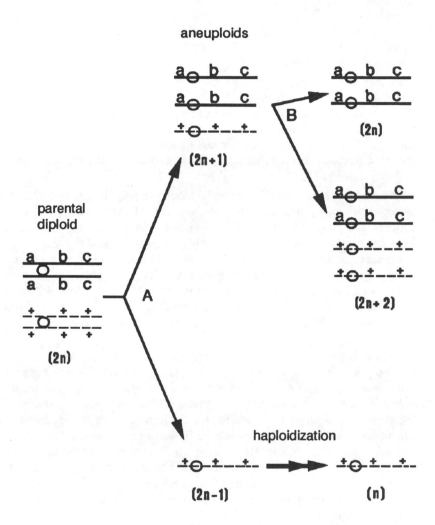

Fig. 11.3. Genetic outcome of non-disjunction during mitotic division in a heterozygous diploid nucleus. A = primary non-disjunction; B = secondary non-disjunction.

grossly unbalanced with one nuclear type predominating and little or no apparent recombination (Weeden, Stasz & Harman, 1987). In *Verticillium lecanii*, incompatibility between some strain combinations could be overcome using protoplast fusion (Jackson & Heale, 1987). The stability of the diploids produced varied considerably, but in most cases rare recombinants could be obtained.

In *Gibberella zeae*, a species which normally shows a mosaic heterokaryon organization, heterokaryotic growth between two incompatible strains cannot be sustained because the hyphal bridges, essential for the maintenance of the heterokaryon, do not form. The slow growing, hybrid colonies obtained following protoplast fusion were considered to be the result of nuclear fusion at an early stage and, consequently, polyploid (Adams *et al.*, 1987).

In many cases, the genetic events following fusion are not well understood, particularly if a considerable amount of incompatibility exists between the parent strains. The frequency of cytoplasmic recombinants from such a cross, for example, is likely greatly to exceed the frequency of nuclear recombination (Croft, 1985). In addition, many fusion experiments reported in the literature involve strains with only a few genetic markers and interpretation of the results needs to be guarded. With the advent of pulse-field gel electrophoresis, enabling the separation of whole chromosomes, the analysis of such crosses may be far simpler in the future.

Wider crosses between totally incompatible isolates or distinct genera may be possible using a dead donor strategy developed for intergeneric plant fusions (Dudits *et al.*, 1987). Protoplasts of the donor parent are irradiated with γ-radiation prior to fusion. This treatment results in a transient hybrid phase from which the chromosomes of the inactivated parent are preferentially lost during the first nuclear divisions. The hybrid condition survives sufficiently long to allow genetic recombination to occur and the transfer of part of the donor genome to that of the recipient. Dudits *et al.* (1987) used this strategy to transfer selected and non-selected markers from carrot to tobacco. This procedure may have value for fungal hybridization.

Applications and future prospects

In addition to its use in genetic analysis, protoplast fusion is an invaluable aid for strain improvement and breeding. Strains with suitable mycoparasitic or antagonistic attributes for potential as biocontrol organisms, often lack features which are necessary if they are to be fully exploited. These may be characteristics enabling the mass production of a suitable inoculum or involve host specificity or rhizosphere colonising ability (Zimmerman, 1986). Greater tolerance of environmental stress (e.g. high or low temperatures, desiccation, solar radiation) may increase the effectiveness of the organism and extend its useful range to new crop situations (Wilson & Pusey, 1985). Improvements in mass production might include utilization of a cheaper substrate and enhanced sporulation in culture. Increased spore size in *Verticillium lecanii* has been correlated with rapid germination and insect control potential in the field (Hall, 1984).

The natural populations of these fungi are a valuable source of potentially useful variation which a breeding programme can exploit (Riba, Lourdes de Aquino & Alves Ribeiro, 1987; Jalata 1986). Crosses between isolates with particular desirable characters can be made to construct strains in which these features are combined. A breeding programme can be used to enhance a particular characteristic, for example enzyme or antibiotic production and to improve field performance (Stasz & Harman, 1986, 1987; Cullen & Andrews, 1984). In certain circumstances it may be necessary to introduce particular beneficial mutations into commercial strains. If biocontrol is to be used within an integrated approach to pest management, then pesticide resistance in the biocontrol organism would be an advantage (Abd-el Moity, Papavizas & Shatla, 1982; Papavizas & Lewis, 1981). Since the direct use of mutagenic treatments is not always desirable, the introduction of such characters may be best achieved by crossing with strains with the appropriate resistance phenotype.

To be effective a breeding programme needs to have definite aims. It is important to know which features are desirable in a commercial strain, both for the selection of the most appropriate parental isolates, and for the assessment of performance of the recombinant progeny. For this reason an understanding of the basis of the biocontrol interaction is vital (Howell, 1987). In addition, a relatively simple method of assessment is essential to allow a sufficient number of progeny to be screened for improved phenotypes (Davet, 1986).

Protoplast isolation, fusion and regeneration systems have been developed for several important biocontrol fungi including *Trichoderma harzianum* (Stasz & Harman, 1986, 1987; Weeden *et al*., 1987), *Metarhizium anisopliae* (Silveira & Azevedo, 1987), *Beauveria bassiana* (Kawamoto & Aizawa, 1986) and *Verticillium lecanii* (Jackson & Heale, 1987). These advances have been made relatively recently and so far few strains with improved biocontrol performance have been generated. Stasz & Harman (1987) crossed auxotrophically marked strains of *T. harzianum* by protoplast fusion. Recombinant progeny, as determined by colony appearance, biological characteristics and isozyme phenotype, were assessed for biocontrol of *Pythium ultimum*. Several recombinants proved to be more effective than the original, prototrophic progenitor isolates (See Baker, chapter 10 for further discussion).

The use of protoplast fusion to overcome incompatibility barriers allows characteristics from even distantly related isolates to be combined. Interspecies hybridization would seem to be a worthwhile approach to strain improvement in *Trichoderma* (Stasz, Harman & Matteson, 1987; Park *et al.*, 1986). Competitive saprophytic ability and rhizosphere competence, in *T. harzianum*, have been correlated with cellulolytic activity (Ahmad & Baker, 1986, 1987). Several other species of *Trichoderma* are known to be extremely effective producers of cellulolytic enzymes, characteristics that could usefully be transferred to the biocontrol species. *Gliocladium virens*, which also has potential for biocontrol, can suppress its target fungi through antibiosis, mycoparasitism or competition (Howell, 1987). Increased cellulolytic and mycolytic activity could enhance the effectiveness of this organism for biocontrol. It is no coincidence that some of the most effective mycolytic enzymes used in protoplast isolation are derived from strains of *Trichoderma*. Alternatively, the transfer of genes for the production of the *G. virens* antibiotic into *T. harzianum* strains by hybridization may be desirable.

As the basis of biocontrol interactions becomes better understood, breeding programmes to improve strains will become more common. Our experience with pesticides has clearly shown how adaptable pest and pathogen populations can be. It is not unlikely, therefore, that with the increasing use of biocontrol organisms, even in an integrated approach, there will be a continual need to 'update' the strains in use, as the pathogen populations change to meet the challenge. Protoplast fusion will undoubtedly play an important part in the exploitation of biocontrol fungi.

References

Abd-el Moity, T. H., Papavizas, G. C. & Shatla, M. N. (1982). Induction of new isolates of *Trichoderma harzianum* tolerant to fungicides and their experimental use for control of white rot of onion. *Phytopathology*, **72**, 396-400.

Adams, G., Johnson, N., Leslie, J. F. & Hart, L. P. (1987). Heterokaryons of *Gibberella zeae* formed following hyphal anastomosis of protoplast fusion. *Experimental Mycology*, **11**, 339- 353.

Ahmad, J. S. & Baker, R. (1986). Competitive saprophytic ability and cellulolytic activity of rhizosphere competent mutants of *Trichoderma harzianum*. *Phytopathology*, **76**, 1104.

Ahmad, J. S. & Baker, R. (1987) Rhizosphere competence of *Trichoderma harzianum*. *Phytopathology*, **77**, 182-189.

Alikhanian, S. I., Kameneva, S. V. & Krylov, V. N. (1960). Experimental increase in the frequency of nuclear diploidization in the mycelium of heterokaryons of *Penicillium janchewskii*. *Mikrobiologia*, **29**, 821-825.

Anne, J. & Peberdy, J. F. (1975). Conditions for induced fusions of fungal protoplasts in polyethylene glycol solutions. *Archives of Microbiology*, **105**, 201-205.

Anne, J. & Peberdy, J. F. (1976). Induced fusion of fungal protoplasts following treatment with polyethylene glycol. *Journal of General Microbiology*, **92**, 413-417.

Apirion, D. (1962). A general system for the automatic selection of auxotrophs from prototrophs and vice versa in micro-organisms. *Nature*, **195**, 959-961.

Apirion, D. (1965). The two-way selection of mutants and revertants in respect of acetate utilization and resistance to fluoroacetate in *Aspergillus nidulans*. *Genetical Research*, **6**, 317-329.

Arst, H. N. jr. (1968). Genetic analysis of the first steps of sulphate metabolism in *Aspergillus nidulans*. *Nature*, **219**, 268-270.

Barron, G. L. (1962). The parasexual cycle and linkage relationships in the storage rot fungus *Penicillium expansum*. *Canadian Journal of Botany*, **40**, 1603-1613.

Beccari, E., Modigliani, P. & Morpurgo, G. (1967). Induction of inter- and intragenic mitotic recombination by fluorodeoxyuridine and fluorouracil in *Aspergillus nidulans*. *Genetics*, **56**, 7-12.

Bignami, M., Morpurgo, G., Pagliano, R., Carene, A., Conti, G. & di Guiseppe, G. (1974). Non-disjunction and crossing-over induced by pharmaceutical drugs in *Aspergillus nidulans*. *Mutation Research*, **26**, 159-170.

Boeke, J. D., La Croute, F. & Fink, G. R. (1984). A positive selection for mutants lacking orotidine-5'-phosphate decarboxylase in yeasts: 5-fluoro-orotic acid resistance. *Molecular and General Genetics*, **197**, 345-346.

Bos, C. J. (1985). Protoplasts from fungal spores. In *Fungal Protoplasts: Applications in Biochemistry and Genetics*, ed. J. F. Peberdy & L. Ferenczy, pp. 73-85. Marcel Dekker Inc.: New York.

Bradley, S. G. (1962). Parasexual phenomena in micro-organisms. *Annual Review of Microbiology*, **16**, 35-52.

Bradshaw, R. E. & Peberdy, J. F. (1984). Protoplast fusion in *Aspergillus*: selection of interspecific heterokaryons using antifungal inhibitors. *Journal of Microbiological Methods*, **3**, 27-32.

Broda, H. G., Schnettler, R. & Zimmerman, U. (1987). Parameters controlling yeast hybrid yield in electrofusion the relevance of pre-incubation and the skewness of the size distribution of both fusion partners. *Biochimica et Biophysica Acta*, **899**, 25-34.

Caten, C. E. (1981). Parasexual processes in fungi. In *The Fungal Nucleus*, ed. K. Gull & S. G. Oliver, pp. 191-214. Cambridge University Press.

Caten, C. E. & Day, A. W. (1977). Diploidy in plant pathogenic fungi. *Annual Review of Phytopathology*, **15**, 295-318.

Caten, C. E. & Jinks, J. L. (1966). Heterokaryosis: its significance in wild homothallic ascomycetes and fungi imperfecti. *Transactions of the British Mycological Society*, **49**, 81-93.

Coudray, M-R. & Canevascini, G. (1980). Isolation and regeneration of protoplasts from the cellulolytic fungus *Sporotrichum thermophile*. *Berichte der Schweizerischen Botanischen Gesellschaft*, **90**, 108-117.

Cove, D. J. (1976a). Chlorate toxicity in *Aspergillus nidulans*: studies of mutants altered in nitrate assimilation. *Molecular and General Genetics*, **146**, 147-159.

Cove, D. J. (1976b). Chlorate toxicity in *Aspergillus nidulans*: the selection and characterisation of chlorate resistant mutants. *Heredity*, **36**, 191-203.

Croft, J. H. (1985). Protoplast fusion and incompatibility in *Aspergillus*. In *Fungal Protoplasts: Applications in Biochemistry and Genetics*, ed. J. F. Peberdy & L. Ferenczy, pp. 225-240. Marcel Dekker Inc.: New York.

Cullen, D. & Andrews, J. H. (1984). Evidence for the role of antibiosis in the antagonism of *Chaetomium globosum* to the apple scab pathogen *Venturia inaequalis*. *Canadian Journal of Botany*, **68**, 1819-1823.

Dales, R. B. G. & Croft, J. H. (1977). Protoplast fusion and the isolation of heterokaryons and diploids from vegetatively incompatible strains of *Aspergillus nidulans*. *FEMS Microbiology Letters*, **1**, 201-204.

Davet, P. (1986). Activité parasitaire des *Trichoderma* vis-à-vis des champignons à sclérotes; corrélation avec l'aptitude à la competition dans un sol non stérile. *Agronomie* (Paris), **6**, 863-868.

Davis, B. (1985). Factors affecting protoplast isolation. In *Fungal Protoplasts: Applications in Biochemistry and Genetics*, ed. J. F. Peberdy & L. Ferenczy, pp. 45-71. Marcel Dekker Inc.: New York.

Day, A. W. & Day, L. L. (1974). The control of karyogamy in somatic cells of *Ustilago violacea*. *Journal of Cell Science*, **15**, 619-632.

Day, A. W. & Jones, J. K. (1968). The production and characters of diploids in *Ustilago violacea*. *Genetical Research*, **11**, 63-81.

De Bertoldi, M. & Caten, C. E. (1975). Isolation and haploidization of heterozygous diploid strains in a species of *Humicola*. *Journal of General Microbiology*, **91**, 63-73.

De Vries, O. M. H. & Wessels, J. G. H. (1972). Release of protoplasts from *Schizophyllum commune* by combined action of purified β-1,3-glucanase and chitinase derived from *Trichoderma viride*. *Journal of General Microbiology*, **76**, 319-330.

Dudits, D., Maroy, E., Praznovszky, T., Olah, Z., Gyorgyey, J. & Cella, R. (1987). Transfer of resistance traits from carrot into tobacco by asymmetric somatic hybridisation: regeneration of fertile plants. *Proceedings of the National Academy of Sciences of the USA*, **84**, 8434-8438.

Emeis, C. C. (1987). Intergeneric hybridization of yeasts by electrofusion. *Studies in Biophysics*, **119**, 31-34.

Elander, R. P. (1974). Genetic aspects of cephalosporin- and cephamycin-producing micro-organisms. *Developments in Industrial Microbiology*, **16**, 356-374.

Esposito, R. E. & Holliday, R. (1964). The effect of 5-fluorodeoxyuridine on genetic replication and somatic recombination in synchronously dividing cultures of *Ustilago maydis*. *Genetics*, **50**, 1009-1017.

Fargasova, A., Sipiczki, M. & Betina, V. (1985). Morphological and colour mutants of *Trichoderma viride*: characterisation and complementation. *Folia Microbiologia*, **30**, 433-442.

Ferenczy, L. (1981). Microbial protoplast fusion. In *Genetics as a Tool in Microbiology*, Thirty-first Symposium of the Society for General Microbiology, ed. S. W. Glover & D. A. Hopwood, pp. 1-34. Cambridge University Press.

Ferenczy, L., Bradshaw, R. E., Kevei, F. & Peberdy, J. F. (1987). Biochemical alternative to mutagenesis: model experiments with auxotrophic strains of *Aspergillus nidulans*. *Proceedings of the Forth European Congress on Biotechnology*, vol. 1, Abstracts/Extended Abstracts, ed. O. M. Neijssel, R. R. van der Meer, & K. Ch. A. M. Luyben, p. 505. Elsevier Science Publishers, B.V.: Amsterdam.

Ferenczy, L., Kevei, F. & Szegedi, M. (1975). High-frequency fusion of fungal protoplasts. *Experientia*, **31**, 1028-1030.

Ferenczy, L., Kevei, F. & Szegedi, M. (1976). Fusion of fungal protoplasts induced by polyethylene glycol. In *Microbial and Plant Protoplasts*, ed. J. F. Peberdy, A. H. Rose, H. J. Rogers & E. C. Cocking, pp. 177-187. Academic Press: London.

Fincham, J. R. S., Day, P. R. & Radford, A. (1979). *Fungal Genetics*, 4th Edition. Blackwell Scientific Publications: Oxford.

Forster, E. & Emeis, C. C. (1986). Enhanced frequency of karyogamy in electrofusion of yeast protoplasts by means of preceeding G1 arrest. *FEMS Microbiology Letters*, **34**, 69-72.

Groth, I., Jacob, H. E., Kuenkel, W. & Berg, H. (1987). Electrofusion of *Penicillium* protoplasts after dielectrophoresis. *Journal of Basic Microbiology*, **26**, 341-344.

Hall, R. A. (1984). Epizootic potential for aphids of different isolates of the fungus *Verticillium lecanii*. *Entomophaga*, **29**, 311-321.

Hamlyn, P. F. (1982). Protoplast fusion and genetic analysis in *Cephalosporium acremonium*. Ph.D. Thesis, University of Nottingham, UK.

Hamlyn, P. F. & Ball, C. (1979). Recombination studies with *Cephalosporium acremonium*. In *Third International Symposium on the Genetics of Industrial Microorganisms*, ed. O. K. Sebek & A. I. Laskin, pp. 185-191. American Society for Microbiology: Washington, D.C.

Hamlyn, P. F., Birkett, J. A., Perez-Martinez, G. & Peberdy, J. F. (1985). Protoplast fusion as a tool for genetic analysis in *Cephalosporium acremonium*. *Journal of General Microbiology*, **131**, 2813-2823.

Hamlyn, P. F., Bradshaw, R. E., Mellon, F. M., Santiago, C. M., Wilson, J. M. & Peberdy, J. F. (1981). Efficient protoplast isolation using commercial enzymes. *Enzyme and Microbial Technology*, **3**, 321-325.

Hashiba, T. & Yamada, M. (1982). Formation and purification of protoplasts from *Rhizoctonia solani*. *Phytopathology*, **72**, 849-853.

Hastie, A. C. (1970). Benlate-induced instability of *Aspergillus* diploids. *Nature*, **226**, 771.

Hastie, A. C. (1981). The genetics of conidial fungi. In *Biology of Conidial Fungi*, vol. 2, ed. G. T. Cole & B. Kendrick, pp. 357-393. Academic Press: London.

Hastie, A. C. & Heale, J. B. (1984). Genetics of *Verticillium*. *Phytopathologia Mediterranea*, **23**, 130-162.

Hocart, M. J., Lucas, J. A. & Peberdy, J. F. (1987). Production and regeneration of protoplasts from *Pseudocercosporella herpotrichoides* (Fron) Deighton. *Journal of Phytopathology*, **119**, 193-205.

Holliday, R. (1961). Induced mitotic crossing-over in *Ustilago maydis*. *Genetical Research*, **2**, 231-248.

Holliday, R. (1964). The induction of mitotic recombination by mitomycin C in *Ustilago maydis* and *Saccharomyces*. *Genetics*, **50**, 323-335.

Howell, C. R. (1987). Relevance of mycoparasitism in the biological control of *Rhizoctonia solani* by *Gliocladium virens*. *Phytopathology*, **77**, 992-994.

Jackson, C. W. & Heale J. B. (1987). Parasexual crosses by hyphal anastomosis and protoplast fusion in the entomopathogen *Verticillium lecanii*. *Journal of General Microbiology*, **133**, 3537-3548.

Jalata, P. (1986). Biological control of plant-parasitic nematodes. *Annual Review of Phytopathology*, **24**, 453-489.

Kafer, E., Scott, B. R. & Kappas, A. (1986). Systems and results of tests for chemical induction of mitotic malsegregation and aneuploidy in *Aspergillus nidulans*. *Mutation Research*, 167, 9-34.

Kao, K. N. & Michayluk, M. R. (1974). A method for high- frequency intergeneric fusion of plant protoplasts. *Planta*, 115, 355-367.

Kawamoto, H. & Aizawa, K. (1986). Fusion conditions for protoplasts of an entomogenous fungus *Beauveria bassiana*. *Applied and Entomological Zoology*, 21, 624-626.

Kokontis, J. M. & Garber, E. D. (1983). Spontaneous and induced mitotic recombination in *Ustilago violacea* detected at the cellular level. *Current Genetics*, 7, 465-471.

Leach, J. B. & Yoder, O. C. (1983). Heterokaryon incompatibility in the plant pathogenic fungus, *Cochliobolus heterostrophus*. *Journal of Heredity*, 74, 149-152.

Lhoas, P. (1961). Mitotic haploidisation by treatment of *Aspergillus niger* diploids with *p*-fluorophenylalanine. *Nature*, 190, 744.

Lhotakova, M. & Vondrejs, V. (1984). Fusion of protoplasts of UV- irradiated yeasts. *Folia Microbiologia*, 29, 395-396.

Messias, C. L. & Azevedo, J. L. (1980). Parasexuality in the deuteromycete *Metarhizium anisopliae*. *Transactions of the British Mycological Society*, 75, 473-477.

Molnar, A., Pesti, M. & Hornok, L. (1985). Isolation, regeneration and fusion of *Fusarium oxysporum* protoplasts. *Acta Phytopathologica Academia Scientiarum Hungaricae*, 20, 175-182.

Morpurgo, G. (1961). Somatic segregation induced by *p*-fluorophenylalanine. *Aspergillus Newsletter*, 2, 10.

Nolan, R. A. (1985). Protoplasts from Entomophthorales. In *Fungal Protoplasts: Applications in Biochemistry and Genetics*, ed. J. F. Peberdy & L. Ferenczy, pp. 87-112. Marcel Dekker Inc.: New York.

Nolan, R. A. (1988). A simplified defined medium for growth of *Entomophaga aulicae* protoplasts. *Canadian Journal of Microbiology*, 34, 45-51.

Norman, E. (1988). Biochemical genetics of cephalosporin C production. Ph.D. Thesis, University of Nottingham, Nottingham, UK.

Nuesch, J., Treichler, H. J. & Liersch, M. (1973). The biosynthesis of cephalosporin C. In *Genetics of Industrial Microorganisms, vol. 2, Actinomycetes and Fungi*, ed. Z. Vanek, Z. Hostalek & J. Cudlin, pp. 309-334. Elsevier Science Publishers, B.V.: Amsterdam.

Ogawa, K., Brown, J. A. & Wood, T. M. (1987). Intraspecific hybridization of *Trichoderma reesei* QM-9414 by protoplast fusion using colour mutants. *Enzyme and Microbial Technology*, 9, 229-232.

Papavizas, G. C. & Lewis, J. A. (1981). Introduction of new biotypes of *Trichoderma harzianum* resistant to benomyl and other fungicides. *Phytopathology*, 71, 247-248.

Park, H. M., Jeong, J. M., Hong, S. W., Hah, Y. C. & Seong, C. N. (1986). Interspecific protoplast fusion of *Trichoderma koningii* and *Trichoderma reesei*. *Korean Journal of Microbiology*, 24, 91-97.

Peberdy, J. F. (1979). Fungal protoplasts: isolation, reversion and fusion. *Annual Review of Microbiology*, 33, 21-39.

Peberdy, J. F. (1985). Mycolytic enzymes. In *Fungal Protoplasts: Applications in Biochemistry and Genetics*, ed. J. F. Peberdy & L. Ferenczy, pp. 31-44. Marcel Dekker Inc.: New York.

Pontecorvo, G. (1956). The parasexual cycle. *Annual Review of Microbiology*, 10, 393-400.

Pontecorvo, G. & Kafer, E. (1958). Genetic analysis based on mitotic recombination. *Advances in Genetics*, 9, 71-104.

Pontecorvo, G., Roper, J. A., Hemmons, L. M., MacDonald, K. D. & Bufton, A. W. J. (1953). The genetics of *Aspergillus nidulans*. *Advances in Genetics*, 5, 142-238.

Puhalla, J. E. & Mayfield, J. E. (1974). The mechanism of heterokaryotic growth in *Verticillium dahliae*. *Genetics*, 76, 411- 422.

Puhalla, J. E. & Speith, P. T. (1983). Heterokaryosis in *Fusarium moniliforme*. *Experimental Mycology*, 7, 328-335.

Reyes, F., Perez-Leblic, M. I., Martinez, M. J. & Lahoz, R. (1984). Protoplast production from filamentous fungi with their own autolytic enzymes. *FEMS Microbiology Letters*, 24, 281-283.

Riba, G., Lourdes De Aquino, M. & Alves Ribeiro, S. (1987). Polymorphisme chez des souches brésiliennes de l'Hyphomycète *Metarhizium anisopliae* (Metsch.) Sorokini, inféodées à des Cercopidae. *Agronomie* (Paris), 7, 763-768.

Rowlands, R. T. (1984). Industrial strain improvements: mutagenesis and random screening procedures. *Enzyme and Microbial Technology*, 6, 3-10.

Silveira, W. D. & Azevedo, J. L. (1987). Protoplast fusion and genetic recombination in *Metarhizium anisopliae*. *Enzyme and Microbial Technology*, 9, 149-152.

Singh, A. & Sherman, F. (1974). Methionine auxotrophs by isolating methylmercury resistance mutants. *Nature*, 274, 227-229.

Sonnenberg, A. S. M. & Wessels, J. G. H. (1987). Heterokaryon formation in the basidiomycete *Schizophyllum commune* by electrofusion of protoplasts. *Theoretical and Applied Genetics*, 74, 654-658.

Stasz, T. E. & Harman G. E. (1986). Preparation of nonparental strains of *Trichoderma harzianum* by protoplast fusion. *Phytopathology*, 76, 1104.

Stasz, T. E. & Harman G. E. (1987). Improved biocontrol strains of *Trichoderma harzianum* developed by protoplast fusion. *Phytopathology*, 77, 1771.

Stasz, T. E., Harman, G. E. & Matteson, M. C. (1987). Intraspecific and interspecific hybridization of *Trichoderma* strains by protoplast fusion. *Phytopathology*, 77, 1619.

Tanabe, S. M. & Garber E. D. (1980). Genetics of *Ustilago violacea* IX. Spontaneous and induced mitotic recombination. *Botanical Gazette*, 141, 483-489.

Tinline, R. D. & MacNeill, B. H. (1969) Parasexuality in plant pathogenic fungi. *Annual Review of Phytopathology*, 7, 147-170.

Toyama, H., Shinmyo, A. & Okada, H. (1983). Protoplast formation from conidia of *Trichoderma reesei* by cell wall lytic enzymes of a strain of *Trichoderma viride*. *Journal of Fermentation Technology*, 61, 409-411.

Typas, M. A. (1983) Heterokaryon incompatibility and interspecific hybridisation between *Verticillium albo-atrum* and *V. dahliae* following protoplast fusion and microinjection. *Journal of General Microbiology*, 129, 3043-3056.

Typas, M. A. & Heale, J. B. (1976). Heterokaryosis and the role of cytoplasmic inheritance in the dark resting structure formation in *Verticillium* spp. *Molecular and General Genetics*, **146**, 17-26.

Tyrrell, D. (1977). Occurrence of protoplasts in the natural life cycle of *Entomophthora egressa*. *Experimental Mycology*, **1**, 259-263.

Wallin, A., Glimelius, K. & Eriksson, T. (1974). The induction and aggregation and fusion of *Daucus carota* protoplasts by polyethylene glycol. *Zeitschrift fur Pflanzenphysiologie*, **74**, 64-80.

Weeden, N. F., Stasz, T. E. & Harman, G. E. (1987). Lack of expression of a parental genome in *Trichoderma* colonies derived from fusion of protoplasts. *Genetics*, **116**, S59 (abstract 13·51).

Wilson, C. L. & Pusey, P. L. (1985). Potential for biological control of postharvest plant diseases. *Plant Disease*, **69**, 375-378.

Yost, H. T., Chaleff, R. S. & Finerty, J. P. (1967). Induction of mitotic recombination in *Saccharomyces cerevisiae* by ethyl methane sulphonate. *Nature*, **215**, 660.

Zhemchuzhina, N. S., Kazakevich, G. D., Voinova, T. M., Shchurov, M. N. & Javakhiya, V. G. (1985). Overcoming vegetative incompatibility barriers in *Pyricularia oryzae* Cav. with the fusion of mycelial protoplasts. *Mikologiya I Fitopatologiya*, **19**, 462-466.

Zimmerman, E. K. (1971). Induction of mitotic gene conversion by mutagens. *Mutation Research*, **11**, 327-337.

Zimmerman, G. (1986). Insect pathogenic fungi as pest control agents. *Progress in Zoology*, **32**, 217-232.

Zimmerman, V. & Vienken, J. (1982). Electric field induced cell to cell fusion. *Journal of Membrane Biology*, **67**, 165-82.

Chapter 12

Commercial approaches to the use of biological control agents

K. A. Powell[1] & J. L. Faull[2]

[1]ICI Agrochemicals, Jealott's Hill Research Station, Bracknell,
Berks. RG12 6EY & [2]Biology Department, Birkbeck College, Malet Street,
London WC1E 7HX, UK

Introduction

The success of affluent countries in meeting the increasing food demands of their populace has largely been achieved by increased agricultural production and the control of losses during storage and distribution. New chemicals and new technologies have played an indispensable role in achieving this agronomic 'miracle' (Walker, 1987) and this is reflected in the 1986 global figures for the agrochemical market which amounted to some $17·4 billion. Of this figure, 44% of the market share was for herbicides, 31% for insecticides, 19% fungicides and 6% for plant growth regulators (Wood Mackenzie, 1987). Over the last 20 years, as the demand for food has risen, the global market for pesticides has also steadily increased. This market expansion shows signs of slowing as the developed world's population has stabilised and the agrochemical products discovered have satisfied most major niches in the market. As the development of each new product has been estimated to take over 8 years and cost $80 million (Wood Mackenzie, 1987) the price of continuous search and registration of new pesticides can be appreciated.

Currently the only reliable and relatively cheap solution to problems of pests and diseases in large scale agriculture is chemical treatment. Environmental and organic foods groups have lobbied for change but there is little alternative to chemical treatment if large scale agriculture is to continue to provide stable food supplies in the future. Agriculture, therefore, has a problem. High yields depend upon effective use of agrochemicals but the costs of registration of new products are becoming increasingly expensive.

Looking to the future, biological control agents (BCAs) offer a potential solution to the dilemma from both the environmental and the industrial viewpoint. The environmental impact of BCAs is assumed to be less than that of agrochemicals. Their usage is presumed to reduce the global consumption of chemical pesticides, and lead to fewer resistance problems in target organisms. BCAs can also be used in integrated pest management (IPM) schemes, where in combination with good crop husbandry and hygiene, a high quality product can be produced with low chemical input. IPM requires the acceptance by the public of a degree of damage on a product rather than the perfection normally expected (Soper & Ward, 1981). On the industrial side, attractions include cheaper discovery and registration of BCAs relative to chemical pesticides, with a consequently shorter lead time. The effect of these factors on the financial viability of a market can be quite dramatic.

No chemical pesticide would be considered for development unless its projected market could cover costs of discovery, development and registration plus make a reasonable profit to recycle into the system. Normal lead time for development of a new chemical, from discovery to the market place, is in excess of eight years, and patents run for 20 years from the date of application. Costs, therefore, have to be recouped in 12 years. As registration can cost $20 million, a minimum annual profit of $1·6 million is necessary merely to cover these costs alone. Obviously, the lower registration costs of BCAs estimated at about $140,000 in 1983 (Markle, 1983) and the shorter lead time (estimated at 3 years) mean that smaller markets can be targeted or larger ones addressed by a series of products or mixtures. The number of pests, diseases and weeds that can be targeted by chemicals is limited by their potential market size, and a list of those which the agrochemical industry considers as justifiable targets can be found in Jutsum (1988). The list contains only six insect species, five plant diseases (all foliar pathogens) and seven weeds. The market niches that can also be defined for BCA targeting include pests, diseases and weeds which have no current effective chemical control agents, have chemical controls that are too expensive or toxic to be considered, or where targets have become resistant to the existing chemical control. For example, a situation now exists in Canada where legislation in some Provinces has precluded the use of chemical pesticides in forestry. This has lead to the widespread use of *Bacillus thuringiensis* as an insecticide in this environment where previously it would not have been able to compete with chemicals on a cost/efficiency basis. It is currently sold at roughly twice the cost of the alternative chemical pesticide (Jutsum, 1988). Another potential market niche for BCAs are the so-called protected environments where environmental parameters are much more controlled than in the field, and a certain degree of manipulation of growing conditions can give better results for BCAs than that achievable in the field.

At present, the total global market for BCAs and other microbial based products amounts to about $48 million per annum, of which about 50% is accounted for by *Bacillus thuringiensis* and 18% by *Rhizobium*

(Lethbridge, 1989). The bulk of the remaining section of the market is accounted for by *Agrobacterium radiobacter* and silage additives. Even with the relatively reduced costs of developing BCAs permitting the exploitation of smaller markets there is still a lower limit for commercial investment. For example, the production by Upjohn and the USDA of Collego® (*Colletotrichum gloeosporioides* f.sp. *aeschynomene*) to control northern jointvetch, will never recover development costs of approximately $2 million (Templeton, 1986) as the market for Collego is estimated at $132,000 per annum. This illustrates a problem common to many of the smaller targets, that in the absence of public funding many of them cannot be addressed by the agrochemical industry.

Requirements for the successful discovery of a BCA which include the standard types of methodology for isolation, screening and mode of activity studies are well documented (Singh & Faull, 1988). In the scientific literature there are perhaps thousands of papers on the isolation and *in vitro* screening of micro-organisms against a plethora of plant pathogens (see Cook & Baker, 1983 for references). It is a sad fact that very few of these potential BCAs have actually proceeded to commercialization. It seems it is relatively simple to select for antagonism between two organisms in an *in vitro* system, but it is much harder to develop a BCA to be effective in the field. Very rarely in academic papers is any consideration given to the suitability of a microbe to fermentation, formulation and persistence in the field.

To date, only 5% of deliberate releases of BCAs have actually achieved their aim (Jutsum, 1988). There is obviously a serious gap between the scientific observations in the laboratory and the practicalities of mass production and field usage of BCAs. The absence of real dialogue between the discoverer of the BCA phenomenon and the company commercializing it could be the reason for these poor success rates. We intend to examine some of the factors that might improve the development of reported BCAs through to commercialization. Wherever possible, reference will be made to the development of fungal BCAs or to the control of fungal diseases but in many cases the only commercial BCAs available are bacteria or viruses.

It may be instructive to consider some of the features of the successful BCAs that are currently in the marketplace. Like agrochemicals, BCAs fall into the categories of insecticides, herbicides and fungicides. Different aspects of each class will be examined to illustrate the various important features of successful BCAs. For instance, details of formulation technology will be dealt with in the fungal insecticides section, as much has been published in this area, but the comments apply equally to all of the categories of BCA considered.

Features of successful biological control agents

Bacteria, viruses and fungi have all been used as microbial insecticides and the greatest commercial impacts of BCAs have been made in the insecticide markets. Consequently, their success deserves closer consideration since the lessons learned in this area may serve to focus our approach to the development of other classes of BCA.

Bacterial insecticides

To date, the most successful BCA has been the insecticidal bacterium *Bacillus thuringiensis*. Sales of *B. thuringiensis* in forestry, agriculture and public health amounted to some $30 million (Jutsum, 1988). It is the only BCA that stimulates a considerable level of interest from the agrochemical industry. This bacterium is also well researched in terms of fermentation microbiology and, therefore, will serve as a good example of the features of a commercially viable BCA.

Much information is available concerning its metabolism during vegetative growth and sporulation, and on the control of fermentation to maximise yield of endotoxin and spores (Luthy, Cordier & Fischer, 1982). The current modest success of *B. thuringiensis* comes after a long period of stagnation owing to inconsistent field results (due to inadequate standardization), and competition from cheaper chemicals. The turning point for *B. thuringiensis* came with the increased potency and reliability of the product, the occurrence of resistance to chemical pesticides in some potential target organisms and the changes in Canadian forestry practice (Luthy *et al.*, 1982). There has also been a change in the attitudes of the forestry users, who take great pains to make the product work. The advantages of *B. thuringiensis* are the low mammalian toxicity, lack of resistance problems, and its specificity. However, specificity may also be a disadvantage, in that some of the *B. thuringiensis* strains have extremely narrow host ranges. *B. thuringiensis* also requires a greater sophistication in usage relative to chemical pesticides, and can be expensive to produce. At the moment the advantages outweigh the disadvantages because of the primary importance of environmental safety.

B. thuringiensis is a suitable organism for large scale fermentation as it grows easily in submerged culture using standard fermentation equipment, and its nutrient requirements are simple (Faust, 1974; Ignoffo & Anderson, 1979). The ultimate aim of the fermentation is to produce the maximum yield of insecticidally active parasporal endotoxin crystal (Scherrer, Luthy & Trumpi, 1973). *B. thuringiensis* is formulated either directly or after centrifugation followed by spray or freeze drying. The latter is expensive and not strictly necessary for *B. thuringiensis* as the crystal endotoxin is heat stable and will withstand the other methods of harvest.

A major problem encountered with the mass production of *Bacillus thuringiensis* comes with trying to standardise the active ingredient which is mixed with cell debris, spores and the remains of the media in which it was grown. Furthermore, it is not possible to compare batch to batch activity solely on the basis of protein content, and consequently its insecticidal effect on its target host has to be recorded as a bioassay. Therefore, it has been necessary to devise an international standard. Samples of defined International Units (IU) are available from the Pasteur Institute. The problems of standardization of bioassays are numerous (Burgerjon & Dulmage, 1977). In general, the results vary with the physiological state of the insect and the heterogeneity of food and environment. Comparative bioassays are time consuming and expensive and the IU achieved on standard insects may not be relevant to the target pest. Other direct methods for determining activity, such as immunoelectrophoresis or ELISA, are currently under development and these may improve matters.

Formulation of *Bacillus thuringiensis* is very difficult, but is essential for field efficacy. There are constraints that may vary with the target insect. For instance, it is important that the insect ingests sufficient toxin to cause death, rather than debilitation followed by recovery. This entails ensuring that the toxin, which prevents feeding within a few minutes, is suitably sized, and perhaps mixed with baits or feeding stimulants. Bulla & Yousten (1979) have listed different *B. thuringiensis* formulations and potencies to give optimum control of certain targets in the US, and Angus & Luthy (1971) list some of the additives that can be used. UV light and rain tolerance are important characters of formulation.

The location of genes for toxin production varies between strains from the chromosome to plasmids or to a combination of both (Held *et al.*, 1982; Kronstat, Schnepf & Whiteley, 1983). This has opened the way for genetic manipulation of toxin genes between strains, so that more than one toxin is produced by one *Bacillus thuringiensis* strain, thereby increasing the host range (Karamata & Piot, 1987). Other developments include the insertion and expression of the *B. thuringiensis* toxins in higher plants which has already led to some success (Vaeck *et al.*, 1987). However, if toxin genes are incorporated into plants it may spell the end of some *B. thuringiensis* markets although it is unlikely to have any immediate effect in forestry. The expression of *Bacillus thuringiensis* toxin genes in higher plants may lead to sufficient selective pressure on insect pests that toxin resistance may develop in some of them. There are already reports of resistance to *B. thuringiensis* toxin developing in some insect pests (McGaughey, 1985).

Viral insecticides

In the case of insect viruses, an initial problem in mass production was the obligate nature of their parasitism, which meant that they had to be produced in living insects. Despite this, several of the insect nucleopolyhedrosis viruses (NPVs) have been commercially produced *in vivo*, in large numbers of host insects which are either collected from the wild, or more satisfactorily, are reared specially. In 1982 three insect viruses had been registered for use in the US: the NPVs of Cotton Bollworm, Douglas fir Tussock Moth and the Gypsy Moth. During mass production of insect viruses, host tissue is used as efficiently as possible to maximise yield of biologically active virus per insect (Ignoffo, 1966). The details of such methodology are given in Shapiro (1982) but factors that are important include the state of the host (its general health, density, sex, age and growth stage), the conditions under which it is reared (the type of container, its temperature, humidity, photoperiod, diet), the source of the viral inoculum (its purity, activity and size), and the length of incubation and the point of harvest.

Given these highly exacting requirements for *in vivo* production it is barely feasible to produce NPVs on a commercial basis in this way. Thus, another possibility is to use suspension culture of insect cells for production. This offers greater possibilities of process control, the virus should be free of microbial contaminants and production should be less labour intensive. There are however, tremendous problems in scaling-up insect cell cultures from shake flasks to fermenter culture. To some extent these have been partially overcome by increasing the viscosity of culture media using compounds like methylcellulose to reduce shearing effects in highly stirred vessels. Semicontinuous production of NPV has been achieved and costs have been much reduced by recycling the medium, but this method is still far too expensive and not competitive with *in vivo* production (Hink, 1982).

Downstream processing of the viruses is also beset by problems. Harvest from the host insect involves some form of grinding or blending followed by concentration and drying. These stages are points at which problems may arise, and contamination by other microbes may occur. The chaff from the insects, particularly their setae, are allergenic which further complicates this stage of mass production. Once extracted from their insect host the viruses must be formulated. The most common carrier is water, with the addition of a number of different stickers, spreaders and most importantly a UV light protectant (Cunningham, 1982).

Although there are considerable advantages of viral insecticides over chemical pesticides, such as their host specificity and absence of toxic effects on mammals, non-target insects, birds and fish, these features rarely outweigh the high cost of production (Summers *et al.*, 1975; Cunningham, 1982). At the moment only small specialist companies seem to be interested in commercialization of these minor products in niche markets. Major improvements in cost of production could be made if a satisfactory formulation could be found for the application of NPVs. At present the formulated product is highly sensitive to UV light and applications could be much reduced if the virus could be made to exist in a stable form on the leaf surface.

Fungal insecticides

The early evaluations of mycoinsecticides considered them to be impracticable because of the adverse effects of the environment on the activity of the fungus (Snow, 1890, 1895; Billings & Glen, 1911). At that time, the basis of epizootics was not understood. However, more recent research has revealed the principles behind the establishment and maintenance of epizootics and at present there are several entomopathogenic fungi registered for use (see Gillespie & Moorhouse, Chapter 4).

The general strategy for use is to start epizootics early by the introduction of a large initial source of inoculum, to gear application to the insect infestation rates and to be willing to use chemicals if the population gets too high. Therefore, agrochemical compatibility is of great importance. The entomopathogenic fungi fall into two nutritional categories, those which can be grown in culture and those that are obligate biotrophs and require a living host. As with the insect viruses and with other biotrophic pathogens, great doubts currently exist as to whether mass production of obligately biotrophic fungi will ever be an economic proposition. To date, *in vitro* production is the only commercially viable method for producing mycoinsecticides. Once again the methods for the successful production of these fungi serve as useful indicators of the likely successful methods for other categories and types of BCA.

For successful *in vitro* production of fungi several important factors should be considered, including strain selection, which involves the use of bioassays to indicate improved performance; strain stability, which requires liquid nitrogen storage of stock cultures; single spore isolation and media development. The optimum C/N ratio, O_2 requirements, temperature and pH optima for both vegetative growth and sporulation have to be ascertained, and there are occasions when the optimum environmental conditions for growth and sporulation do not coincide. There can also be problems in the scale-up of process from flasks to small fermenters and then to large fermenters, but these can be alleviated by judicious changes in the culture medium. Details of media development and laboratory and pilot scale production of fungal insecticides can be found in Soper & Ward (1981).

Generally, there are four options for mass production of fungi: surface cultivation on solid or semi-solid media, biphasic fermentation and submerged fermentation. Solid medium production is the most expensive option as it requires specialist equipment, but it has the advantage that it is the simplest, and harvesting can be achieved by vacuuming spores from the surface of the culture. Semi-solid media involve the impregnation of solid granules with culture medium, followed by fermentation with tumbling. This type of fermentation has in the past been plagued with sterility problems, and requires the use of specialised fermenters. In this situation, the substrate is milled with the fungus. With biphasic fermentation, the fungus is grown to the end of log phase in a fermenter and then the mycelium is harvested, spread onto trays and allowed to sporulate. Again, this can be harvested by vacuuming. The production method of choice remains submerged fermentation as this method utilises existing equipment, and similar downstream processing for harvesting and drying as other fermented products.

A number of criteria have to be met during formulation of fungal insecticides. When applied to an area, the microbe should spread uniformly and quickly and remain in that site. These are the so-called spreader-sticker qualities. The formulation must not inhibit the infection process and, if possible, should enhance disease transmission. All components of the formulation must be compatible to ensure product stability. Finally, all the formulation costs, application procedures and ingredients must result in a product that is competitively priced.

There are various types of formulation that are currently in use. Dry formulations include dusts and granules, where the active ingredient is formulated and stored dry until use. Dusts are formulated from the active ingredient and a diluent or filler, normally an inert material of low absorbent capacity. Clays are often first choice. Dusts contain 5-10% spores by weight and 90-95% diluent. They are mixed dry, and the exact proportions of the components varied to ensure a free flowing product. Factors like particle size, bulk density and flowability of a formulation are extremely important. The aim is to produce a free flowing powder which does not allow separation of the spores from the filler during transport, storage and application. The bulk density should be between 320 kg m^{-3} and 800 kg m^{-3} (Polon, 1973). The application of dusts has considerable advantages over other formulations as they give effective cover of the underside of leaves (Rivers, 1967). This targets feeding insects and gives protection from UV light by leaf shade (Soper & Ward, 1981). Dusts are also easy to store. Their disadvantage is that small particle size causes difficulty in applying the formulation under windy conditions, and there can also be an inhalation problem.

In contrast, granular formulations contain 5-20% spores, 80-95% carriers and 1-5% binder on a weight basis. These materials are usually dry blended with the granule carrier which is an absorbent material in the 150 μm to 1 mm range. It is important that the ratio between the largest and smallest particle be kept to 2:1 otherwise particle separation will occur. The number of spores held on a carrier will depend on the surface area available, the roughness of the surface and the electrostatic attraction of the carrier. In turn, particle size needs to be determined, with a lower level at a weight that ensures settling through a canopy. Commonly used carriers include kaolinite, montmorillonite and other clays but other carriers can be used including organic substrates like ground corn cobs.

Wettable powders can be stored dry until use, when they are mixed with water and sprayed. Dry powder storage minimises interactions between spores and the other components. However, there are increased problems of safety with powders as they are easily inhaled. Formulation is also more difficult than with dry powders because surface active agents have to be added to ensure dispersal when the wettable powder is mixed with water in an applicator tank. Most wettable powders contain 50-80% spores by weight, 15-45% filler, 1-10% dispersant and 3-5% wetting agent. The ratio of spores to filler is critical. The presence of the filler maintains flowability and prevents agglomeration of spores during storage which would reduce wettability and clog spray nozzles (Couch & Ignoffo, 1981). The filler can be the same as that in dry powders but it must be hydrophilic and disperse quickly in a uniform suspension. The filler must also be physically and chemically inert towards the spores.

When a dry powder is introduced into a liquid it must penetrate the surface by overcoming the surface tension of the liquid/solid interface, and the presence of a surfactant in the formulation helps to reduce the tension and allow displacement of air around the spores by liquid (Heistland, 1964). Some fungal spores are highly hydrophobic and have to be covered with a layer of surfactant if they are to wet readily. However, this reduces shelf life (Soper & Ward, 1981). There are many hundreds of wetters available but little is known of their effects on fungal spores. The dispersant ensures that the particles do not attract each other and, therefore, that spores will distribute uniformly in a water column without settling. A list of dispersants can be found in Polon (1973).

Liquid formulations are potentially the most useful and are currently the least developed. A liquid formulation is essentially a suspension of an insoluble particle (spores) in a liquid. The active ingredients by weight account for some 10-40%, the suspending agent for 1-3%, the surfactant 3-8%, the dispersant 1-5% and the solvent vehicle, normally water or oil, 35-65%. The general concept is to have a liquid phase which has a viscosity roughly equal to that of the settling rate of the active ingredient. This is achieved by the use of colloidal clays, polysaccharide gums, cellulose or synthetic polymers, and details have been summarised by Battista (1975) and Theng (1979). The desired formulation is one that will resist a small force like spore settling, but will flow freely when a greater force is applied.

During the fermentation and formulation of a microbial pesticide a number of other crucial requirements must also be considered. The release of a BCA from a formulation must be rapid. The formulation must swell and crack in response to water, and the BCA must recover and germinate quickly. The formulation of the BCA has a fundamental role to play here in protecting the organism during storage and application, but it must also aid escape and growth when the formulated product is applied to the target.

Targeting of the BCA to its site of action is also crucial. In some cases the BCA is motile, and it can actively move to its site of action, but in others hyphal growth has to occur for a BCA to leave the point of inoculation and arrive at its site of action. Controlled delivery over time is highly desirable in some cases and this can in part be achieved by formulation (Rhodes *et al.*, 1989). The metabolic state of the BCA once it has germinated is also important. If the organism has been subjected to excessive stress during fermentation, formulation and application then there may be a period of metabolic repair necessary before outgrowth can occur. This phenomenon has been observed in dormant seeds, bacteria and yeasts (Osborne, Dell'Aquila & Elder, 1984; Elder *et al.*, 1987). This is obviously highly undesirable, and efforts must be made to minimise these deleterious effects during product development.

Microbial herbicides

At present, there are very few plant pathogens that are being considered as microbial herbicides, and as yet there are no commercially available viral or bacterial preparations. All interest is currently on the use of fungi in this respect. This reflects the fundamental differences in the physiologies of the other groups, in that viruses tend not to be specific and they require a vector, and that plant pathogenic bacteria normally require wounds to enter their hosts. In contrast, fungi can actively penetrate a healthy plant. An extensive list of projects has been compiled by Scheepens & Van Zon (1982) but to date there are only three commercial mycoherbicide preparations available: Collego (*Colletotrichum gloeosporioides* f. sp. *aeschynomene*) against *Aeschynomene virginica* in rice and soybean; *Cercospora rodmanii* against water hyacinth; and Devine® (*Phytophthora palmivora*) against milkweed vine in citrus groves (see also Templeton & Heiny, Chapter 6). There is a clear preference for the facultative pathogens for commercialization, but obligate pathogens have been used in a 'release once–control forever' system associated with classical biological control. There have been undoubted technical successes using biotrophic fungi as mycoherbicides. For example, *Puccinia chondrillina* has been used as a BCA to control skeleton weed in Australia-with an estimated saving of US$26 million at final equilibrium. The use of Devine to control milkweed vine falls into this category too, only one application is required to establish this BCA in soils for 8 years (Kenney, 1986). Obviously no company could afford too many products such as these as repeat sales are needed to recover the costs of development and production.

The three existing commercial preparations are wettable powder formulations, and as such have the same characteristics as the fungal insecticides already discussed.

Microbial control of plant pathogens

Biocontrol of soilborne plant pathogens with antagonistic fungi and bacteria has been under intensive investigation for many years, and much of the background can be found in Cook & Baker (1983) and Lumsden & Lewis (Chapter 8). Despite this, there is at present only one registered and commercialised product in the USA (Dagger G® for the control of *Pythium* and *Rhizoctonia* damping-off complex on cotton), and only two in Europe (*Trichoderma viride* for the treatment of timber and trees against *Chondrostereum purpureum* and *Armillaria mellea*, and *Peniophora (Phlebia) gigantea* for the protection of tree stumps against *Heterobasidion (Fomes) annosum*). The control of plant viruses with attenuated strains or avirulent strains of virus has achieved some degree of success. For example, the use of avirulent strains of tobacco mosaic virus (TMV) to protect against infection of tomato plants with virulent TMV has become widespread (Channon *et al.*, 1978). The use of the bacterium *Agrobacterium radiobacter* to protect plants against crown gall caused by *A. tumefaciens* is the only effective way of controlling this disease, and *A. radiobacter* K84 is also a commercial product. Apart from these few minor products, the success of plant disease control using other microbes is limited to the laboratory, to the greenhouse and to some very special soils that are essentially artificial (Elad, Chet & Henis, 1981). However, there are similarities between the current situation in this area and that of microbial insecticides such as *Bacillus thuringiensis* twenty years ago. With both bacterial and fungal insecticides, early results in the field were disappointing for a number of reasons. The interactions were not understood and this lead to errors in formulation and application of early products. There were also problems with the strains not being sufficiently virulent. This was overcome by strain improvements using natural isolates and conventional mutagenesis. Jutsum (1988) has listed the technical shortcomings of microbes as BCAs for plant pathogen control.

Scientific and commercial requirements for biological control agents

Requirements for successful pest/pathogen control

Consideration of the currently successful BCAs that are available should give us some idea of the sorts of characteristics that are required to develop new commercial BCAs:

- *The speed of action* of the BCA relative to that of the pest or host plant is critical. The pathogen may outgrow the BCA effectively leaving it behind whilst the pathogen exploits new niches. This seems to be the case with the use of some soil bacteria against soilborne plant pathogens, where plant root growth rate exceeds the rate of bacterial growth leaving the niche open to pathogen attack (Maplestone, 1986). A BCA may be added to the environment with a nutritive carrier. If the BCA is slow growing and the pathogen is fast growing then the pathogen may utilise any extra substrate and exacerbate disease. For example, Kelley (1976) reported that *Trichoderma harzianum*-impregnated clay granules provided nutrients for the growth of *Phytophthora cinnamomi* which worsened rather than controlled the disease.

- *The persistence* of a BCA in the relevant environment is important. Host plant vulnerability to some pests and diseases can be defined very closely as a 'window' of time, before and after which susceptibility does not occur. In other cases the host is continuously susceptible to attack throughout its lifetime. It is thus important that the BCA exists long enough to protect the plant during its vulnerable period, whatever that may be. It may also be necessary for a BCA to colonise new substrates and multiply in the environment. For example, many of the commercialised entomopathogenic fungi and viruses rely on multiplication in the environment to establish epidemics.

- *The environmental tolerance* to moisture, temperature, UV light, and pH of the BCA must be known, and if they do not match that of the target niche, the formulation of the BCA can be used to try and modify the niche. Garrett's (1956) concept of competitive saprophytic ability is useful when considering whether or not the BCA is a good competitor in a given environment. Information on how its performance is affected by the host plant species or soil types is vital for success of the BCA.

- *The mechanism of action* of the BCA will affect the type of organism chosen for a niche and its application rates. The BCA effect may be mediated via a highly active volatile antibiotic or by close physical contact between the BCA and its target, as is the case with hyperparasitism and hyphal interference (see Lewis *et al.*, Chapter 9). Thus the distance between the target and BCA can be greater with a volatile product than if there has to be close physical contact between the two. The need for close proximity between the BCA and the target will affect application methodology, rates and formulation. In some cases if very close contact is required between a BCA and a moving target in a complex matrix like soil, then use of that particular BCA may be precluded. Information about the various mechanisms involved in BCA/target interactions is contained in Mukerji & Garg (1988a, b)

Requirements for successful registration

After discovery, a potential BCA must be produced and formulated so that it is stable in storage and disperses well at application. Subsequently, delivery systems must be developed so that the BCA arrives at the correct site at the right time in sufficient quantity to do its job on a reliable basis. Finally, it is necessary to register it as a product.

Like chemicals, there are a number of regulatory requirements that need to be satisfied before a product will be granted a licence. At present in the UK and France, the use of indigenous microorganisms as BCAs in small field trials requires no special permission, and this is also the case in the USA for field trials of under 10 acres. For non-indigenous microorganisms the legislation is currently under discussion. For further development there is a registration requirement that includes toxicity studies and persistence studies which in the USA are listed by the United States Environmental Protection Agency (USEPA) in the Federal Register (USEPA, 1984).

For genetically engineered organisms the situation is different and currently under discussion in Europe. In the USA, the EPA has issued guidelines for obtaining an experimental use permit (EUP) for the release of these organisms, and these requirements are reviewed in Milewski (1988). At present it is very much more complex, and therefore more expensive, to bring genetically engineered organisms into the market place, so the market niches that can be targeted must be very valuable ones to recoup expenditure.

Requirements for commercialization

From the preceding discussions it can be seen that there are several characteristics that make for success of a microorganism as a BCA:

- *A viable market size* and customer demand for BCAs. It is preferable that there is broad spectrum activity of a BCA against a number of targets. This makes the market size larger and improves the likelihood of cost effective manufacture.

- *High performance and consistency* of effect of the BCA has to be equal to that of chemical pesticides.

- *Persistence of effect* must be acceptable to both the user and the producer, i.e. the BCA should persist long enough to do the job.

- *Safety*; the product has to be free of all toxicological problems, i.e. have low mammalian toxicity and have little or no effect on non-target species and be environmentally compatible.

- *Stability*; the product must have a two year shelf life at -5 to $+30°C$ but this could be modified.

- *Product differentiation*; this is necessary where there are competing chemicals.

- *Indigenous microorganism*; at present, unless aimed at a very valuable market, BCAs should be indigenous microorganisms that are not genetically engineered. This criterion may change within the foreseeable future.

- *Saprotrophic microorganism*; the BCA should be a saprotroph.

- *Minimal capital costs*; BCA mass production should be by submerged fermentation utilizing currently existing fermentation equipment.

- *Cost and practicality of production*; substrates should be cheap and simple. The BCA should be easily harvested, stable when dry, and be easily formulated, i.e. have naturally wettable spores of regular size and shape with long term viability and compatibility with other commonly used ingredients of formulations. They should be relatively insensitive to UV light and desiccation, but they should germinate and grow readily when moisture is available. They should be insensitive to other commonly used agrochemicals.

- *Application*; they should be applied using conventional technology and their application should not involve major changes in current agricultural practices.

When looking at this list it may seem that it will take nothing short of a miracle to obtain a successful BCA. However, it is worth remembering that there are a number of BCAs already available as commercial products that satisfy, at least in part, most of these criteria. In the past there has been no evidence of a coordinated approach to these requirements by research scientists. It may be that by addressing these preceding points in a systematic manner during all stages of research and development the passage of BCAs through to commercialization may improve. There is also a need for realistic requirements to be set by regulatory bodies so that costs of registration do not escalate dramatically.

The experience of both academic and industrial scientists in this field has increased knowledge of microbial physiology and genetics, and should lead us away from the low input, crude screening systems that to date have given false hopes of BCAs. It is now up to these scientists to provide a sound base from which these potentially valuable microbes can be successfully commercialised, and for the industrialists to keep an open mind on the prospects for their success.

This chapter was written whilst JLF was on sabbatical leave at ICI Jealott's Hill, funded by a Royal Society/SERC Industrial Fellowship.

References

Angus, T. A. & Luthy, P. (1971). Formulation of microbial insecticides. In *Microbial Control of Insects and Mites*, ed. H. H. Burges & N. W. Hussey, pp. 623-638. Academic Press: New York.

Battista, O. A. (1975). *Microcrystal polymer Science*, McGraw Hill: New York.

Billings, F. H. & Glen, P. A. (1911). Results of the artificial use of white fungus disease in Kansas. *USDA Agricultural Bureau Entomological Bulletin 107*, ed. L. O. Howard. USDA: Washington, DC.

Bulla, L. A. & Yousten, A. A. (1979). Bacterial insecticides. In *Microbial Biomass, Economic Microbiology*, vol. 4, ed. A. H. Rose, pp. 91-114. Academic Press: New York.

Burgerjon, A. & Dulmage, H. T. (1977). Industrial and international standardisation of microbial pesticides. 1. *Bacillus thuringiensis*. *Entomophaga*, 22, 121-129.

Channon, A. G., Cheffins, N. J., Hitchon, G. M. & Barker, J. (1978). The effect of inoculation with attenuated mutant strains of Tobacco Mosaic Virus on the yield of early glasshouse tomatoes. *Annals of Applied Biology*, 88, 121-129.

Cook, R. J. & Baker, K. F. (1983). *The Nature and Practice of Biological Control of Plant Pathogens*. American Phytopathology Society: St Paul, Minnesota, USA.

Couch, T. L. & Ignoffo, C. M. (1981). Formulation of insect pathogens. In *Microbial Control of Pests and Plant Diseases 1970-1980*, ed. H. D. Burges, pp. 621-634. Academic Press: New York.

Cunningham, J. C. (1982). Field trials with baculoviruses in control of forest insect pests. In *Microbial and Viral Pesticides*, ed. E. Kurstack, pp 335-386. Marcel Dekker: New York.

Elad, Y., Chet, I. & Henis, Y. (1981). Biological control of *Rhizoctonia solani* in strawberry fields with *Trichoderma harzianum*. *Plant and Soil*, 60, 245-254.

Elder, R. H., Dell'Aquila, A., Mezzina, M., Sarasin, A. & Osbourne, D. J. (1987). DNA ligase in repair and replication in the embryos of rye, *Secale cereale*. *Mutation Research*, 181, 61-71.

Faust, R. M. (1974). Bacterial Diseases. In *Insect Diseases*, ed. G. E. Cantwell, pp. 87-183. Marcel Dekker: New York.

Garrett, S. D. (1956). *Biology of Root Infecting Fungi*. Cambridge University Press.

Held, G. A., Bulla, L. A., Ferrari, E., Hoch, J., Aronson, A. L., & Minnich, S. A. (1982). Cloning and localisation of the lepidopteran protoxin gene of *Bacillus thuringiensis* subsp. *kurstaki*. Proceedings of the National Academy of Sciences of the U. S. A., **79**, 6065-6069.

Heistland, E. N. (1964). Theory of coarse suspension formulation. *Journal of Pharmacological Science*, **54**, 1-18.

Hink, W. F. (1982). Production of *Autographa californica* nuclear polyhedrosis virus in cells from large scale suspension cultures. In *Microbial and Viral Pesticides*, ed. E. Kurstack, pp. 493-506. Marcel Dekker: New York.

Ignoffo, C. M. (1966). Insect viruses. In *Insect Colonisation and Mass Production*, ed. C. N. Smith, pp. 501-530. Academic Press: New York.

Ignoffo, C. M. & Anderson, R. F. (1979). Bioinsecticides. In *Microbial Technology*, vol. 1, ed. H. J. Peppler & D. Perlman, pp. 1-28. Academic Press: New York.

Jutsum, A. R. (1988). Commercial application of biological control: status and prospects. *Philosophical Transactions of the Royal Society of London, series B*, **318**, 357-373.

Kararmata, D. & Piot, J. C. (1987). Hybrid *Bacillus thuringiensis* cells useful for biological control of pests with synergistic insecticidal activity and wider spectrum of activity than parent strains. European Patent no. 221024.

Kelley, W. D. (1976). Evaluation of *Trichoderma harzianum* impregnated clay granules as a biocontrol for *Phytophthora cinnamomi* causing damping off of pine seedlings. *Phytopathology*, **66**, 1023-1027.

Kenney, D. S. (1986). Devine, the way it was developed. An overview. *Weed Science*, **34** (Suppl. 1), 15-16.

Kronstat, J. W., Schnepf, H. E. & Whiteley, H. R. (1983). Diversity of locations for *Bacillus thuringiensis* crystal protein genes. *Journal of Bacteriology*, **154**, 419-428.

Lethbridge, G. (1989). An industrial view of microbial inoculants for crop plants. In *Microbial Inoculation of Crop Plants*, Society for General Microbiology special publication 25, ed. R. C. Campbell & R. M. Macdonald, in press. IRL Press: Oxford.

Luthy, P., Cordier, J.-L. & Fischer, H.-M. (1982). *Bacillus thuringiensis* as a bacterial insecticide. In *Microbial and Viral Pesticides*, ed. E. Kurstak, pp. 35-74. Marcel Dekker: New York.

Maplestone, P. A. (1986). Interactions between soil bacteria and the take-all fungus: root colonisation and potential for biological control. Ph.D. thesis, University of Bristol.

Markle, G. M. (1983). Registration of biorationals. In *Proceedings of the National Interdisciplinary Biological Control Conference*, ed. G. E. Allen & M. C. Nelson, p. 106. USDA: Washington, DC.

McGaughey, W. H. (1985). Insect resistance to the biological insecticide *Bacillus thuringiensis*. *Science*, **229**, 193-195.

Milewski, E. A. (1988). The approach of the US Environmental Protection Agency in regulating certain biotechnology products. In *Risk Assessment for Deliberate Release*, ed. W. Klingmuller, pp. 184-190. Springer-Verlag: Berlin.

Mukerji, K. C. & Garg, K. L. (1988a). Biocontrol of Plant Pathogens. 1. CRC Press: Boca Raton, Florida.

Mukerji, K. C. & Garg, K. L. (1988b). Biocontrol of Plant Pathogens. 2. CRC Press: Boca Raton, Florida.

Osborne, D. J, Dell'Aquila, A. & Elder, R. H. (1984). DNA repair in plant cells. An essential part of early embryo germination in seeds. *Folio Biologica* (Praha), Special Publication, 155-170.

Polon, J. A. (1973). Formulation of pesticidal dusts, wettable powders and granules. In *Pesticide Formulations*, ed. W. V. Walkenburg, pp. 143-234. Marcel Dekker: New York.

Rhodes, D., Powell, K.A., MacQueen, M. D. & Greaves, M. P. (1989). Controlled release of biological control agents. In *Controlled Delivery and Pest Management*, ed. R. M. Wilkins, in press. Taylor & Francis: London.

Rivers, C. F. (1967). The natural and artificial dispersion of pathogens. In *Insect Pathology and Microbial Control*, ed. P. A. VanDerLaan, pp. 252-263. North Holland Publishing Company: Amsterdam.

Scheepens, P. C. & Van Zon, H. C. J. (1982). Microbial Herbicides. In *Microbial and Viral Pesticides*, ed. E. Kurstack, pp. 623-641. Marcel Dekker: New York.

Scherrer, P., Luthy, P. & Trumpi, B. (1973). Production of endotoxin by *Bacillus thuringiensis* as a function of glucose concentration. *Applied Microbiology*, **25**, 644-646.

Shapiro, M. (1982). *In vivo* mass production of insect viruses for use as pesticides. In *Microbial and Viral Pesticides*, ed. E. Kurstack, pp. 463-492. Marcel Dekker: New York.

Singh, J. & Faull, J. L. (1988). Hyperparasitism and biological control. In *Biocontrol of Plant Pathogens*, vol. 2, ed K. G. Mukerji & K. L. Garg, pp. 167-179. CRC Press: Boca Raton, Florida.

Snow, F. H. (1890). Experiments for the destruction of chinch bugs. *21st Report of the Entomological Society of Ontario*, pp. 93-97.

Snow, F. H. (1895). Contagious diseases of the chinch bug. *4th Annual Report of the Directorate of the University of Kansas*.

Soper, R. S. & Ward, M. G. (1981). Production, formulation and application of fungi for insect control. In *Biological Control in Crop Production*, Beltsville Agricultural Research Center symposium 5, ed. G. Papavizas, pp. 161-180. Allanheld Osmun & Co.: Totowa, New Jersey.

Summers, M., Engler, R., Falcon, L. A. & Vail, P. (1975). *Baculovirus for Insect Pest Control: Safety Considerations*. American Society for Microbiology: Washington, DC.

Templeton, G. E. (1986). Mycoherbicide research at the University of Arkansas, past, present and future. *Weed Science*, **34** (suppl. 1), 35-37.

Theng, B. K. G. (1979). *Formulations and Properties of Clay Polymer Complexes*. Elsevier: New York.

USEPA (1984). Microbial pesticides: interim policy on small scale testing. Notice of interim policy. *Federal Register*, **49**, 202.

Vaek, M., Raynaerts, A., Holte, H., Jansens, S., DeBeuckeleer, M., Dean, C., Zabeau, A. & Leemans, L. (1987). Transgenic plants protected from insect attacks. *Nature*, **238**, 33-37.

Walker, R. (1987). Foreword. In *Toxicological Aspects of Food*, ed. K. Miller, pp. v-ix. Elsevier Applied Science: London.

Wood Mackenzie, (1987). Agrochemical Service, March 1986. Wood Mackenzie: Edinburgh.

Chapter 13

The environmental challenge to biological control of plant pathogens

Annabel Renwick & Nigel Poole

ICI Agrochemicals, Jealott's Hill Research Station, Bracknell, Berkshire RG12 6EY, UK

Introduction

The biological control of plant pathogens is not yet a commercial reality (Powell & Faull, Chapter 12) and in general, the development of commercial biological control agents (BCAs) of plant diseases is still at the exploratory stage. The major challenge facing the commercialization of BCAs is to obtain effective, reproducible and environmentally acceptable disease control without changing standard agronomic practices. At present, the introduction of microorganisms into the environment as 'pesticides' is being carefully considered for their 'environmental impact' by many governmental authorities and research institutions (Domsch *et al.*, 1988). There is, however, much less consideration and research effort being given to obtaining effective and reproducible disease control. Yet, without reliable and reproducible disease control there are unlikely to be any worthwhile products to be released into the environment.

The lack of consistent plant disease control by microorganisms is often ascribed to 'environmental factors', which are sometimes difficult to define. In non-controlled environments, fluctuations in temperature and ultraviolet (UV) light, pH, water and nutrient availability, or changes in a competitive microflora can all play a major role in restricting activity of the introduced microbial antagonists used to control plant pathogens (Cook & Baker, 1983). The effect of such environmental factors on the biological interactions which take place between the host, pathogen and BCA are complex, and this is further compounded as the underlying mechanisms of disease control are rarely understood.

In the main, most BCAs have been empirically isolated and screened for disease control of plant pathogens. Considering the importance of environmental factors in biological control, it is surprising that there have been so few published studies fully evaluating the robustness (i.e. maintenance of efficacy) of microbial antagonists under a wide range of environmental conditions, particularly in the field. Therefore, in this chapter we will consider the design and development of a programme for the isolation and selection of a commercial BCA for disease control with particular emphasis on the effects of environmental factors.

Interactions between biological control agents and pathogens

Antagonistic microbes have frequently been selected on the basis of their mode of action (Geels & Schippers, 1983; Weller, Zhang & Cook, 1985; Toyodo *et al.*, 1988). However, *in vitro* antagonistic activity rarely correlates with disease control in the field (Andrews, 1985; Rovira, 1985). Nevertheless, Weller, Howie & Cook (1988) have recently demonstrated a relationship between the *in vitro* inhibition of *Gaeumannomyces graminis* and suppression of take-all by fluorescent pseudomonads. Similarly, microbes selected for degradation of fusaric acid have shown good control of *Fusarium* wilt of tomatoes in a field soil (Toyodo & Utsumi, 1987). In order to improve the reliability of biological control it will be necessary to understand the mechanisms of antagonism and those environmental factors which effect changes. Most of the recognized modes of action fall into the following catagories: 1) competition for nutrients, 2) antibiosis, and 3) mycoparasitism (see also Lewis, Whipps & Cooke, Chapter 9).

Competition for nutrients

Experimentation on the competitive abilities of microorganisms for different nutrients can be very difficult but correlations between competition for nutrients and disease control have been shown (Alabouvette, Couteaudier & Louvet, 1985; Elad & Chet 1987). Chen *et al.* (1988) demonstrated that competition for nutrients between the indigenous microbial population in compost, resulted in reduced damping-off in cucumber-seedlings caused by *Pythium*. The suppressive nature of the compost was lost with the addition of nutrients.

Competition between microbes for available Fe^{3+} involving the production of Fe^{3+} chelators (siderophores) has been well investigated as a mechanism for disease control (Swinburne, 1986; de Weger, Schippers & Lugtenberg, 1987). One of the major restrictions with this particular mechanism of activity is that siderophores are only produced in environments where the availability of Fe^{3+} is low (normally at high pH). This will undoubtedly reduce the value of microbial antagonists with siderophore production as a major mechanism of disease control when applied to soil with a pH below neutral.

Antibiosis

Antibiotics have been isolated from many antagonistic microbes and have often been indirectly implicated in disease control. Sometimes, antibiotics produced by BCAs are inactivated by soil (Howell & Stipanovic, 1979, 1980). For example, a phenazine-1-carboxylic acid produced by the fluorescent pseudomonad 2.79 (NRRLB-15132) in soil was found to have a role in the suppression of take-all (Gurusiddaiah *et al.*, 1986). However, Brisbane & Rovira (1988) have shown that the value of this antibiotic may be restricted to acid soil as under alkaline conditions it is inactivated.

Mycoparasitism

Mycoparasitism of fungal pathogens requires initial growth of the antagonist, 'recognition' between antagonist and pathogen and the production of cell-wall degrading enzymes (Elad, Chet & Henis, 1979; Elad, 1986). All of these biological processes will be influenced by changes in environmental conditions, resulting in changed biocontrol activity. However, Howell (1987) found that after mutagenesis of the mycoparasite *Gliocladium virens*, the mycoparasitic deficient strains continued to control *Rhizoctonia solani* on cottonseed in soil. Mycoparasitism was, therefore, not seen to be the major mechanism of disease control in this system.

Glasshouse and field screening systems

Appropriate glasshouse and field screens for the selection and evaluation of biocontrol agents may follow *in vitro* primary screening or form a primary screen itself. If reproducible and effective control of target pathogens cannot be shown in controlled glasshouse and field screens, it will be difficult to produce a commercial product. The main objectives of glasshouse and field screens for biocontrol agents are not unlike those for fungicide screening:

- To select active, 'lead' antagonistic microbes, which control a spectrum of plant pathogens.
- To evaluate the biocontrol activity of lead microbes under a range of environmental conditions.
- To evaluate test formulations.
- To develop and evaluate application methods.

Differences in effectiveness and reliability of strains do occur between field trials and investigators, even for those microbial antagonists which have been well studied. Hornby, (1987) found that the fluorescent pseudomonad 2.79 (NRRLB-15132) and selected *Bacillus* strains previously shown to have activity against take-all (*Gaeumannomyces graminis* var. *tritici*) of wheat (Weller, 1983; Capper & Campbell, 1986), gave no control of take-all in three years of field trials. Hornby (1987) discussed the weaknesses of screening methods used to evaluate biological control agents of take-all. At present there are no standard methods for evaluating the activity of BCAs in glasshouse or field screens. In contrast, the registration and commercialization of fungicides requires reputable, standard operating procedures to evaluate the effectiveness and safety of test chemicals. This will also have to be the requirement for BCAs in order for them to be registered.

Screening systems for fungicide evaluation have been well developed for most of the economically important plant pathogens (Shephard, 1987). The challenge for biocontrol is to adopt the expertise from fungicide evaluation and use it to improve both glasshouse and field techniques for evaluating biocontrol agents. Agrochemical companies have this transferrable expertise. However, it is recognised that biocontrol agents are more sensitive to changes in environmental conditions than fungicides, and this will have to be taken into consideration in redesigning the screens.

There are major experimental criteria which have to be considered prior to designing glasshouse and field screens for biocontrol agents. In practice these can only be chosen when the objectives of the screening programme are clearly established. Examples of these criteria are:

- Use of soils naturally infested with specific plant pathogens, or use of soils artificially inoculated with the required plant pathogens.
- Quantity and pathogenicity of the plant pathogens in the soil.
- Evaluation against complexes or single species of plant pathogens.
- Choice of chemical and/or BCA standards.
- Environmental factors e.g. radiation, water potential, pH, fertilizers, other pesticides.
- Choice of crops/cultivars.
- Biological nature of the BCA, e.g. colonizing activity and its response to prevailing environmental conditions.
- Methods for scale-up fermentation of the BCA.
- Type of formulation of the BCA.
- Method of application of the BCA; placement and timing.
- Method of disease assessment; timing of disease assessment.

Efficient glasshouse and field screens should be aimed at specific plant pathogens and the test microbial antagonists should be evaluated against a range of pathogen concentrations, representing a range of disease pressures. In glasshouse screens at ICI Agrochemicals, we have found the threshold level of disease control to be very clear cut. This information is very important to differentiate biocontrol activity between strains. In many published experiments, BCAs are evaluated against natural levels of inocula. The main advantages of using soils or sites where the pathogen is naturally present, are that the BCA will challenge the target pathogen(s) under realistic disease pressures, and in competition with the natural microflora. In addition to these factors, complexes of pathogens often cause disease, and therefore in naturally infested soils the biocontrol agents will be evaluated against such pathogen complexes. There are, however, several serious drawbacks to the use of naturally infested sites. These are:

- In practice it is important to have a reliable screen of known disease pressure which naturally infested sites do not provide.
- Finding potential trial sites containing natural inoculum is time consuming. Satisfactory fields are often located many miles from the research station, which can cause extra difficulties in managing the trial, especially when the trial requires regular attention, such as irrigation for inducing damping-off diseases.
- Disease control is only evaluated at one disease pressure at each site, and against a complex of pathogens.
- Variability between sites and years is often high since the level of disease expressed is a result of the inoculum pressure and prevailing environmental conditions.
- Naturally infested sites can sometimes result in patchy distribution of disease symptoms, e.g. take-all. Artificial inoculation of soil with the pathogen can sometimes overcome this problem (Simon, Rovira & Foster, 1987).
- A wide range of chemicals (when available) have to be used as standards to determine which pathogen is causing the disease. This results in large trials and complex statistical analyses.

Our own experience with field trials using *Gaeumannomyces graminis* naturally infested sites took place during a research project on the biological control of take-all in collaboration with Dr R. Campbell at Bristol University. In spring 1984, fields in their first or second year of wheat or barley, showing serious symptoms of take-all, were identified (Feest & Campbell, 1986). The most severe sites were used in the autumn for evaluating several lead microbial antagonists. However, the development of take-all was poor during the following growing season. This exercise was repeated in 1985, but again we found low levels of take-all disease developed in the autumn-sown wheat crop. After two years of field trials inconclusive, evaluation results were obtained because of unreliable disease pressures. This clearly demonstrated that in the process of commercializing a BCA, reliable field trial techniques for evaluation of lead antagonistic microorganisms were needed.

Improvements to screening systems

Improved field trials involving take-all have been developed by producing suitable forms of pathogen inoculum in sufficient quantities to give reproducible levels of disease. Work has shown that the nature, size and placement of the pathogen propagule are important in the development of realistic levels of disease under field conditions (Wilkinson, Cook & Alldredge, 1985; Simon, Rovira & Foster, 1987). However, in the case of *Pythium* spp., water availability and soil temperature are more important than the quantity and distribution of the pathogen propagules in limiting the development of damping-off (Stanghellini *et al.*, 1983).

The sensitivity and speed of screens could be increased if disease assessment methods were capable of detecting the presence of the pathogen in plant tissue prior to the production of visible symptoms. Using the ELISA technique, *Gaeumannomyces graminis* was detected in wheat roots as early as 6 to 10 days after inoculation (El-Nashaar, Moore & George, 1986). A direct relationship was shown between the inoculum concentration of *G. graminis* and detection of pathogen antigens in the roots. Using this technique, reduced levels of *G. graminis* were found in the roots 10 days after sowing seed inoculated with microbial antagonists. This technique could be used for a series of diseases where a pathogen infects the young seedlings but the symptoms do not develop until the plants reach maturity.

Environmental factors

Many of the environmental factors which can influence microbial activity in field screens can first be tested in the glasshouse. For example, Nelson (1988) showed that *Enterobacter cloacae* and *Erwinia herbicola* controlled *Pythium* disease more effectively at 25° than at 15°C. It may therefore be possible to select microbial antagonists which control disease under specified conditions in the field. The same may hold for a range of other microbial inoculants. For instance, the percentage occupancy of nodules, by two *Rhizobium trifolii* strains (CP3B and R4) on two white clover cultivars (Milkanova and S.184) was influenced by liming (Renwick & Jones, 1986). Overall, strain CP3B was the more competitive for nodule occupancy, but by selection of the cultivar Milkanova, previously shown to have a 'preference' for R4 at high pH (Jones & Morley, 1981), and by increasing the soil pH by liming, it was possible to increase the proportion of nodules occupied by the less competitive strain R4. This work was supported by *in vitro* and glasshouse studies and illustrates that by selection of strains and cultivars for particular environmental conditions manipulation of microbial activity in the field is feasible.

Establishment and detection of biological control agents

Colonization

The initial establishment of microbial antagonists and subsequent colonization of the plant surface are believed, in some circumstances, to be crucial elements to disease control (Van Vuurde & Schippers, 1980; Suslow, 1982; Schroth, Loper & Hildebrand, 1984; Bahme & Schroth, 1987). This two-stage process is dependent on the physiology of the microbial antagonists, the formulation used, the rate and timing of application, and the prevailing environmental conditions. The dynamics of colonization by antagonistic microflora and those environmental factors which affect this are little understood, although studies are now being published which address this (Bahme & Schroth, 1987; Bahme *et al.* 1988)

Detection and quantification of microbes in the environment

In order to determine microbial colonization of the plant, the introduced microbes have to be detected. A range of techniques are available (Andrews, 1987). Quantitative and qualitative information about microbial communities on plant surfaces has been obtained using microscopical methods and plating techniques (Foster, 1985; Morris & Rouse, 1987). However, in order to quantify the changes in the population of an introduced microorganism, specific markers or labels are required. Marker techniques must be sensitive enough to enable detection in low numbers and, importantly, the methods should be easy to use and be relatively cheap, as large numbers of microbes will have to be screened. In general bacteria are easier to quantify in soil and on plant tissue than fungi, although techniques for the estimation of fungal colonization are developing (Mendgen, 1987). Antibiotic resistance is the most common marker used for the detection of bacteria but there are problems related with this method because since 'mutant' colonies are selected, the strain may lose other wild-type properties including biocontrol activities.

The ability of bacteria and fungi to colonize plant surfaces actively is often of prime importance during the selection of BCAs. Scher *et al.*,(1988) investigated several characteristics of *Pseudomonas* and *Serratia* spp. which may influence colonizing ability, such as cell motility, chemotaxis and generation rates. However, no positive correlation was found between any of the measured characteristics and the colonizing ability of the bacteria. There is increasing evidence, though, that agglutination of bacteria to roots may play a role in their ability to colonize root surfaces (Anderson, Habibzadegah-tari & Tepper, 1988). Vesper, (1987) found a good correlation between pili produced by a fluorescent pseudomonad and the proportion of attachment to corn roots, and Tari & Anderson (1988) correlated the loss of agglutination of a bacterium with the loss of *Fusarium* wilt control. As with many biological mechanisms, pilus production and agglutination are also affected by environmental factors.

Formulation and application of biological control agents

The competitive nature of BCA strains can be influenced by developing suitable formulations, and by adjusting the growing conditions of the plants to encourage microbial colonization. Fungal and bacterial antagonists are introduced into the environment for control of plant pathogens as seed treatments (Harman, Chet & Baker, 1981; Weller, 1983; Lewis & Papavizas, 1987; Harman & Taylor, 1988), granules (Dagger G®) or as liquid suspensions in the form of sprays or drenches (Collego® and Devine®). Although it is not the case with Devine and Collego, most commercial products require a shelf life of a minimum of 12 months. Therefore, BCAs will need to be formulated as metabolically inactive viable cells. Fungal spores are adapted for survival under these conditions, but many potential biocontrol agents are gram-negative, non-spore forming bacteria. This obviously poses a difficult challenge for the development of suitable formulations for these bacteria.

Formulations can be developed to aid the competitive ability of microbial antagonists on application to the plant or soil. The addition of organic acids to the formulation of *Trichoderma koningii*, and the addition of polysaccharides and polyhydroxyl alcohols to *T. harzianum* increased the biocontrol activity of these strains (Nelson, Harman & Nash, 1988). It was suggested that certain fatty acids stimulated fungal spore germination, but also that the compounds could also have affected the pH of the formulation and thus indirectly affected the growth and biocontrol activity of the *Trichoderma* strains. The application of BCAs to seeds is appropriate for the control of seedling diseases, such as those caused by *Pythium*, *Phytophthora*, *Fusarium* and *Rhizoctonia* spp. There has been moderate success in controlling some of these pathogens (Howell & Stipanovic, 1979, 1980; Windels, 1981; Hadar *et al.*, 1983). Plant diseases such as take-all of wheat and barley and the *Fusarium* wilts are more difficult to control with either biocontrol agents or fungicides, since these pathogens can infect susceptible plant roots throughout the growing season. To control these diseases microorganisms would need to colonize the whole root system. In general, bacteria are more accomplished at colonizing plant roots than fungi (Weller, 1984; Chao *et al.*, 1986). However, Weller (1984) found that a *Pseudomonas* strain colonized the seminal roots, but few nodal roots, when applied as a seed treatment for the control of take-all on wheat.

In collaboration with R. Campbell at Bristol University, we compared the colonization of wheat roots by a *Pseudomonas fluorescens* strain applied as dry granules and a liquid drench to the furrows at sowing. Assessments of root colonization by the bacteria were made at weekly intervals for two months after seedling emergence. Fewer bacteria (a tenfold difference) were detected on the roots of wheat plants when the bacteria were applied as granules compared to the liquid drench. More importantly, closer investigation of the roots revealed very patchy distribution of the bacteria on the roots when they were applied as granules. In addition to this, bacteria were detected in soil samples at 10 cm depth, but no bacteria were detected in the soil when applied in granules. This information poses questions about the efficacy of bacterial transfer to the root system from dry granules applied to soil. Fungal antagonists can grow between soil particles under drier soil conditions and hence granules may be more acceptable for fungal agents (Campbell & Clor, 1983).

A wide range of environmental factors can influence the ability of bacterial and fungal antagonists to colonize plant surfaces throughout a growing season. Fluctuations in environmental conditions can greatly alter microbial colonization or even survival. In the same colonization study at Bristol University, the daily changes in air and soil temperature (10 cm below surface) were also monitored throughout the growing season of winter wheat. Data of this nature can be of value in understanding any loss of or change in colonizing ability of the microbial antagonists on the roots (Bahme & Schroth, 1987). The downward movement of water from irrigation can influence the downward transfer of bacteria on roots (Bahme & Schroth, 1987; Parke et al,, 1986; Chao et al., 1986). In most cases it is believed that the bacteria are carried down the root profile by passive movement of water rather than by active motility of the bacteria. Several studies have shown that non-motile mutants retain their ability to colonize roots (Howie, Cook & Weller, 1987; Scher et al., 1988).

Environmental and regulatory aspects of biological control

A chemical pesticide will be subjected to a long and expensive evaluation of its environmental impact before it can be marketed. This evaluation can take 7 to 8 years from discovery and cost £10 million for safety testing alone. This cost is obviously a major factor in determining which pests and diseases are selected as targets suitable for research and, eventually, which products are commercialized. The environmental impact of a BCA will also have to be evaluated before it can be commercialized. It is clear from the literature (Domsch et al., 1988), that the research requirements for environmental impact studies overlap with those for obtaining efficient and reliable systems for evaluation of biological control.

The cost of this evaluation will be a major factor in selecting targets for BCA research and commercialization of lead strains. The ideal regulatory system for determining the environmental impact of a BCA should have a range of functions:

- Prevent the release of anything into the environment which is likely to cause 'unacceptable' damage to the environment.

- Permit the release of everything else.
- Should not impose unnecessary expense or delay upon applicants.
- Should have the full and justified trust of the public.
- Be flexible and evolve as experience builds up.

The standards should not be unrealistically high. Absolute certainty and definitions of 'unacceptable' damage are unobtainable and funds to answer such imponderable questions are limited.

A chemical pesticide and a BCA have the same aim, i.e. control of a pest, with concomitant increase in the value of a crop; a result for which the farmer is prepared to pay. However, their mode of action is different and this fact should be recognized in the design of an environmental impact programme. The BCA, for example, will probably be selective for the target pest but also be capable of multiplying in the environment. This difference is at present recognized in the approach to the registration of a BCA. The main factors determining how detailed the environmental impact study of a BCA depend on whether the BCA is:

- indigenous or non-indigenous;
- a known pathogen or non-pathogen of crops or man;
- specific or non-specific to the target pest or disease;
- genetically engineered or natural.

At present, for an indigenous, natural BCA the registration process is both considerably cheaper and quicker than that for a chemical. However, testing of non-indigenous microbes in the UK may be complex.

The debate on the release of genetically engineered organisms is an emotive issue which was initiated in the mid 1970s. In the USA there is a regulatory framework now in existence and the release of genetically manipulated microbes is taking place. Similar regulatory issues are being discussed at the moment in Europe, but unlike the USA, Europe is a collection of independent countries with very different cultures, political views, legal systems and approaches to risk assessment. In Denmark at the moment there is a legal ban on release experiments, while in the Federal Republic of Germany, a recommendation is being discussed that there should be a five year moratorium on the release of genetically engineered microbes, while the release of viruses should be prohibited. In the UK, Netherlands and France, a rigorous vetting system is being developed on a case by case approach and some releases have already taken place. The European Community in its move towards a single market without trade barriers in 1992 is producing directives in this area. Without a doubt, the final regulatory system must be constructed so that it can evolve to take account of new knowledge.

References

Alabouvette, C., Couteaudier, Y. & Louvet, J. (1985). Soils suppressive to *Fusarium* wilt: mechanisms and management of suppression. In *Ecology and Management of Soilborne Plant Pathogens*, ed. C. A. Parker, A. D. Rovira, K. J. Moore, P. T. W. Wong & J. F. Kollmorgen, pp. 101 - 105. The American Phytopathological Society: St. Paul, Minnesota.

Anderson, A. J., Habibzadegah-tari, P. & Tepper, C. S. (1988). Molecular studies on the role of root surface agglutinin adherence and colonisation by *Pseudomonas putida. Applied and Environmental Microbiology*, **54**, 375-380.

Andrews, J. H. (1985). Strategies for selecting antagonistic microorganisms from the phylloplane. In *Biological Control on the Phylloplane*, ed. C. E. Windels & S. E. Lindow, pp. 31-44. American Phytopathological Society: St Paul, Minnesota.

Andrews, J. H. (1987). How to track a microbe. In *Microbiology of the Phyllosphere*, pp. 14-35. Cambridge University Press.

Bahme, J. B. & Schroth, M. N. (1987). Spatial-temporal colonization patterns of a rhizobacterium on underground organs of potato. *Phytopathology*, **77**, 1093-1100.

Bahme, J. B., Schroth, M. N., Van Gundy, S. D., Weinhold, A. R. & Tolentino, D. M. (1988). Effect of inocula delivery systems on rhizobacterial colonization of underground organs of potato. *Phytopathology*, **78**, 535-542.

Brisbane, P. G. & Rovira, A. D. (1988). Mechanism of inhibition of *Gaeumannomyces graminis* var. *tritici* by fluorescent pseudomonads. *Plant Pathology*, **37**, 104-111.

Campbell, R. & Clor, A. (1985). Soil moisture affects the interaction between *Gaeumannomyces graminis* var. *tritici* and antagonistic bacteria. *Soil Biology and Biochemistry*, **17**, 441-446.

Capper A. L. & Campbell R. (1986). The effect of artificially inoculated antagonistic bacteria on the prevalence of take-all disease of wheat in field experiments. *Journal of Applied Bacteriology*, **60**, 155-160.

Chao, W. L., Nelson, E. B., Harman, G. E. & Hoch, H. C. (1986). Colonization of biological control agents applied to the seeds. *Phytopathology*, **76**, 60-65.

Chen, W., Hoitink, H. A., Schmitthener, A. F. & Tuovinen, O. H. (1988). The role of microbial activity in suppression of damping-off caused by *Pythium ultimum*. *Phytopathology*, **78**, 314-322.

Cook, R. J. & Baker, K. F. (1983). *The Nature and Practice of Biological Control of Plant Pathogens*. American Phytopathological Society: St. Paul, Minnesota.

de Weger, L. A., Schippers, B. S Lugtenberg, B, (1987). Plant growth stimulation by biological interference in iron metabolism in the rhizosphere. In *Iron Transport in Microbes, Plants and Animals*, ed. G. Winkelmann, D. van der Helm & J. B. Nielands, pp. 387-400. VCH Weinheim.

Domsch, K. H., Driesel, A. J., Goebel, W., Andersch, W., Lindenmaier, W., Lotz, W., Reber, H. & Schmidt, F. (1988). Considerations on release of gene-technologically engineered microorganisms into the environment. *FEMS Microbiology Ecology*, **53**, 261-272.

Elad, Y. (1986). Mechanisms of interactions between rhizosphere microorganisms and soilborne plant pathogens. In *Microbial Communities in Soil*, FEMS symposium 33, ed. V. Jensen, A. Kjoller & L. H. Sorensen, pp. 49-60. Elsevier: Amsterdam.

Elad, Y. & Chet, I. (1987). Possible role of competition for nutrients in biocontrol of *Pythium* damping off by bacteria. *Phytopathology*, **77**, 190-195.

Elad, Y., Chet, I. & Henis, Y. (1979). Degradation of plant pathogenic fungi by *Trichoderma harzianum*. *Canadian Journal of Microbiology*, **28**, 719-725.

El-Nashaar, H. M., Moore, L. W. & George, R. A. (1986). ELISA quantification of initial infection of wheat by *Gaeumannomyces graminis* var. *tritici* as moderated by biocontrol agents. *Phytopathology*, **76**, 1319-1322.

Feest, A. & Campbell, R. (1986). The microbiology of soils under successive wheat crops in relation to take-all disease. *FEMS Microbiology Ecology*, **38**, 99-111.

Foster, R. C. (1985). The biology of the rhizosphere. In *Ecology and Management of Soilborne Plant Pathogens*, ed. C. A. Parker, A. D. Rovira, K. J. Moore, P. T. W. Wong & J. F. Kollmorgen. pp. 75-79. The American Phytopathological Society: St. Paul, Minnesota.

Geels, F. P. & Schippers, B. (1983). Selection of antagonistic fluorescent *Pseudomonas* spp. and their root colonization and persistence following treatment of seed pieces. *Phytopathologische Zeitschrift*, **108**, 193-206.

Gurusiddaiah, S., Weller, D. M., Sarkar, A. & Cook, R. J. (1986). Characterization of an antibiotic produced by a strain of *Pseudomonas fluorescens* inhibitory to *Gaeumannomyces graminis* var. *tritici* and *Pythium* spp. *Antimicrobial Agents and Chemotherapy*, **29**, 488-495.

Hadar, Y., Harman, G. E., Taylor, A. G. & Norton, J. M. (1983). Effects of pregermination of pea and cucumber seeds and of seed treatment with *Enterobacter cloacae* on rots caused by *Pythium* spp. *Phytopathology*, **73**, 1322-1325.

Harman, G. E., Chet, I. & Baker, R. (1981). Factors affecting *Trichoderma hamatum* applied to seeds as a biocontrol agent. *Phytopathology*, **71**, 569-572.

Harman, G. E. & Taylor, A. G. (1988). Improved seedling performance by integration of biological control agents at favourable pH levels with solid matrix priming. *Phytopathology*, **78**, 520-525.

Hornby, D. (1987). Field testing putative biological controls of take-all: rationale and results. *EPPO Bulletin*, **17**, 615-623.

Howell, C. R. (1987). Relevance of mycoparasitism in the biological control of *Rhizoctonia solani* by *Gliocladium virens*. *Phytopathology*, **77**, 992-994.

Howell, C. R. & Stipanovic, R. D. (1979). Control of *Rhizoctonia solani* on cotton seedlings with *Pseudomonas fluorescens* and with an antibiotic produced by the bacterium. *Phytopathology*, **69**, 480-482.

Howell, C. R. & Stipanovic, R. D. (1980). Suppression of *Pythium ultimum* induced damping off of cotton seedlings by *Pseudomonas fluorescens* and its antibiotic pyoluteorin. *Phytopathology*, **70**, 712-715.

Howie, W. J., Cook R. J. & Weller, D. M. (1987). Effects of the soil matric potential and cell motility on wheat root colonization by fluorescent pseudomonads suppressive to take-all. *Phytopathology*, **77**, 286-292.

Jones, D. G. & Morley, S. J. (1981). The effect of pH on host plant 'preference' for strains of *Rhizobium trifolii* using the fluorescent ELISA technique for *Rhizobium* strains for identification. *Annals of Applied Biology*, **88**, 448-450.

Lewis, J. A. & Papavizas, G. C. (1987). Application of *Trichoderma* and *Gliocladium* in alginate pellets for control of *Rhizoctonia* damping off. *Plant Pathology*, **36**, 438-446.

Mendgen, K. (1987). Quantitative serological estimation of fungal colonization. In *Microbiology of the Phyllosphere*, ed. N. J. Fokkema & J. Van Den Heuvel, pp. 50-63. Cambridge University Press.

Morris C. E. & Rouse D. I. (1987). Microbiological and sampling considerations for quantification of epiphytic microbial community structure. In *Microbiology of the Phyllosphere*, ed. N. J. Fokkema & J. Van Den Heuvel. pp. 3-14. Cambridge University Press.

Nelson, E. B. (1988). Biological control of *Pythium* seed rot and pre-emergence damping off of cotton with *Enterobacter cloacae* and *Erwinia herbicola* applied as seed treatments. *Plant Disease*, **72**, 140-142.

Nelson, E. B., Harman, G. E. & Nash, G. T. (1988). Enhancement of *Trichoderma* induced biological control of *Pythium* seed rot and pre-emergence damping off of peas. *Soil Biology and Biochemistry*, **20**, 145-150.

Parke, J. L., Moen, R., Rovira, A. D. & Bowen G. D. (1986). Soil water flow affects the rhizosphere distribution of a seed-borne biological control agent *Pseudomonas fluorescens*. *Soil Biology and Biochemistry*, **18**, 583-588.

Renwick, A. & Jones D. G. (1986). The manipulation of white clover 'host preference' for strains of *Rhizobium trifolii* in an upland soil. *Annals of Applied Biology*, **108**, 291-302.

Rovira, A. D. (1985). Manipulation of the rhizosphere microflora to increase plant production. In *Reviews of Rural Science, 6 Biotechnology and Recombinant DNA Technology in the Animal Production Industries*, ed. R. A. Leng, J. S. F. Barker, D. B. Adams & K. J. Hutchinson. pp. 185-197. University of New England: Armidale, Australia.

Scher, F. M., Kloepper, J. W., Singleton, C., Zaleska, I. & Laliberte, M. (1988). Colonization of soyabean roots by *Pseudomonas* and *Serratia* species. Relationship to bacterial motility, chemotaxis and generation time. *Phytopathology*, **78**, 1055-1059.

Schroth, M. N., Loper, J. E. & Hildebrand D. C. (1984). Bacteria as biocontrol agents of plant disease. In *Current Perspectives in Microbial Ecology*, eds. M. J. Klug & C. A. Reddy, pp. 362-369. American Society for Microbiology: Washington, DC.

Shephard, W. C. (1987). Screening for fungicides. *Annual Review of Phytopathology*, **25**, 389-206.

Simon, A., Rovira, A. D. & Foster, R. C. (1987). Inocula of Gaeumannomyces graminis var. tritici for field and glasshouse studies. *Soil Biology and Biochemistry*, **19**, 363-370.

Stanghellini, M. E., Stowell, L. J., Kronland, W. C. & Van Bretzel, P. (1983). Distributions of *Pythium aphanidermatum* in rhizosphere soil and factors affecting the expression of the absolute inoculum potential. *Phytopathology*, **73**, 1463-1466.

Suslow, T. V. (1982). Role of root colonizing bacteria in plant growth. In *Phytopathogenic Prokaryotes*, vol 1, ed. M. S. Marks & G. H. Lacy, pp. 187-223. Academic Press: London.

Swinburne, T. R. (1986). *Iron. Siderophores and Plant Diseases* NATO ASI Series, Life Sciences vol. 117. Plenum Press: New York & London.

Tari, P. H. & Anderson, A. J. (1988). *Fusarium* wilt suppression and agglutinability of *Pseudomonas putida*. *Applied and Environmental Microbiology*, 54, 2037-2041.

Toyodo, H., Hashimoto, H., Utsumi, R., Kobayashi, H. & Ouchi S. (1988). Detoxification of fusaric acid by a fusaric acid-resistant mutant of *Pseudomonas solanacearum* and its application to biological control of fusarium wilt of tomato. *Phytopathology*, 78, 1307-1311.

Toyodo, H. & Utsumi, R. (1987). Prevention of fusarium diseases and microorganisms therefore. European Patent Application 87306103.0.

Van Vuurde, J. W. L. & Schippers B. (1980). Bacterial colonization of seminal wheat roots. *Soil Biology and Biochemistry*, 12, 559-565.

Vesper, S. J. (1987). Production of pili (fimbriae) by *Pseudomonas fluorescens* and correlation with attachment to corn roots. *Applied and Environmental Microbiology*, 53, 1397-1405.

Weller, D. M. (1983). Colonization of wheat roots by a fluorescent pseudomonad suppressive to take-all. *Phytopathology*, 73, 1548-1553.

Weller, D. M. (1984). Distribution of a take-all suppressive strain of *Pseudomonas fluorescens* on seminal roots of winter wheat. *Applied Environmental Microbiology*, 48, 897-899.

Weller, D. M., Zhang, B. X. & Cook, R. J. (1985). Application of a rapid screening test for selection of bacteria suppressive to take-all of wheat. *Plant Disease*, 69, 710-713.

Weller, D. M., Howie, W. J. & Cook, R. J. (1988). Relationship between *in vitro* inhibition of *Gaeumannomyces graminis* var. *tritici* and suppression of take-all of wheat by fluorescent pseudomonads. *Phytopathology*, 78, 1094-1100.

Wilkinson, H. T., Cook, R. J. & Alldredge, J. R. (1985). Relation of inoculum size and concentration to infection of wheat roots by *Gaeumannomyces graminis* var. *tritici*. *Phytopathology*, 75, 98-103.

Windels, C. E. (1981). Growth of *Penicillium oxalicum* as a biological seed treatment on peas seed in soil. *Phytopathology*, 71, 929-933.

Index

A

F

Q

T